Atmospheric Re-Entry Vehi

Patrick Gallais

Atmospheric Re-Entry Vehicle Mechanics

With 193 Figures and 23 Tables

 Springer

Patrick Gallais
Civil Engineer ENST
CEA/CESTA, BP 2,
33114 Le Barp
France

ISBN 978-3-642-09281-7   e-ISBN 978-3-540-73647-9

This work is subject to copyright. All rights are reserved, whether the whole or part of the material is concerned, specifically the rights of translation, reprinting, reuse of illustrations, recitation, broadcasting, reproduction on microfilm or in any other way, and storage in data banks. Duplication of this publication or parts thereof is permitted only under the provisions of the German Copyright Law of September 9, 1965, in its current version, and permission for use must always be obtained from Springer. Violations are liable for prosecution under the German Copyright Law.

Springer is a part of Springer Science+Business Media
springer.com
© Springer-Verlag Berlin Heidelberg 2010

The use of general descriptive names, registered names, trademarks, etc. in this publication does not imply, even in the absence of a specific statement, that such names are exempt from the relevant protective laws and regulations and therefore free for general use.

Cover design: eStudio Calamar, Girona, Spain

TEMPLE EXPIATORI
SAGRADA FAMÍLIA

AUDIOGUÍA

DESCOBREIX LA
SAGRADA FAMÍLIA!

¡DESCUBRE LA
SAGRADA FAMILIA!

DISCOVER
SAGRADA FAMILIA!

4 €

Preface

This textbook is derived from a document originally written for colleagues of my company (CEA) who are concerned with ballistic reentry vehicles. This free and open document was published in French in 1998. It dealt with reentry topics restricted to aerodynamics and flight mechanics. It was submitted to my publisher, Springer, who expressed interest in publishing it as a textbook for universities, engineering schools, aerospace companies, and government agencies. We then proceeded to enlarge and modify the original French manuscript and translate it to English.

The final product is not purely academic, but derives from the experiences and experiments during my long career in this field. In the original text, for rigorous concerns, we provided demonstration of the majority of theoretical results, which seemed more suited to the purpose of transferring knowledge. Relating to the first edition, the present text has been corrected from some errors and we greatly appreciate reader feedback upon discovery of additional errors.

New chapters have been added; particularly those relating to flight qualities, instabilities and dispersions, and other chapters have been modified. The subject of planetary entry capsules has been included, following recent collaboration with Alcatel Space Company and French Spatial Agency CNES (MSR/Netlander program). At the end of the textbook, we included a set of exercises as applying principles of each chapter.

I thank CEA for sponsoring my career in this very special field and providing the opportunity to write this book. I owe particular thanks to my colleague Georges Duffa, who initially encouraged me to write the different editions and kindly reviewed the French text. I also thank Springer for their support throughout the evolution of the textbook.

Patrick Gallais

Contents

1 Classical Mechanics 1
1.1 Classical Point Mass Mechanics 1
    1.1.1 Fundamental Principles 1
    1.1.2 Noninertial Frames 2
    1.1.3 Linear and Angular Momentum 3
    1.1.4 Modeling the Forces 4
    1.1.5 Conservation of Energy 4
    1.1.6 Isolated System 6
1.2 Mechanics of Rigid Bodies 7
    1.2.1 Linear Momentum and Angular Momentum 7
    1.2.2 Center of Mass and Equations of Movement 8
    1.2.3 Eulerian Frames 9

2 Topography and Gravitation 15
2.1 The Geodetic Frame of Reference 15
    2.1.1 Coordinates of a Point Relating to the Earth 16
    2.1.2 The Geodetic Systems 17
    2.1.3 Calculation of the Geographical Latitude and Height 18
2.2 The Terrestrial Field of Gravitation 19

3 Models of Atmosphere 21
3.1 Main Parameters and Hypotheses 21
3.2 The Isothermal Exponential Model 22
3.3 Standard Models of Earth's Atmosphere 23
3.4 Martian Models 26

4 Aerodynamics 31
4.1 Aerodynamic Coefficients 33
    4.1.1 Static Coefficients 34
    4.1.2 Dynamic Derivatives 36
    4.1.3 Axisymmetric Vehicles 37
4.2 Modes of Flow 38
    4.2.1 Parameters of Similarity 38
    4.2.2 Characteristics of the Main Flow Modes 39

4.3		Continuous Mode	40
	4.3.1	Experimental	40
	4.3.2	Numerical [AND]	40
	4.3.3	Approximate Analytical Method [TRU]	41
	4.3.4	Continuous Coefficients by Newton's Method	43
4.4		Rarefied Mode	52
	4.4.1	Free Molecular Flow	52
	4.4.2	Intermediate Flow	56
4.5		Qualities of Flight	58
	4.5.1	Static Stability	59
	4.5.2	Gyroscopic Stability	64
4.6		Characteristics of a Family of Sphere Cones	65
	4.6.1	Mach Number Influence	66
	4.6.2	Influence of Angle of Attack	70
	4.6.3	Aerodynamic Modeling for Trajectory Codes	74
4.7		Planetary Entry Capsule	76

5 Inertial Models 85
5.1 Moments of Inertia 85
5.2 CG Offset and Principal Axis Misalignment 86

6 Changing of Reference Frame
6.1 Direction Cosine Matrices 89
  6.1.1 Angular Velocity 91
  6.1.2 Composition of Angular Velocities 94
  6.1.3 Evolution of the Direction Cosine Matrices 95
6.2 Euler Angles 96
  6.2.1 Euler Rotation Matrix 96
  6.2.2 Evolution of Euler Angles 99
6.3 Representations with Four Parameters 103
  6.3.1 Vectorial Representation 103
  6.3.2 Quaternion 104

7 Exoatmospheric Phase
7.1 Movement of the Center of Mass 109
  7.1.1 Keplerian Trajectories 109
  7.1.2 Ballistic Trajectories 118
  7.1.3 Influence of Earth Rotation 128
7.2 Movement Around Mass Center 136
  7.2.1 Rotation Around a Principal Axis 136
  7.2.2 Coning Motion 138

8 Six Degree-of-Freedom Reentry
8.1 General Equations of Motion 143
8.2 Solutions of General Equations 146

9 Zero Angle of Attack Reentry 149
9.1 Allen's Reentry Results 149
    9.1.1 Axial Load Factor and Dynamic Pressure 151
    9.1.2 Heat Flux 151
    9.1.3 Thermal Energy at Stagnation Point 152
    9.1.4 Duration of Reentry 154
9.2 Influence of Ballistic Coefficient and Flight Path Angle 156
9.3 Influence of Range 157

10 Decay of Initial Incidence 163
10.1 Zero Spin Rate 165
    10.1.1 First Approximate Solution 166
    10.1.2 Second Approximate Solution 168
10.2 Nonzero Spin 172

11 End of the Convergence of the Incidence 179
11.1 Linear Equations 180
11.2 Instantaneous Angular Movement 185
    11.2.1 Epicyclic Movement 186
    11.2.2 Tricyclic Movement 190
11.3 Real Angular Motion 198

12 Roll-lock-in Phenomenon 201
12.1 Association of Aerodynamic Asymmetry and CG Offset 202
    12.1.1 Equilibrium on Critical Frequency 204
    12.1.2 Lock-in Near Resonance 206
    12.1.3 Variable Critical Frequency 209
    12.1.4 Criterion for Out-of-plane Asymmetries 209
12.2 Isolated Center of Gravity Offset 216
12.3 Isolated Principal Axis Misalignment 217
12.4 Combined CG Offset and Principal Axis Misalignment 220
    12.4.1 Out-of-plane Misalignment 221
    12.4.2 In-plane Misalignment 226

13 Instabilities 231
13.1 Static Instabilities 231
13.2 Dynamic Instabilities 238
    13.2.1 Approximate Study at Low Angle of Attack 238
    13.2.2 Examples of Unstable Dynamic Behaviors 241

14 Reentry Errors 245
14.1 Zero Angle-of-attack Dispersions 245
    14.1.1 Initial Conditions 245
    14.1.2 Consequences of Drag Dispersions 246
    14.1.3 Sensitivity to Initial Conditions 250
14.2 Nonzero Angle of Attack 251

		14.2.1 Effects of Incidence on Aerodynamic Loads 251
		14.2.2 Effects of Initial Angle-of-attack. 253

Epilog 281

Exercises 285

Solutions 297

1	Equations of Motion Relating to a Reference Frame in Rotation 297
2	Accelerometer Measurements 299
3	Vertical and Apparent Gravity 300
4	Coriolis Force. 301
5	Quaternions 302
6	Quaternions and Euler Angles 303
7	Equivalence of Changing Frame and Rotation Quaternion Operators for Euler Angle Sequence 304
8	Pendulum of Foucault 305
9	Motion of the Earth's Mass Center 311
10	Launch Windows to Mars 311
11	Energy of a Solid in Free Rotation 311
12	Total Angular Momentum of a Deformable System in Gravitation 312
13	Lunar Motion 315
14	Angular Stabilization with Inertia Wheels 315
15	Balancing Machines 316
16	Aerodynamics of Apollo Reentry Capsule in Continuous Flow 318
17	Aerodynamics of Apollo Reentry Capsule in Free Molecular Flow. 320
18	Aerodynamics of Viking Reentry Capsule in Continuous Flow 321
19	Aerodynamics of Viking Reentry Capsule in Free Molecular Flow 324
20	Aerodynamics of Pathfinder Entry Capsule in Intermediate Flow 325
21	Aerodynamics of a Biconic Reentry Vehicle in Continuous Mode 327
22	Targeting Errors of Ballistic Trajectories 330
23	Stability of Free Rotational Motion of a Satellite with Spin 332
24	Allen and Eggers Reentry 333
25	Mars Atmosphere Measurement 334
26	Entry of Meteorites 336
27	Normal Load Factor Related to Incidence. 338
28	Artefact in Computer Codes 339
29	Gyroscopic Stabilization of an Entry Capsule 340
30	Effect of Equilibrium Lift on CG Motion 341
31	Effects of Drag Dispersions 344
32	Wind Effects. 345
33	Effect of a Sharp Stability Variation, Case of a Plane Oscillation 348
34	Mars Skip Out Trajectories. 349

Abbreviations

AAAF	Association Aeronautique et Astronautique de France
AIAA	American Institute of Aeronautic and Atonautic
AoA	Angle of Attack
CG	Center of mass
CP	Aerodynamic center of pressure
DSMC	Direct Simulation Monte-Carlo
EMCD	European Martian Climate Database
FPA	Flight Path Angle
HABP	Hypersonic Arbitrary Body Program
NACA	National Advisory Committee for Aeronautics
NAVSTAR	US satellites navigation system
PAM	Principal Axis Misalignment
PNS	Parabolized Navier Stokes Code
NF	French Standard
NS	Navier Stokes Code
WGS	World Geodetic System

Symbols

Latin Letters

a	Half length of ellipse major axis, sound velocity	m, m/s
A	Aerodynamic component axial or trim amplification factor, directors cosine matrix	N or no dimension
b	Half length of ellipse minor axis	m
B	Inverse matrix of directors cosine	no dim
$\bar{c}_{2,0}$	Constant of gravitational model	no dim
D	Diameter, distance to the axis of poles or damping parameter	m, no dim
$e, \vec{e}$	Eccentricity of the ellipsoid, blunting parameter, basic vector unit	no dim, m
E	Indicate an Eulerian frame related to the vehicle	
$\vec{f}, f$	Elementary force or ellipsoid flatness	N, no dim
$\vec{F}, F$	Force, coefficient of the pulsations equation	N, rad/s
$g, \vec{g}$	Terrestrial gravity field	$m.s^{-2}$
G	Amplification factor of rolling acceleration, coefficient of the pulsations equation	no dim $(rad/s)^2$
$H, \vec{H}$	Height relating to the ellipsoid, angular momentum	m, N.m.s
i	Base of imaginary numbers	
I	Moment of inertia	$m^2.kg$
k	Wave number	m^{-1}
K	Indicate an inertial or Galilean reference frame	
$L, \vec{L}$	Length, aerodynamic roll moment or angular momentum	m, N.m, N.m.s
m	Mass	kg
$M, \vec{M}$	Mass or specific weight, pitching aerodynamic moment, total moment	kg, N.m
N	Yawing aerodynamic moment, length of the great normal to ellipsoid	N.m, m
p	Pressure, roll angular rate, linear momentum	Pa, rad/s, N. s
$\vec{P}$	Linear momentum vector	N. s

xiii

Symbol	Description	Units
q, q̄	Pitch angular rate, quaternion component, dynamic pressure	rad/s, no dim, Pa
Q	Quaternion	
$\vec{r}$, r	Position, geocentric radius vector, specific constant of a gas, yaw angular rate	m, J/kg/K, rad/s
R	Location vector or universal constant of gases	m, J/mole/K
sgn	Sign function (± 1)	no dim
S	Area	m^2
T	Kinetic energy, absolute temperature	W, K
t	Time	s
u	Component along X of speed relating to the air	m/s
v	Component along Y of speed relating to the air	m/s
V_c	Speed of circular orbit	m/s
$\vec{V}$	Velocity	m/s
w	Component along Z of speed relating to the air	m/s
x, y, z	Geocentric co-ordinates	m
$\vec{X}$	Vector	m
Z	Altitude relating to sea level	m
ΔX	Static margin	m

Greek Letters

Symbol	Description	Units
$\bar{\alpha}$	Total aerodynamic incidence, angle of attack	radian
α	Component of incidence in the symmetry plane, angular range	radian
β	Yaw component of incidence	radian
$\vec{\Delta}, \Delta$	Axis of rotation or difference operator	
ε	Half angle of coning motion, first order small number	radian or arbitrary unit
φ	Geographical latitude, Euler angle	radian
$\vec{\varphi}$	Gravitation field vector	m.s^{-2}
Φ	Diameter, angle	m, radian
$\vec{\gamma}, \gamma$	Acceleration, polytropic gas constant	m.s^{-2} or no dim.
$\vec{\Gamma}$	Acceleration	m.s^{-2}
λ	Geocentric longitude, mean free path, pulsation (complex)	rad, m, rad.s^{-1}
Λ	Real part of pulsation	rad.s^{-1}

Symbols

μ	Gravitational constant, inertia ratio, mass	$m^3.s^{-2}$, no dim, kg
θ	Nutation Euler angle, principal inertia axis misalignment	rad
ω	Angular rate, aerodynamic pulsation	rad / s
$\vec{\Omega}, \Omega$	Angular rate, complex angular rate $(q + i\,r)$	rad / s
ξ	Complex incidence $(\beta + i\,\alpha)$	rad
τ	Initial angle between kinetic momentum and linear velocity	rad
ψ	Geocentric latitude, Euler angle	rad

Indices

a	Aerodynamic, apex, absolute
A	Aerodynamic, axial
$stag$	Stagnation point value
b	Relating to vehicle base plane
c	Relating to, relative to base plane
CP, cp	Relating to pressure center
cr	Relating to critical conditions (resonance)
D	Relating to initial re-entry point
e	Relating to the movement of a reference frame, static equilibrium value
E	Value relating to a non Galilean or Eulerian frame, at static trim
G	Relating to center of gravity, gyrometric
i	Indicate a point of a solid
l, m, n	Relating to roll, pitch and yaw axis, relating to vehicle nose tip
N	Normal, relating to vehicle nose tip
O	Relating to origin of aerodynamic reference frame
p, q, r	Relating to pressure (p), relating to rolling or to resonance (r), gradient relating to components of angular rate
Ref, R	Reference value
T	Transposed, transverse
x, y, z	Relating to x, y, z axis

Y	Side component
w	Relating to wind
α, β	Gradient relating to components of angle of attack
θ	Relating to principal axis misalignment
0	Value at zero pitch and yaw angles of attack, indices of the real part of a quaternion, relating to initial time
$1, 2, 3$	Relating to axis number
$\dot{X}, \ddot{X}$	Derivative operators relating to time
∞	Relating to undisturbed incoming conditions

Constants[CRC],[GPS]

Avogadro number	$N = 6.02213674 \times 10^{23}$ mole^{-1}
Boltzmann constant	$k = 1.3806581 \times 10^{-23}$ J/K
Perfect gas constant	$R = Nk = 8.3145107$ J/mole/K

WGS84 Ellipsoid

Geometrical Model

Equatorial semi major axis	$a = 6378137.$m
Flatness parameter	$f = (a - b)/a = 1/298.257223563$

Model of gravity

Angular velocity of rotation	$\omega = 7.292115167 \cdot 10^{-5} rad/s$
Geocentric gravitational constant	$\mu = 3.986005 \cdot 10^{14} m^3 \cdot s^{-2}$
Coefficient of the first non-spherical term of the potential	$\bar{c}_{2,0} = -0.48416685 \cdot 10^{-3}$

List of Reference

[ABR] Abramowitz, M., Stegun, J.E.: Handbook of Mathematical Functions. US Department of Commerce, National Bureau of Standard Applied Mathematical Series (1964), Dover, New York (1970)
[ALL] Allen, H.J., Eggers, A.J. Jr.: A study of the Motion and Aerodynamic Heating of Missiles Entering the Earth's Atmosphere at High Supersonic Speeds, NACA TN 4047, October, (1957)
[ALP] Lin, T.C., et al.: Flight-Dynamics Instability Induced by Heat-Shield Ablation Lag Phenomenon, TRW Systems. J. Spacecr. Rockets, **40**, 5, (2003)
[AND] Anderson, J.D.: Hypersonic and High Temperature Gas Dynamics. McGraw Hill, New York (1989)
[ARF] Arfken, G.: Mathematical Methods for Physicists. Academic Press, London (1970)
[BAS] Bass, J.: Mathematics, Part 1, 3rd edition. Masson and Co., Paris (1964)
[BGK] Bathnagar, P.L., Gross, E.P., Krook, M.: A model for collision process in gazes. Phys. Rev., **94**, 511–A-524 (1954)
[CRC] CRC Handbook of Chemistry and Physics, 74th edition. CRC Press, Cleveland (1993)
[DIR] Dirling, R.B. Jr: Asymmetric Nosetip Shape Change During Atmospheric Entry. AIAA 12th Thermophysics Conference, Albuquerque (1977)
[DUF] Ablation, M.C., Duffa, G.: Commissariat à l'Energie Atomique. CESTA, France (1996)
[DYN] Platus, D.H.: Ballistic re-entry vehicle flight dynamics. J. Guid. Control (1982)
[EMC] Collins, M., Lewis, S.R., Read, P.L., Thomas, N.P.J., Talagrand, O., Forget, F., Fournier, R., Hourdin, F., Huot, J.P.: A climate database for the Martian atmosphere. In: Environment Modelling for Space Based Applications. European Space Agency SP- 392, pp. 323–327 (1996)
[ERI] Ericsson, L.E.: Hyperballistic vehicle dynamics. J. Spacecr. Rockets, **19**(6), AIAA (1982)
[GAL] Gallais, P.: Netlander Aerodynamic Design. 3rd International Symposium Atmospheric Re-entry Vehicles and Systems. Arcachon (France), AAAF, March (2003)
[GLO] Glover, L.S., Hagen, J.C.: The Motion of Ballistic Missiles. The Johns Hopkins University, Applied Physics Laboratory, Silver Spring, MD (1971)
[GPS] Leick, A.: GPS Satellite Surveying. John Wiley & Sons, New York (1990)
[GRA] Justus, C.G., Alyca, F.N., Jeffries, W.R. III, Johnson, D.L.: The NASA/MFSC Global Reference Atmosphere Model—1995 Version (GRAM 95). NASA Technical Memorandum (1995)
[HAB] Gregoire, J.E. et al.: Aerodynamic Prediction Rationale for Advanced Arbitrarily Shaped Missile Concepts. AIAA 18th Aerospace Sciences Meeting, Pasadena (1980)
[HBK] Handbook of Astronautical and Engineering, New York & London, Mc Graw-Hill (1961)
[KAR] Karatekin, Ö.: Aerodynamics of a planetary entry capsule at low speed, Université Libre de Bruxelles, Février (2002)
[LIN] Lin, T.C., Grabowsky, W.R.: Ballistic re-entry vehicle dispersion due to precession stoppage. TRW electronics and defence sector. J. Spacecr. Rockets, **21**(4), AIAA (1984)
[NFX] NF X02-115, Symboles et Vocabulaire de la Mécanique du vol, AFNOR (1970)
[PAL] Palmer, R.H. et al.: A Phenomenological Framework for Reentry Dispersion Source Modelling. Space and Missile Systems Organization, Los Angeles (1977)

[PAR] Moss, J.N., Blanchard, R.C., Wilmoth, R.G., Braun, R.D.: Mars Pathfinder Rarefied Aerodynamics: Computation and Measurements. NASA Langley Research Center, AIAA 98–0298 (1998)

[PAT] Gnoffo, P.A. et al.: Effect of Sonic Line Transition on the Aerodynamics of the Mars Pathfinder Probe. NASA Langley Research Center, AIAA (1995)

[PLA] Platus, D.H.: Re-Entry Vehicle Dispersion from Entry Angular Misalignment. The Aerospace Corporation, El Segundo, California. J. Guid., Control, **2**(4), (1978)

[RAT] Wilmoth, R.G. et al.: Rarefied Transitional Bridging Functions of Blunt Body Aerodynamics. NASA Langley Research Center, 1st International Symposium on Rarefied Gas Dynamics, Marseille, France, July (1998)

[RYH] Ryhming, I.L.: Dynamique des fluides. Presses Polytechniques Romandes, Lausannes, Suisse (1985)

[SUT] Sutton, K., Graves, R.A.: A general stagnation point convective heating equation for arbitrary gas mixtures. NASA TR- 376 November (1971)

[TRU] Truitt, R.W.: Hypersonic Aerodynamics. The Ronald Press Company, New York (1959)

[USS] U.S. Standard Atmosphere 1966. U.S. Government Printing Office, Washington D.C. (1966)

[VAU] Vaughn, H.R.: A Detailed Development of the Tricyclic Theory. Sandia Laboratories, Albuquerque (1968)

[VFL] Lin, T.C., Rubin, S.G.: Viscous flow over spinning cones at angle of attack. AIAA J., **12**(7), (1974)

[VIK] Experimental Aerodynamic Characteristics of the Viking Entry Vehicle Over the Mach Range 1.5–10. NASA – CR—15225. Martin Marietta Corporation (1971)

[WAT] Waterfall, A.P.: Effect of ablation on the dynamics of spinning re-entry vehicles. J. Spacecrafts and Rockets, **6**, (1969)

Introduction

The mission of a ballistic reentry vehicle (RV) consists of a trajectory having three phases of very different characteristics and duration: launch, vacuum, and atmospheric reentry to the target.

The launch phase is dominated by engine thrust, which is typically designed for nearly uniform acceleration. Duration is short and, assuming a protective shroud or fairing is provided, the primary concerns for the RV are the mechanical loads induced by shock and vibration.

Final stage thrust terminates at a very high altitude, and separation of a deployment module (or perhaps only the RV) follows in a virtual vacuum environment. Using reaction jets, the deployment module may exercise maneuvers to dispense RVs and other objects designed to subsequently reenter with specific mission objectives. The ballistic phase is governed by the kinematics conditions of separation, which are then subjected to gravitational and inertial forces. The trajectory, which is very close to elliptical, usually has a long duration of tens of minutes and contributes the major part of the range to the target zone.

Following deployments, the RV accelerates under gravitational force until the first perceptible aerodynamic effects of the atmosphere on motion occur between 120 and 90 km, depending on the mass and drag parameter of the body, known as the "ballistic coefficient (β)." Nonmilitary missions generally require high altitude deceleration (low β), whereas most military RVs need to retain velocity by achieving high β's, which results in endoatmospheric reentry times of less than a minute.

During reentry, the RV is subjected to gravitational, as well as both inertial and aerodynamic forces and moments. The aerodynamic effects become increasingly important as the altitude decreases and usually become dominant by 40 km, a result of the exponential increase of the air density.

Velocities may decrease from their initial reentry values of approximately Mach 20 to nearly subsonic at impact. Maximum axial loads typically vary from 50 to 100 g with stagnation pressures in excess of 100 atmospheres and thermal flux of 100 MW/m^2. The thermal protection system for this nosetip environment must endure several seconds, which is a formidable materials engineering problem. This problem also extends to the frustum where fluxes reach about one-tenth the nosetip values and require sophisticated insulation materials that rejects much of the incident energy.

Transverse load depends on the convergence of the initial angle of attack at reentry and generally does not exceed a few g's maximum value at about 20 km.

However, ablation of the nosetip may occur nonuniformly and results in asymmetric shapes that induce small sustained angles of attack ("trim" angles) at low altitude. When coupled with inertial asymmetries, these trim angles may be amplified through complex dynamic effects to much higher angles, having lateral loads that may threaten vehicle survival.

The above discussion shows the major importance that the combined aerodynamics and flight mechanics subject realms have in the successful design, accuracy, and evaluation of RV. As design tools, they provide the necessary aeroshell shape and mechanical tolerances on dimensions to control mass and inertial properties that ultimately result in the accuracy achieved by the configuration. Evaluation of this process follows applications to specific ground and flight tests, which yield the empirical data to further improve the elements of the analyses. Thus, the overall mission is vitally dependant on sound execution in these reentry disciplines, which are treated in this book.

Additionally, the case of planetary probes entry will be developed in this new edition. These missions have very long durations of several months to several years and initial conditions vary greatly, with different atmospheric conditions and gravitational properties. Aeroshell shapes are tailored to optimize velocity and path angles in these environments in order to achieve favorable experimental conditions. Although environmental levels may be different from typical Earth conditions, the technical formulations and approaches are quite similar and the methodology is virtually identical.

The following subjects are treated in separate chapters: classical mechanics; topography and gravitation; atmosphere models; aerodynamics; inertial model; changing of reference frames; ballistic phase; six degrees of freedom reentry equations; zero angle of attack reentry; initial angle of attack convergence; final angle of attack convergence; roll lock in; instabilities; and dispersions.

Chapter 1
Classical Mechanics

Lagrange mechanics, based on variational calculus and minimum action principle, is a beautiful intellectual construction but most frequently[1] of little use in dealing with our applications, which are characterized by nonconservative forces. All we need is contained in the "old" mechanics of Galileo and Newton and its consequences, deduced (among others) by Euler and d'Alembert. Excellent textbooks are dedicated to classical mechanics [BRK][LAN]. Our goal is neither to play the scientist nor to compete with them but only to remind the reader basic hypothesis or results he may have forgotten, especially the rigid body mechanics, which are used throughout this book. We assume the reader has a sound knowledge of basic kinematics notions.

1.1 Classical Point Mass Mechanics

Let us first recall that the fundamental principles were entirely deduced from the observation of movement of planets of solar system, in order to correlate and explain the measured phenomena with predictions.

Like any theory in physics, they were only assumptions, set up in principles, which proved for a long time to predict the physical phenomena with great exactitude. It is quite clear that these principles required modification in some domains by electromagnetic and relativity theory. However, traditional mechanics applies perfectly to the usual macroscopic objects at velocities well below the speed of light.

1.1.1 Fundamental Principles

1. The universe bathes in a medium at rest relative to which the objects move, "absolute space" (historically defined using stars assumed "fixed"). The measurement of time is the same for all the observers, whatever their movement relating to this absolute space.

---

[1] However, I found they can be very useful in some special case such as nonrigid body effects.

P. Gallais, *Atmospheric Re-Entry Vehicle Mechanics.*
© Springer 2007

2. The principles of mechanics apply to privileged observation systems, Galileo or inertial reference frame (K), in uniform translation relating to absolute space.

   a. Principle of inertia: "When no external forces are applied on it, the "free" movement of a point mass is uniform relating to any inertial reference frame:"

   $$\vec{v}_K = \vec{v}_K \, (t=0)$$

   b. Fundamental principle: The acceleration of a point mass "m" subjected to a force is the same relating to any inertial reference frame:

   $$\frac{d\vec{v}_K}{dt} = \frac{\vec{f}}{m} = \vec{\gamma}_a \qquad (1.1)$$

   c. Principle of equality of the action and reaction: For any set of two point masses interacting,

   $$\vec{f}_{1\to 2} + \vec{f}_{2\to 1} = 0$$

1.1.1.1 Remarks

- The laws of mechanics are invariant through Galilean transform:

$$\vec{x}' = \vec{x} + \vec{v}t; \quad t' = t \quad \leftrightarrow \quad \vec{x} = \vec{x}' - \vec{v}t'; \quad t = t'$$

- The fundamental principle implies assumptions of existence of mass invariant relating to change of inertial reference frame. Forces and accelerations are also invariant through Galilean transform:

$$\vec{x}' = \vec{x} + \vec{v}t; \quad t' = t \quad \Rightarrow \quad f'\left(\vec{x}', t'\right) = f\left(\vec{x}, t\right)$$

- The principle of inertia is a consequence of the fundamental principle; however, it can be used to define the inertial reference frames.
- Then the second principle can be used to define forces.

1.1.2 Noninertial Frames

The corollary of the principle of inertia is that relating to a noninertial reference frame,

1. the free movement (zero applied forces) of a point mass is not uniform and
2. the second principle is not valid.

A noninertial reference frame has a rotation and/or nonuniform translation movement relating to inertial frames. To determine the acceleration of a point mass

1.1 Classical Point Mass Mechanics

relating to a noninertial frame, we must account for latent forces, fictitious forces, only related to the movement of the noninertial observation frame.

- Let us label O the origin of axes of such a reference frame (E), $\vec{V}_e(t)$ the linear velocity of O, $\vec{\Omega}_e(t)$ angular velocity of (E) around O relating to an inertial reference frame (K), and $\vec{r}_E(t)$ the instantaneous location of a point mass P relating to (E).
- Let us develop the movement $\vec{r}(t)$ of P relating to (K) while composing the movement of [E] and the relative movement $\dot{\vec{r}}_E$ of P for an observer fixed to [E], located at the origin O of [E]. We will demonstrate in Chap. 6 that:

$$\dot{\vec{r}} = \vec{V}_e + \dot{\vec{r}}_E + \vec{\Omega}_e \wedge \vec{r} \tag{1.2}$$

$$\ddot{\vec{r}} = \dot{\vec{V}}_e + \ddot{\vec{r}}_E + 2\,\vec{\Omega}_e \wedge \dot{\vec{r}}_E + \dot{\vec{\Omega}}_e \wedge \vec{r} + \vec{\Omega}_e \wedge (\vec{\Omega}_e \wedge \vec{r}) \tag{1.3}$$

- We thus obtain the expression of the second derivative of the apparent movement by replacing the left term of (1.3) with its value from (1.1), $\ddot{\vec{r}} = \vec{\gamma}_a = \frac{\vec{f}}{m}$, which yields:

$$\ddot{\vec{r}}_E = \vec{\gamma}_E = \frac{\vec{f}}{m} - \left[\vec{\Gamma}_e + 2\vec{\Omega}_e \wedge \vec{v}_E + \dot{\vec{\Omega}}_e \wedge \vec{r} + \vec{\Omega}_e \wedge (\vec{\Omega}_e \wedge \vec{r})\right] \tag{1.4}$$

with $\vec{\Gamma}_e = \dot{\vec{V}}_e$ (acceleration of the origin of E relating to K) and $\vec{v}_E = \dot{\vec{r}}_E$ (apparent velocity of P for the observer fixed to E, obtained by derivation of the components of $\vec{r}_E$ relating to E).

It must be noted that the bracketed term of acceleration does not correspond to real forces but has an entirely kinematics origin in the movement of the observer. This acceleration is independent of the mass, which is also the case for gravitational fields. For an observer fixed to such a reference frame, the effect is equivalent to a gravitational field[2]. The virtual force appears as a gravity field in which all the masses interact. However, unlike an ordinary gravitational field, this one depends not only on the location but also on the relative velocity (Coriolis acceleration term). Another difference is that the gravitational forces vanish ad infinitum, which is not the case of virtual forces.

Finally, we observe that the real forces, invariant through changes of inertial reference frames, are also invariant with respect to changes of noninertial reference frame.

1.1.3 Linear and Angular Momentum

These physical entities are relative to the selected inertial frame (K):

---

[2] This equivalence is the foundation of general theory of relativity.

$$\vec{p} = m\vec{v} \text{ (linear momentum)} \tag{1.5}$$

$$\vec{l} = \vec{r} \wedge \vec{p} \text{ (angular momentum)} \tag{1.6}$$

The fundamental principle (1.1) stated by Newton is equivalent to:

$$\dot{\vec{p}} = \vec{f} \tag{1.7}$$

Therefore:

$$\dot{\vec{l}} = \vec{r} \wedge \vec{f} \tag{1.8}$$

1.1.4 Modeling the Forces

The most general model of force is such that the instantaneous value of $\vec{f}$ depends on the entire past of the point mass. Such a force cannot be represented a priori by a functional relationship. Fortunately, there is a class of simpler models that represent most situations such that the instantaneous force corresponds to a state function. A state function depends only on kinematics state parameters $(\vec{r}, \dot{\vec{r}})$ and time:

$$\vec{f}(t) = \vec{f}(t, \vec{r}, \dot{\vec{r}})$$

We will use this exclusively in all that follows, assuming the function is piecewise continuous.

1.1.5 Conservation of Energy

1.1.5.1 General Case

Let us consider a point mass subject to a force defined by a state function $\vec{f}(t) = \vec{f}(t, \vec{r}, \dot{\vec{r}})$. For an inertial observer, we obtain:

$$\int_{t1}^{t2} \dot{\vec{p}} \cdot \vec{v} dt = \int_{t1}^{t2} \vec{f} \cdot \vec{v} dt \Rightarrow \int_{t1}^{t2} m\dot{\vec{v}} \cdot \vec{v} dt = \oint_{s_{1 \to 2}} \vec{f} \cdot d\vec{r}$$

$$\frac{1}{2}mv_2^2 - \frac{1}{2}mv_1^2 = W_{12} \tag{1.9}$$

Increase of the state function "kinetic energy" T of the point mass equals the work done by the force:

$$T = \frac{1}{2}mv^2 = \frac{p^2}{2m} \tag{1.10}$$

1.1 Classical Point Mass Mechanics

The relationship $T_2 - T_1 = W_{12} \Leftrightarrow dT = \Delta W$ corresponds to the expression of the conservation of energy.

The kinetic energy and work are depending on the inertial reference frame.

Indeed, let us consider two inertial frames K and K' such that $\vec{V}_e$ is the velocity of K' relating to K, $\vec{r}$, and $\vec{r}'$, the instantaneous relative positions of the point mass respective to K and K'. From the Galileo relativity principle, we know $t = t'$, $\vec{f} = \vec{f}'$, $\vec{v} = \vec{v}' + \vec{V}_e$, and $\vec{r} = \vec{r}' + \vec{V}_e \cdot t$.

We obtain from (1.9) and (1.10):

$$dT' = dT - \vec{V}_e \cdot md\vec{v}$$
$$\Delta W' = \Delta W - \vec{V}_e \cdot \vec{f} dt$$

The inequality of the kinetic energy in the various reference frames is obviously related to different relative velocities.

The inequality of work in different inertial frames is related to differences in relative displacement, as a consequence of differences of displacement of these frames.

According to the fundamental principle, $md\vec{v} = \vec{f} dt$, it is checked that the expressions are in agreement with the conservation of energy in any reference frame K'.

$$dT' = \Delta W' \Leftrightarrow T'_2 - T'_1 = W'_{12}$$

1.1.5.2 Particular Case of Conservative Forces

In the case of a "conservative" force in the inertial reference frame K work, W_{12}, provided by the force during the displacement from location 1 to location 2 is *independent of time and path* $s_{1 \to 2}$ followed. From vector analysis, there exists a "potential energy," $U(\vec{r})$ continuous function of the location such that:

$$\vec{f}(\vec{r}) = -\frac{\partial U}{\partial \vec{r}} \Rightarrow W_{12} = -(U(\vec{r}_2) - U(\vec{r}_1)) = -\Delta U \quad (1.11)$$

Potential energy is defined with an arbitrary additive constant, for only the variation of energy ΔU has a significant physical meaning. In the inertial frame, this force corresponds to a force field, a function of the location.

In this situation, one can define a total mechanical energy function, $E = T + U$. According to the preceding results:

$$E = T_2 + U_2 = T_1 + U_1 = \text{Constant} \quad (1.12)$$

As potential energy U, total mechanical energy is defined with an arbitrary additive constant.

However, the reader should strongly protest at this stage, as the above definition of a conservative force obviously leads to a physical nonsense.

Indeed, except in the particular case of a uniform field force, the conservative character of the most general field of force exists only in the inertial reference frame. Clearly in other inertial frames it depends on time, so it is not conservative, which is not admissible because the conservation of energy is violated. In fact, this nonsense is related to the overly general mathematical definition of the state force given above, as we have seen from the second principle, the force must be independent of the observation frame. As conservative forces are created by mass, electrical charges or any material entities, and applied to other material entities, physical conservative forces depends not directly on the location of the mass point $\vec{r}_i$ but on the relative location of the two material entities interacting $\vec{r}_{ij} = \vec{r}_i - \vec{r}_j$. With this kind of dependence, the force becomes independent of the inertial observation frame. When the force is conservative, namely, $\vec{f}_{j \to i} = -\frac{\partial U}{\partial \vec{r}_{ij}}$, the total energy of the couple of entities is a constant of the movement in any inertial frame:

$$E = T_i + T_j + U = \text{constant}$$

This does not mean that the previous statement cannot be used. We have only to keep in mind it is an approximation, and the force is conservative only in one frame. For example, it is useful in the classical two-body problem in gravitational interaction, when one body has a mass negligible compared to the other (for example a satellite or a RV around the earth). In this case, the center of mass of the system is very close to that of the larger mass, and the approximation is valid to study the movement of the smaller mass.

1.1.6 Isolated System

This is a system in which no forces of external origin are applied.

In the case of a single point mass, we have $\vec{f} = 0$. Thus for any inertial observer:

$$\Rightarrow \vec{p} = \vec{p}_0; \quad \vec{l} = \vec{l}_0 \tag{1.13}$$

Linear momentum and angular momentum are constants in the movement.

For an isolated system of N points subjected to internal forces, by applying the principle of equality of action and reaction to all the combinations of two points, we obtain:

$$\vec{P} = \vec{P}_0 \Leftrightarrow \dot{\vec{P}} = \sum_i \dot{\vec{p}}_i = \sum_i \sum_{j \neq i} \vec{f}_{j \to i} = 0 \tag{1.14}$$

$$\vec{L} = \vec{L}_0 \Leftrightarrow \dot{\vec{L}} = \sum_i \dot{\vec{l}}_i = \sum_i \vec{r}_i \wedge \sum_{j \neq i} \vec{f}_{j \to i} = 0 \tag{1.15}$$

Total linear momentum and total angular momentum are constants in the movement, which is a consequence of the principle of equality of action and reaction.

1.2 Mechanics of Rigid Bodies

1.2.1 Linear Momentum and Angular Momentum

Let us consider an inertial reference frame K centered at O. A rigid body can be modeled as a system of point masses for which the relative distance between any couple of points is uniform (independent of time). We can define the total linear momentum and the total angular momentum of the body relating to K:

$$\vec{P} = \sum_i \vec{p}_i \tag{1.16}$$

$$\vec{L} = \sum_i \vec{l}_i \tag{1.17}$$

We obtain from the point mass mechanics results:

$$\dot{\vec{P}} = \sum_i \vec{f}_i = \sum_i \vec{f}_{ext,i} + \sum_i \sum_{j \neq i} \vec{f}_{j \to i} \tag{1.18}$$

where $\vec{f}_{ext,i}$ are forces external to the system and $\vec{f}_{j \to i}$ is the interior force exerted by point j on point i.

From the principle of equality, the sum of internal forces is null, which leads to:

$$\dot{\vec{P}} = \sum_i \vec{f}_{ext,i} = \vec{F}_{ext} \tag{1.19}$$

Similarly, for angular momentum,

$$\dot{\vec{L}} = \sum_i \dot{\vec{l}}_i = \sum_i \vec{r}_i \wedge \vec{f}_i = \sum_i \left[\vec{r}_i \wedge \vec{f}_{ext,i} + \vec{r}_i \wedge \sum_{j \neq i} \vec{f}_{j \to i} \right] \tag{1.20}$$

The total moment of internal forces vanishes from the principle of equality:

$$\dot{\vec{L}} = \sum_i \vec{r}_i \wedge \vec{f}_{ext,i} = \vec{M}_{ext} \tag{1.21}$$

Let us now develop (1.16) and (1.17) that characterize the inertial properties of the system. We introduce an intermediate point P having at this time an arbitrary location $\vec{R}(t)$ and motion $\vec{V}(t)$ relating to K.

$$\vec{r}_i = \vec{R} + \vec{\mathfrak{R}}_i \Rightarrow \vec{P} = \sum_i m_i \dot{\vec{R}} + \sum_i m_i \dot{\vec{\mathfrak{R}}}_i = M\dot{\vec{R}} + \sum_i m_i \dot{\vec{\mathfrak{R}}}_i \qquad (1.22)$$

$$\vec{L} = \vec{R} \wedge M\dot{\vec{R}} + \vec{R} \wedge \sum_i m_i \dot{\vec{\mathfrak{R}}}_i + \left[\sum_i m_i \vec{\mathfrak{R}}_i\right] \wedge \dot{\vec{R}} + \sum_i m_i \vec{\mathfrak{R}}_i \wedge \dot{\vec{\mathfrak{R}}}_i \qquad (1.23)$$

1.2.2 Center of Mass and Equations of Movement

We now choose the position of point P at the center of mass G of the body such that:

$$\sum_i m_i \vec{\mathfrak{R}}_i = 0 \Leftrightarrow \vec{R} = \frac{1}{M}\sum_i m_i \vec{r}_i \qquad (1.24)$$

Thus we obtain:

$$\vec{P} = M\dot{\vec{R}} \qquad (1.25)$$

$$\vec{L} = \vec{R} \wedge \vec{P} + \sum_i m_i \vec{\mathfrak{R}}_i \wedge \dot{\vec{\mathfrak{R}}}_i = \vec{R} \wedge \vec{P} + \vec{H} \qquad (1.26)$$

This relation allows us to define angular momentum of the body as the angular moment relating to the center of mass:

$$\vec{H} = \sum_i m_i \vec{\mathfrak{R}}_i \wedge \dot{\vec{\mathfrak{R}}}_i \qquad (1.27)$$

Equations (1.19), (1.21), (1.25), and (1.27) result in:

$$\dot{\vec{P}} = \frac{d(M\dot{\vec{R}})}{dt} = \vec{F}_{ext} \qquad (1.28)$$

$$\dot{\vec{L}} = \tfrac{d}{dt}(\vec{R} \wedge \vec{P} + \vec{H}) = \sum_i \vec{r}_i \wedge \vec{f}_{ext,i} = \vec{R} \wedge \vec{F}_{ext} + \sum_i \vec{\mathfrak{R}}_i \wedge \vec{f}_{ext,i} \qquad (1.29)$$

By developing the preceding relation, after simplification:

$$\frac{d\vec{H}}{dt} = \sum_i \vec{\mathfrak{R}}_i \wedge \vec{f}_{ext,i} \qquad (1.30)$$

Finally, the equations of motion of the solid for an inertial observer are:

$$\dot{\vec{P}} = \vec{F}_{ext} \qquad (1.31)$$

$$\dot{\vec{H}} = \vec{M}_{ext/G} \qquad (1.32)$$

1.2 Mechanics of Rigid Bodies

with:

$\vec{P} = M\dot{\vec{R}}$ — Linear momentum relating to K of the center of mass G together with total mass M

$\vec{H} = \sum_i m_i \vec{\Re}_i \wedge \dot{\vec{\Re}}_i$ — Angular momentum of the body relating to its center of mass G

$\vec{F}_{ext} = \sum_i \vec{f}_{ext,i}$ — Sum of external forces applied

$\vec{M}_{ext/P} = \sum_i \vec{\Re}_i \wedge \vec{f}_{ext,i}$ — Moment of exterior forces relating to center of mass G

This well-known theorem simplifies the study of motion of a system of point masses by breaking the movement into two components:

- The movement of the center of mass associated with the total mass, the linear momentum, and the sum of external forces applied.
- The movement of the system around its center of mass associated with the angular momentum and with the sum of external moments applied.

A seldom clarified property of the angular momentum $\vec{H}$ and its derivative is the fact that these quantities are invariant through changes of inertial reference frames and noninertial ones, restricted to *any accelerated translation*. Indeed, let us consider a reference frame having an arbitrary accelerated translation motion, $\vec{\gamma}_e(t)$. For the linear momentum derivative, we must add the inertia forces, $\vec{f}_{I,i} = -m_i\vec{\gamma}_e$, to external forces to take into account the accelerated movement of the observer:

$$\dot{\vec{P}} = \vec{F}_{ext} - \sum_i m_i\vec{\gamma}_e = \vec{F}_{ext} - M\vec{\gamma}_e$$

But in the expression of the derivative of angular momentum about center of mass, the sum of angular moment of the inertia forces vanishes, thanks to the definition of center of mass:

$$\dot{\vec{H}} = \vec{M}_{ext/P} - \left[\sum_i m_i\vec{\Re}_i\right] \wedge \vec{\gamma}_e = \vec{M}_{ext/P}$$

This invariance applies particularly to a nonrotating frame centered at the center of mass G of the body, for which $\dot{\vec{P}} = 0$.

1.2.3 Eulerian Frames

The properties established up to now apply in fact to any system of point masses since we have yet to introduce the assumption of a rigid body.

Let us use this assumption now. We associate the solid with a rigidly fixed reference frame E, whose origin is at the center of mass G. The paternity of this representation has been attributed to Leonard Euler, and they are known as the Eulerian frame. The instantaneous location of the solid relating to the inertial frame K is completely defined by the position of the center of mass G and orientation of the axes of the Eulerian frame E.

The elementary movement of the solid relating to K during the time step "dt" consists of a translation $d\vec{R}$ of the center of mass and of an elementary rotation $d\varphi$ of the frame around an instantaneous axis of unit vector $\vec{n}$. The elementary displacement and the inertial velocity of any point mass i of the solid are:

$$d\vec{r}_i = d\vec{R} + d\varphi \vec{n} \wedge \vec{\Re}_i$$

$$\dot{\vec{r}}_i = \dot{\vec{R}} + \vec{\Omega} \wedge \vec{\Re}_i$$

where $\vec{\Omega} = \frac{d\varphi}{dt}\vec{n}$ is the instantaneous angular velocity.

Under these conditions, the angular moment of the solid is written as:

$$\vec{H} = \sum m_i \vec{\Re}_i \wedge \left[\vec{\Omega} \wedge \vec{\Re}_i \right] \tag{1.33}$$

$$\vec{H} = \vec{\Omega} \sum_i m_i \vec{\Re}_i^2 - \sum_i \left[\vec{\Re}_i \cdot \vec{\Omega} \right] \vec{\Re}_i \tag{1.34}$$

1.2.3.1 Expression of the Angular Momentum

The preceding expression of the angular momentum $\vec{H}$ is very inconvenient to treat in the inertial frame K, because $\vec{\Re}_i$ are rotating vectors. On the other hand, it is definitely easier to develop the components of $\vec{H}$ in an Eulerian frame E, using the components of $\vec{\Omega}$ and the coordinates of points of the solid relating to E (the last being rigidly fixed to E, their components are independent of time). We obtain from expression (1.34):

$$\begin{bmatrix} H_{xE} \\ H_{yE} \\ H_{zE} \end{bmatrix} = \begin{bmatrix} \sum_i m_i \left(y_i^2 + z_i^2 \right) & -\sum_i m_i x_i y_i & -\sum_i m_i x_i z_i \\ -\sum_i m_i x_i y_i & \sum_i m_i \left(x_i^2 + z_i^2 \right) & -\sum_i m_i y_i z_i \\ -\sum_i m_i x_i z_i & -\sum_i m_i y_i z_i & \sum_i m_i \left(x_i^2 + y_i^2 \right) \end{bmatrix} \begin{bmatrix} \Omega_{xE} \\ \Omega_{yE} \\ \Omega_{zE} \end{bmatrix} \tag{1.35}$$

The matrix on the right side is referred to the matrix of inertia of the solid relating to center of mass G and to the Eulerian frame E.

1.2.3.2 Derivative of the Angular Momentum

We have seen that expressions of $\vec{H}$ and $\dot{\vec{H}}$ are independent of the movement of the origin G of the reference frame. On the other hand, the relative derivative

1.2 Mechanics of Rigid Bodies

$\left(\dfrac{d\vec{H}}{dt}\right)_E = \begin{bmatrix} \dot{H}_{xE} \\ \dot{H}_{yE} \\ \dot{H}_{zE} \end{bmatrix}$ of $\vec{H}$ characterizes the apparent movement of $\vec{H}$ relating to E and depends on the rotation movement of E. To connect this relative derivative to the moment of external forces, we must determine the expression of the inertial derivative $\dot{\vec{H}}$ in the E frame, hence to account for the driving rotation movement of E:

$$\dot{\vec{H}} = \left(\dfrac{d\vec{H}}{dt}\right)_E + \vec{\Omega} \wedge \vec{H} = \vec{M}_{ext/G} \Leftrightarrow \left(\dfrac{d\vec{H}}{dt}\right)_E = \vec{M}_{ext/G} - \vec{\Omega} \wedge \vec{H} \quad (1.36)$$

Such is the equation of the rotation motion of the solid in an Eulerian frame E. By noting $\vec{\vec{I}}$, the linear operator corresponding to matrix (1.35), this equation becomes:

$$\vec{\vec{I}}\vec{\Omega} = \vec{M}_{ext/G} - \vec{\Omega} \wedge \vec{\vec{I}}\vec{\Omega} \quad (1.37)$$

Operator $\vec{\vec{I}}$ is represented in E by a symmetrical real matrix:

$$\begin{bmatrix} I_{xx} & I_{xy} & I_{xz} \\ I_{xy} & I_{yy} & I_{yz} \\ I_{zx} & I_{yz} & I_{zz} \end{bmatrix} \quad (1.38)$$

Mathematicians will want to check that it belongs to the hermitical matrix group and that it admits three real characteristic roots (eigenvalues), corresponding to three characteristic vectors mutually orthogonal[BAS]. The expression of the matrix in the Eulerian frame built from these three characteristic vectors is diagonal. The diagonal terms are the preceding characteristic roots and represent the moments of inertia of the solid. The axes are called the principal axes of the solid.

$$\begin{bmatrix} I_x & 0 & 0 \\ 0 & I_y & 0 \\ 0 & 0 & I_z \end{bmatrix} \quad (1.39)$$

The operator $\vec{\vec{I}}$ can also be mathematically represented by a second order symmetrical mixed tensor, called the inertia tensor of the solid. Both representations are mathematically equivalent. Each representation has nine components and gives the same results according to the correspondence (operator) between $\vec{\Omega}$ and $\vec{H}$. In one case, we apply matrix-products rules and in the other case index-contraction rules of tensor product. An important practical aspect is the behavior of the elements or the components of this operator through a change of orthogonal frame E fixed to the solid. Let us consider E′, a new frame derived from E by a rotation around the center of mass. Let us label [R] the corresponding rotation matrix, which transforms the components [$H_{E'}$] and [$\Omega_{E'}$] of $\vec{\Omega}$ and $\vec{H}$ in E′ (new frame) into their components in E (old frame). While [I] and [I′] are the expression of operator $\vec{\vec{I}}$ with respect to E and E′, we obtain:

$$[H_E] = [R][H_{E'}] = [I][\Omega_E] = [I][R][\Omega_{E'}]$$
$$[H_{E'}] = [R^{-1}][I][R][\Omega_{E'}] \tag{1.40}$$

As module of vectors is invariant through a rotation, inverse rotation matrix $[R]^{-1}$ is equal to transpose matrix $[R]^T$ [see (6.7) in Sect. 6.1] and we obtain:

$$[I'] = [R]^T[I][R] \tag{1.41}$$

Such is the transformation rule of the inertia matrix through a change of Eulerian frame (rotation).

Remark: In full rigor $\vec{\Omega}$ is not an ordinary vector but a skew symmetric second order tensor (see Sect. 6.1.1); however, it has the same rule of transformation as ordinary vectors through rotations. It is not the case for any kind of change, for example, symmetries. In that case one must use tensor's change rules.

1.2.3.3 Kinetic Energy of the Solid

Relating to inertial frame, one obtains kinetic energy by addition of kinetic energy of all the point mass constituting the solid:

$$T = \sum_i \frac{1}{2} m_i [\vec{V} + \vec{\Omega} \wedge \vec{\mathfrak{R}}_i]^2 \tag{1.42}$$

The origin of the Eulerian frame is at the center of mass, so we have:

$$T = \frac{1}{2} M \vec{V}^2 + \sum_i \frac{1}{2} m_i \left(\vec{\Omega} \wedge \vec{\mathfrak{R}}_i \right)^2 \tag{1.43}$$

By developing the $\vec{\Omega}$ term in the Eulerian frame E we obtain:

$$T = \frac{1}{2} M V^2 + \frac{1}{2} \vec{\Omega} \cdot \vec{H} \tag{1.44}$$

1.2.3.4 Gyroscopic Moment

The theorem of angular momentum related the angular momentum of a solid to the action of an external moment. From the principle of equality of action and reaction, the solid applies a reaction moment (inertial moment) to the system creating the external moment, equal and opposed to this external moment. From the theorem of angular momentum to impart an inertial rotation rate $\vec{\Omega}_e$ to the angular momentum $\vec{H}$ of a solid, we must apply to it a moment:

1.2 Mechanics of Rigid Bodies

$$\vec{M} = \dot{\vec{H}} = \vec{\Omega}_e \wedge \vec{H}$$

Reciprocally, the solid develops against us a gyroscopic reaction moment:

$$\vec{M}_g = -\vec{\Omega}_e \wedge \vec{H}$$

The gyroscopic reaction moment has a modulus proportional to $\|\vec{H}\|$ and $\|\vec{\Omega}_e\|$, and a direction orthogonal to $\vec{H}$ and $\vec{\Omega}_e$.

A solid with a high angular momentum has a high gyroscopic stiffness. It has a strong capacity of resistance to external moments with respect to rotation of its angular momentum.

Chapter 2
Topography and Gravitation

Thanks to measurements on board satellite and external measurements of their trajectory, terrestrial topography and gravitation field are now modeled with great accuracy (the most exact models are confidential, because they reveal the accuracy of medium- and long-range ballistic missiles). However, open data has more than sufficient accuracy for reentry analyses. The models are derived from knowledge of the distribution of the terrestrial masses and geodesy.

Model also exists for planets, in particular for Mars, for which cartography was recently updated, thanks to measurements from the NASA probe Mars Global Surveyor.

2.1 The Geodetic Frame of Reference

The needs are:

– a universal reference to define the positions of the various points of the trajectory relating to the planet and
– an accurate model of the gravitation field.

Geodetic modeling is appealing:

- with geoids, which is by definition the equipotential surface of the apparent gravity field, which coincides on average with the mean level of the seas (the apparent term indicated for an observer rotating with the planet) and
- with an ellipsoid of revolution, that is, analytical second-order approximation of the geoids (least-square approximate of the geoids). The set including geoids model, ellipsoid, and gravity field constitutes the geodetic frame of reference.

The same standard of modeling exists for planets. From current nomenclature, the geoids for a planet correspond to areoids. In the case of a planet without oceans, areoid must be defined arbitrarily, for example, in the vicinity of the mean level of ground.

2.1.1 Coordinates of a Point Relating to the Earth

The Cartesian geocentric coordinates (x, y, and z) of a point P are defined in a trirectangular reference frame (Cx, Cy, and Cz) (Fig. 2.1) having its origin in the center (C) of the ellipsoid (center of mass of the planet). Axis Cx is defined as the intersection of the equatorial plane with the half-plane meridian origin, and Cz is the geographic polar axis (northbound).

Geocentric spherical coordinates are connected to Cartesian coordinates by the following relationships:

$$x = r \cdot \cos \psi \cdot \cos \lambda$$
$$y = r \cdot \cos \psi \cdot \sin \lambda$$
$$z = r \cdot \sin \psi$$

r module of the radius vector $\vec{r}$ between the origin C and the point P
ψ *geocentric latitude* (angle between the radius vector $\vec{r}$ and the equatorial plane, positive for the northern hemisphere)
λ *longitude* (angle of the meridian half-plane including P with the meridian half-plane origin, positive toward the east).

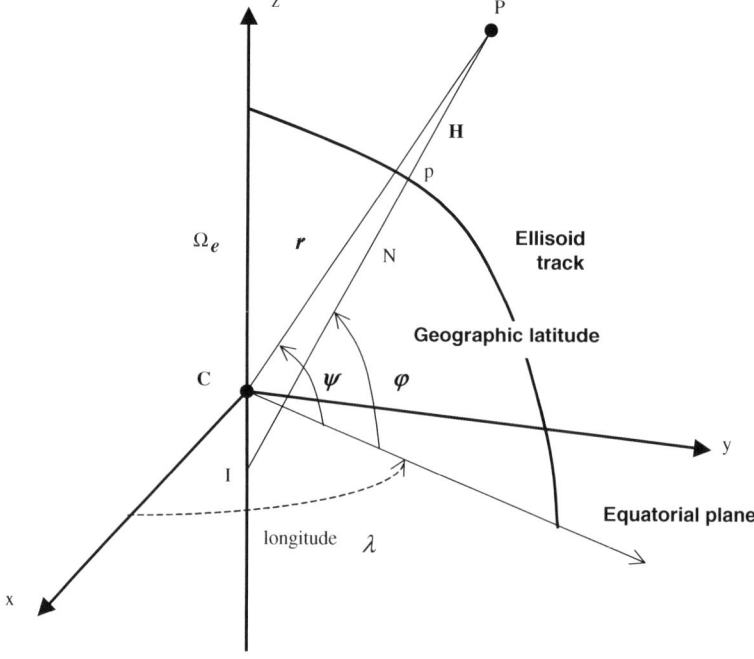

Fig. 2.1 Geocentric and geographical coordinates

2.1 The Geodetic Frame of Reference

The geographical coordinates are defined by:

- longitude, λ, identical to geocentric longitude in the case of an ellipsoid of revolution,
- geographical latitude, φ, angle between the equatorial plane and the normal N to the ellipsoid through the point P, and
- height, H, distance from P to its projection p on the ellipsoid following the normal.

Astronomical coordinates, latitude, longitude, and altitude have definitions similar to the preceding one while replacing the ellipsoid with geoids and the normal with physical direction of vertical (along gravity).

The same definitions hold for planets. According to the current nomenclature, geocentric and geographical terms are replaced by areocentric and aerographic.

2.1.2 The Geodetic Systems

Among the main terrestrial systems used are:

- Hayford ellipsoid, often referred to as "International"(1924),
- Union Astronomique Internationale ellipsoid (UAI),
- Europe 50 ellipsoid, European coverage, and
- currently, however, World Geodetic System 84 (WGS 84) is universally used, which is the reference system used by NAVSTAR.

Characteristics of the system WGS 84 [GPS] are:

- Parameters of the ellipsoid:

    Semimajor axis (equatorial radius), a = 6378137 m
    Flatness, $f = (a - b)/a = 1/298.257223563$
    Semiminor axis (polar radius), $b = a.(1 - f) = 6356752.31425$ m
    Focus location c, $c^2 = a^2 - b^2$
    Eccentricity, $e = \frac{c}{a} = 0.0818191908426$

- Model of gravity (Geodetic Reference System 1980):

    Center of mass G located at origin C of the ellipsoid
    Terrestrial rotation velocity, $\boldsymbol{\omega = 7.292115167 \cdot 10^{-5}}$ **rad/s**
    Geocentric gravitational constant, $\mu = 3.986005 \cdot 10^{14}$ m$^3 \cdot$ s^{-2}
    Coefficient of the first nonspherical term of the potential, $\bar{c}_{2,0} = -0.48416685 \cdot 10^{-3}$

A recent model for the planet Mars uses:

- Parameters of the ellipsoid:

    Semimajor axis (equatorial), a = 3393940 m
    Semiminor axis (polar), b = 3376790 m

- Model of gravity:
  Center of mass G located at the origin C of the ellipsoid
  Rotation velocity, $\omega = 7.08821808 \cdot 10^{-5} rad/s$
  Geocentric gravitational constant, $\mu = 0.428284 \cdot 10^{14} m^3 \cdot s^{-2}$
  Coefficient of the first nonspherical term of the potential, $\bar{c}_{2,0} = -0.8767481910^{-3}$

2.1.3 Calculation of the Geographical Latitude and Height

The numerical calculation of trajectories frequently uses Cartesian coordinates, unlike data on launching points and targets in terms of geographic coordinates. It is thus necessary to convert one set of data into the other; i.e., (λ, φ, H) into (x, y, and z) and reciprocally. The direct problem is easy, and one obtains the relations starting from Fig. 2.1:

$$x = (N + H) \cdot \cos\varphi \cdot \cos\lambda$$
$$y = (N + H) \cdot \cos\varphi \cdot \sin\lambda$$
$$z = \left[N \cdot (1 - e^2) + H\right] \cdot \sin\varphi$$

where N is the length of the great normal in "p" to the ellipsoid (measured between its intersection "I" with the axis Cz and orthogonal projection "p" of P on ellipsoid):

$$N = Ip = \frac{a}{\sqrt{1 - (e \cdot \sin\varphi)^2}}$$

Starting from the preceding results for the geocentric latitude and introducing $D = \sqrt{x^2 + y^2}$, one obtains:

$$\tan\psi = \frac{z}{D}$$

And for the geographical latitude:

$$\tan\varphi = \frac{z}{D} + e^2 \cdot \frac{N}{D} \cdot \sin\varphi$$

This requires solution of a nonlinear equation:

$$\tan\varphi = \tan\psi + \frac{e^2 \cdot a \cdot \sin\varphi}{D \cdot \sqrt{1 - (e \cdot \sin\varphi)^2}}$$

We can calculate φ by successive approximation by replacing φ in the second term with its approximation of order n-1:

2.2 The Terrestrial Field of Gravitation

$$\tan \varphi_n = \tan \psi + \frac{e^2 \cdot a \cdot \sin \varphi_{n-1}}{D \cdot \sqrt{1 - (e \cdot \sin \varphi_{n-1})^2}}$$

In practice, a good approximation is obtained after two iterations by using $\varphi_0 = \psi$ as the initial value.

The calculation of height is determined from the geographical latitude φ:

$$H = \frac{D}{\cos \varphi} - N = \frac{D}{\cos \varphi} - \frac{a}{\sqrt{1 - (e \cdot \sin \varphi)^2}}$$

2.2 The Terrestrial Field of Gravitation

The corrective term in the potential of the terrestrial gravitational field, taking into account the nonspherical mass distribution, was modeled in the form of a development in spherical harmonics,

$$R(\cos\theta, \lambda, r) = \sum_{n=2}^{\infty} \sum_{m=0}^{\infty} \frac{\mu \cdot a^n}{r^{n+1}} \cdot \overline{P}_{n,m}(\cos\theta) \cdot [\overline{c}_{n,m} \cdot \cos(m \cdot \lambda) + \overline{s}_{n,m} \cdot \sin(m \cdot \lambda)]$$

with,

$$\theta = \frac{\pi}{2} - \psi \quad \text{(Geocentric colatitudes)}$$

$$\overline{P}_{n,m}(\cos\theta) = \sqrt{(2n+1)\frac{(n-m)!}{(n+m)!}} \cdot P_{n,m}(\cos\theta)$$

$$P_{n,m}(\cos\theta) = \frac{(1-\cos^2\theta)^{\frac{m}{2}}}{2^n n!} \left(\frac{\partial}{\partial \cos\theta}\right)^{n+m} \left(\cos^2\theta - 1\right)^n$$

    a Semimajor axis of the ellipsoid

In practice, the dominant term is $\overline{c}_{2,0}$, which is approximately 1000 times larger than the following terms. This results in,

$$R \approx \frac{\mu \cdot a^2}{r^3} \cdot \overline{c}_{2,0} \cdot \frac{\sqrt{5}}{2} \cdot (3 \cdot \sin^2 \psi - 1)$$

The potential of the field due to the distribution of the terrestrial masses is thus written as a function of spherical geocentric coordinates,

$$U(r, \psi) = \frac{\mu}{r} + \frac{\mu \cdot a^2}{r^3} \cdot \overline{c}_{2,0} \cdot \frac{\sqrt{5}}{2} \cdot (3 \cdot \sin^2 \psi - 1)$$

From this expression of U, we obtain the radial and orthoradial components of the gravitation field $\vec{\varphi}$,

$$\vec{\varphi} = \vec{\nabla} U = \frac{\partial U}{\partial r} \cdot \frac{\vec{r}}{r} + \frac{1}{r} \cdot \frac{\partial U}{\partial \psi} \cdot \vec{t}$$

where,

$$\vec{t} = -\sin \psi \cdot (\cos \lambda \cdot \vec{i} + \sin \lambda \cdot \vec{j}) + \cos \psi \cdot \vec{k}$$

That is to say,

$$\varphi_r = \frac{\partial U}{\partial r} = -\frac{\mu}{r^2} - \frac{3 \cdot \mu \cdot a^2}{r^4} \cdot \bar{c}_{2,0} \cdot \frac{\sqrt{5}}{2} \cdot (3 \cdot \sin^2 \psi - 1)$$

$$\varphi_\psi = \frac{1}{r} \frac{\partial U}{\partial \psi} = \frac{\mu \cdot a^2}{r^4} \cdot \bar{c}_{2,0} \cdot 3\sqrt{5} \cdot \sin \psi \cdot \cos \psi$$

- This model also applies to planets, by using the specific values of the coefficients.
- It applies to the pure gravitational field. To obtain the field of gravity $\vec{g}$ measured by an observer fixed to revolving planet, it is necessary to add to potential U the centrifugal force $\Phi = \frac{1}{2}\omega^2 (r \cos \psi)^2$. The resulting potential $U + \Phi$ is called the potential of normal gravity as it corresponds in the vicinity of the ellipsoid to the plumb-line direction, normal with the ellipsoid by construction [GPS].

Chapter 3
Models of Atmosphere

The essential problem of dynamic reentry study is related to modeling aerodynamic effects induced by the flow around the vehicle. To approach it, we must first have a model of earth's atmosphere.

3.1 Main Parameters and Hypotheses

A precise model of the ambient gaseous medium as a function of altitude is needed to calculate the flow. The essential parameters are:

- Density ρ
- Speed of sound a
- Dynamic viscosity μ, or kinematics viscosity $\eta = \frac{\mu}{\rho}$
- Mean free path of molecules λ
- Absolute temperature T
- Static pressure p
- Relative humidity, wind
- Specific heat: at constant pressure c_p, at constant volume c_v
- Chemical composition

Geometrical altitude Z, used in the models of atmosphere, is generally the distance, measured along the field line of the normal gravity, between the point considered and the mean sea level (this level corresponds to the surface of geoids). This altitude Z is in practice very close to:

$$Z \approx H - H_g$$

where H and H_g are, respectively, the geographical heights of the point and of the geoids relating to the ellipsoid.

Most of models of standard atmosphere use the following assumptions:

- Standard equation of state of perfect gas $p = \rho \cdot \frac{R}{M} \cdot T$
- Atmosphere in vertical hydrostatic balance $\frac{\partial p}{\partial Z} = -\rho \cdot g$

The preceding equation relates to an observer fixed to the revolving planet and "g" indicates the "normal field of gravity" that includes the influence of the centrifugal force.

3.2 The Isothermal Exponential Model

Simplification of density versus altitude leads to analytical solutions for the equations of the motion during the reentry,

$$\rho = \rho_s \cdot e^{\frac{Z}{H_g}}$$

Parameters ρ_s and H_g can be:

- determined numerically by smoothing a more exact standard atmosphere,
- derived from standard ground values using preceding assumptions.

In the first case, one finds, for example, in the literature $\rho_s \approx 1.39$ kg/m^3, $H_g \approx 7000$ m

In the second case, one obtains by combining the above two assumptions:

$$\frac{dp}{p} \approx -\frac{M \cdot g_s}{R \cdot T_s} \cdot dZ = -H_g$$

$$\rho_s = \frac{M}{R} \cdot \frac{p_s}{T_s}$$

By using the properties of the standard atmosphere at sea level,

$$p_s = 101325 \text{ pa}$$
$$T_s = 288.15 \text{ K}$$
$$M \approx 28.9644 \cdot 10^{-3} \text{ kg}$$
$$g_s = 9.80616 \text{ m.s}^{-2}$$

We obtain the values of the parameters: $\begin{array}{l} \rho_s = 1.225 \text{ kg} \cdot \text{m}^{-3} \\ H_g = 8435 \text{ m} \end{array}$

A comparison of the two models with a standard atmosphere is given in Fig. 3.4.

3.3 Standard Models of Earth's Atmosphere

The US standard model universally used was built from weather compilations in the northern hemisphere. The GRAM model from NASA is now supplanting it. The US standard was updated on several occasions, and we will describe the US66 version.

This version [USS] includes 14 average atmospheres corresponding to points located in the 15°–75° range of northern latitudes, for various seasons (spring/fall, summer, and winter) and for altitudes lower than 120 km. The atmosphere n°6 that corresponds to 45° latitude, spring/fall, is very close to the US62 atmosphere.

The model uses a geopotential altitude "h," which is defined as,

$$g_{ref} \cdot h = \int_0^Z g(z) dz$$

where $g(z)$ represents an approximate value of the normal gravity field at the latitude considered and at geometrical altitude z:

$$g(z) \approx g(0) \cdot \frac{R_t^2}{(R_t + z)^2}$$

In these expressions, R_t and $g(0)$ are functions of the latitude and indicate the terrestrial radius and the apparent field of gravity at the mean sea level, g_{ref} indicates a constant reference value of the normal gravity field.

Atmosphere profile is determined by the ground pressure and the law of temperature as a function of geopotential altitude. Indeed, by definition of geopotential altitude,

$$dp = -\rho \cdot g \cdot dz = -\rho \cdot g_{ref} \cdot dh$$

Then combining with the equation of state,

$$\frac{dp}{p} = -\frac{M \cdot g_{ref}}{R \cdot T} dh$$

We finally obtain:

$$p(h) = p(0) \cdot e^{-\frac{M \cdot g_{ref}}{R} \int_0^h \frac{dh'}{T(h')}}$$

The model uses a continuous law of temperature, linear by section of geopotential altitude:

– Index of section $b \in [1, N]$
– For the section $h_b \leq h < h_{b+1}$, $T = T_b + A_b \cdot (h - h_b)$

This gives for $A_b \neq 0$,

$$p(h) = p(h_b) \cdot \left[\frac{T}{T_b}\right]^{-\frac{M \cdot g_{ref}}{R \cdot A_b}}$$

And for $A_b = 0$

$$p(h) = p(h_b) \cdot e^{-\frac{M \cdot g_{ref}}{R \cdot T_b} \cdot (h - h_b)}$$

Finally, using density from the equation of state:

$$\rho(h) = \frac{M}{R} \cdot \frac{p(h)}{T(h)}$$

Constants of the model are:

$$M = 28.9644 \cdot 10^{-3} \text{ kg} \cdot \text{mole}^{-1}$$
$$g_{ref} = 9.80665 \text{ m} \cdot \text{s}^{-2}$$

And for the atmosphere n°6, 45° north, spring/fall, the parameters are,

$$g(0) = 9.80655 \text{ m} \cdot \text{s}^{-2}$$
$$R_t = 6356360 \text{ m}$$
$$p(0) = 101325. \text{ Pa}$$

Table 3.1 gives geopotential altitudes and the temperatures at beginning of sections.

Table 3.1 Geopotential altitudes and the temperatures at beginning of sections

Index of section (b)	Altitudes h_b (m)	Gradients A_b (°K/m)	Temperatures T_b (°K)
1	0.	$-6.5 \cdot 10^{-3}$	288.15
2	11000.	0.	216.65
3	20000.	$+1 \cdot 10^{-3}$	216.65
4	32000.	$+2.8 \cdot 10^{-3}$	228.65
5	47000.	0.	270.65
6	52000.	$-2 \cdot 10^{-3}$	270.65
7	61000.	$-4 \cdot 10^{-3}$	252.65
8	69000.	$-3 \cdot 10^{-3}$	220.65
9	79000.	0.	190.65
10	90000.	$+2 \cdot 10^{-3}$	190.65
11	100000.	$+4.36 \cdot 10^{-3}$	210.65
12	110000.	$+16.4596 \cdot 10^{-3}$	254.25
13	117776.		382.24

3.3 Standard Models of Earth's Atmosphere

Derived parameters:

- Speed of sound

$$a = \sqrt{\gamma \cdot \frac{R}{M} \cdot T},$$

with $\gamma = \frac{c_p}{c_v} = 1.4$ r $= \frac{R}{M} = 287.053 \, \text{J} \cdot \text{kg}^{-1} \cdot \text{K}^{-1}$

- Mean free path

$$\lambda = \frac{1}{\sqrt{2} \cdot \pi \cdot N_A \cdot \sigma^2} \cdot R \cdot \frac{T}{p},$$

with the average collision diameter for air, $\sigma = 3.65 \cdot 10^{-10}$ m

- Dynamic viscosity (Sutherland formula)

$$\mu = \frac{\beta \cdot T^{\frac{3}{2}}}{T + S}$$

with

$$\beta = 1.458 \cdot 10^{-6} \, \text{kg} \cdot \text{s}^{-1} \cdot \text{m}^{-1} \cdot \text{K}^{-\frac{1}{2}}, \quad S = 110.4 \, \text{K}$$

Figures 3.1–3.6 give the evolution according to geometrical altitude of the principal parameters of the atmosphere US66 n°6, 45° north, spring/fall, together with a comparison between standard density and the exponential models.

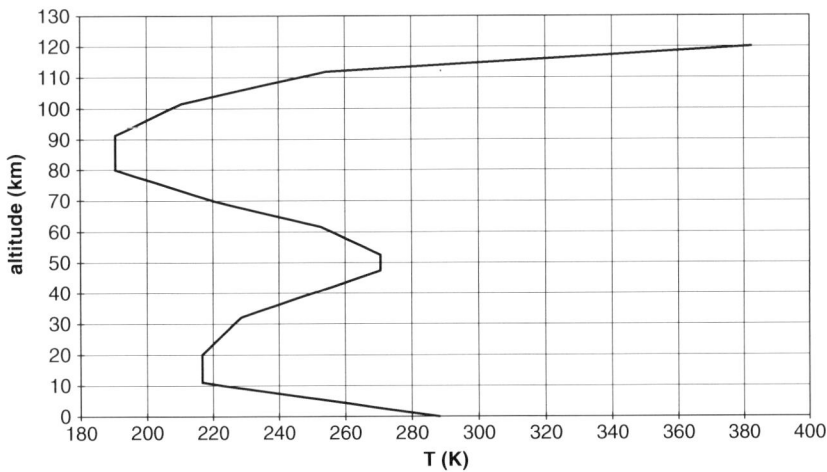

Fig. 3.1 US66 Standard atmosphere, temperature

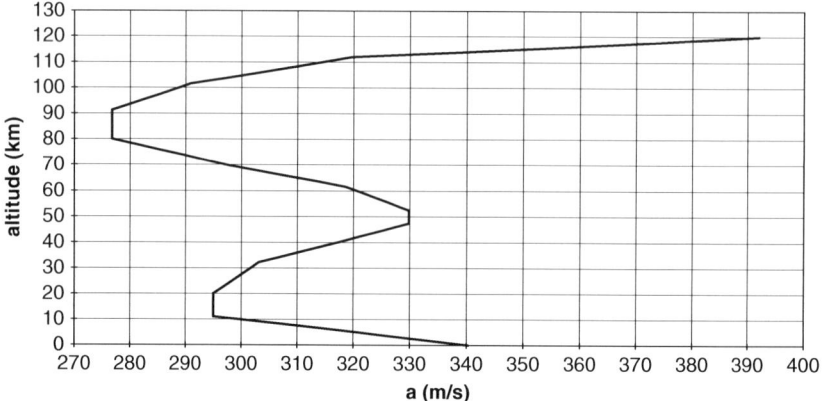

Fig. 3.2 US66 Standard atmosphere, speed of sound

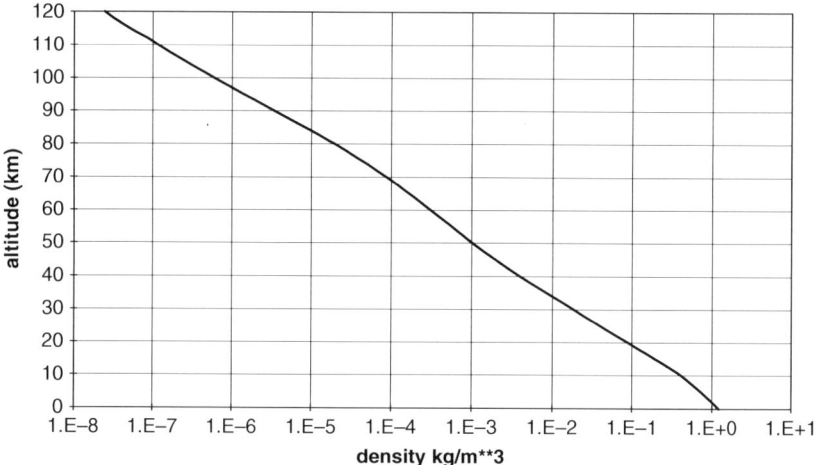

Fig. 3.3 US66 Standard atmosphere, density

3.4 Martian Models

Thanks to quest for extra terrestrial life, our neighbor was visited by numerous probes. It results its topography, and atmosphere are the best known among the solar system planets. It seems it is also the only planet suitable for future human landing.

The equatorial radius of Mars is 3394 km, roughly half the terrestrial radius. The acceleration of gravity at the Martian equator is 3.718 m/s^2, 38% of the terrestrial value 9.798 m/s^2.

3.4 Martian Models

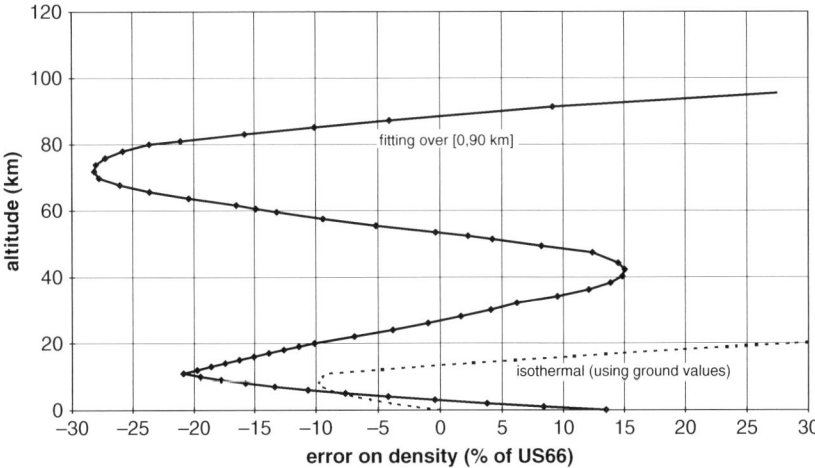

Fig. 3.4 Comparison of exponential approximations and US66

The earth's atmosphere is primarily composed of 78% nitrogen N_2 and 21% oxygen O_2 of average molar mass M ≈ **29 10⁻³ kg** mole. The lower layer of the atmosphere, below 11 km, includes a significant percentage of H_2O as vapor, liquid, or solid. The average profile of the atmosphere in a given place is related primarily to its latitude; it evolves according to the earth's angle relating to the sun during the terrestrial year following the rhythm of seasons. The presence of water in liquid form on surface (oceans) complicates singularly predictive models of the atmosphere.

Mars atmosphere is primarily composed of 95% CO_2 and 3% N_2 of average molar mass M ≈ **43 10⁻³ kg** mole. Taking into account the extremely low tem-

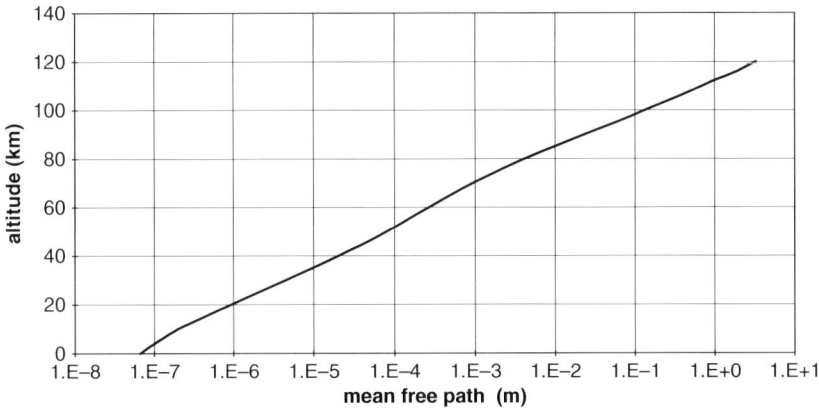

Fig. 3.5 US66 Standard atmosphere, mean free path

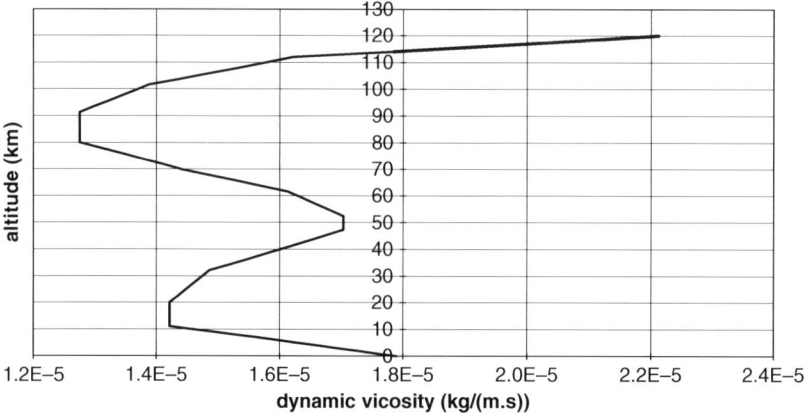

Fig. 3.6 US66 Standard atmosphere, dynamic viscosity

perature, close to 200 K at ground level, the H_2O molecule exists only in solid form. Atmospheric pressure and density at ground level are close to $500\,\text{N/m}^2$ and $0.015\,\textbf{kg/m}^3$, respectively. Like the earth's atmosphere, the Martian mean atmosphere is function of the latitude and longitude, and it evolves according to location relating to the sun during the Martian year, which corresponds to 24 terrestrial months.

The absence of liquid water simplifies heat exchange and modeling of general heat/fluid circulation in the atmosphere, thus allowing prediction of mean values of parameters and dispersions. However, the atmosphere is subjected to violent winds and dust storms that strongly modify the temperature and density profiles compared to the mean model.

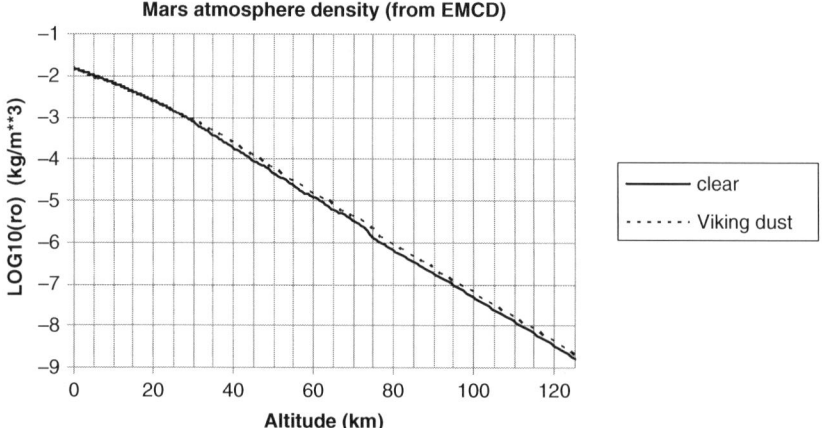

Fig. 3.7 EMCD Mars atmosphere model

3.4 Martian Models

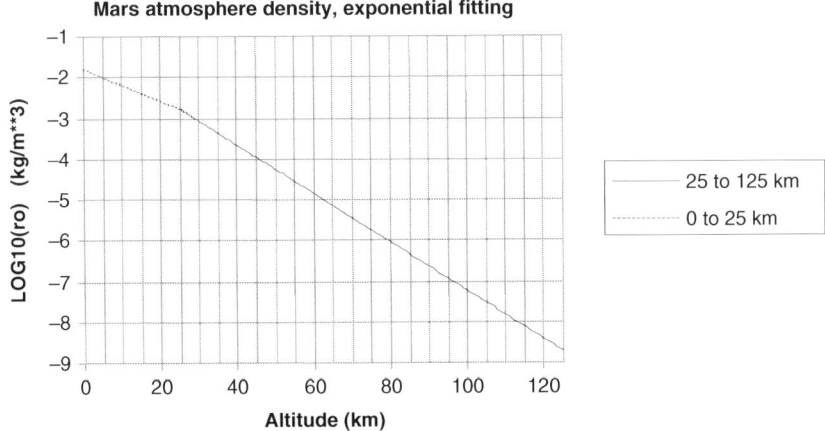

Fig. 3.8 Exponential Mars atmosphere

Table 3.2 The parameters of exponential fittings

Altitudes	ρ_S (kg.m^{-3})	H_g (km)
0–25 km	0.0159	11.049
25–125 km	0.0525	7.295

There are several models used for the prediction of the Martian atmosphere. Explorations from NASA, including the Lander "Mars Pathfinder" (1997) successfully used the Mars GRAM [GRA] model. A more recent model, "European Martian Climate Database" [EMC], was developed by Europeans for their needs. In this model, the annual Martian cycle is divided into 12 seasons of duration equal to two terrestrial months and the Martian day in 12 periods close to two terrestrial hours. It includes four scenarios corresponding to assumptions about dust content: "low dust," "Viking dust," and two dust storm scenarios.

Evolution of the density between 0 and 125 km for two profiles resulting from this model is represented in Fig. 3.7. From these two profiles, Fig. 3.8 shows exponential smoothing for two altitude sections, 0–25 Km and 25–125 Km. The parameters of these exponential fittings are given in Table 3.2.

Chapter 4
Aerodynamics

Reentry vehicles of medium- and long-range ballistic missiles (approximately 3000–12000 km) enter the atmosphere with velocities from 3500 to 7500 m/s. Capsules and space probes have even higher maximum entry velocities, for example, 11 km/s for the Apollo capsule returning from a lunar mission. These speeds are much higher than the local speed of sound (320 m/s at 60 km) and correspond to Mach numbers from 15 to 35 ("hypersonic"). As shown in Fig. 4.1, air disturbances cannot propagate upstream of the vehicle and remain inside a strongly compressed gas layer ("shock layer"). The surface of the shock wave, strongly curved in the vicinity of nosetip, becomes gradually a Mach cone downstream on the vehicle.

In addition to the local properties of air at rest and relative velocity, flow fields and loads are determined by the external geometry and possibly by the properties of the wall (roughness, catalytic capacity, chemical composition, temperature, and pyrolize).

Wind tunnels provide relative velocity by subjecting a motionless vehicle to an air flow, thereby introducing complication of non-uniform ambient conditions.

Characteristics of the incidental flow (or "infinite upstream conditions") are defined by:

1) The velocity relating to air in the vehicle frame: This frame is represented in Fig. 4.2 for the most general configuration of "flyer" admitting a symmetry plane ("pitch" plane). For geometry having an axis of symmetry, the natural choice of axis OX is along this "roll" axis, and orientation of the pitch plane can be chosen arbitrarily around it. In this frame, the components of velocity relative to air $\vec{V}_R = [u, v, w]$ are:

$$u = V_R \cdot \cos \beta \cdot \cos \alpha$$
$$v = V_R \cdot \sin \beta$$
$$w = V_R \cdot \cos \beta \cdot \sin \alpha$$

The incidence angle α is defined between the roll axis and the projection of the relative velocity vector on the pitch plane (positive when w is positive, ranging from

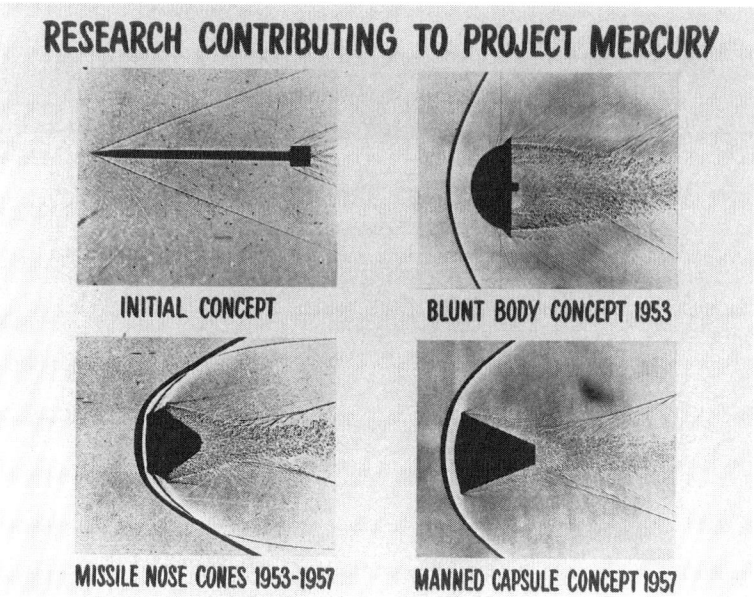

Fig. 4.1 Shock layer of hypersonic vehicle (NASA picture)

−180° to 180°). The sideslip angle β is the angle between the relative velocity vector and its projection on the pitch plane (positive when v is positive, ranging from −90° to 90°).

For bodies of revolution, it is convenient to express a total angle of attack $\bar{\alpha}$ and a windward meridian ϕ_w which can be defined as:

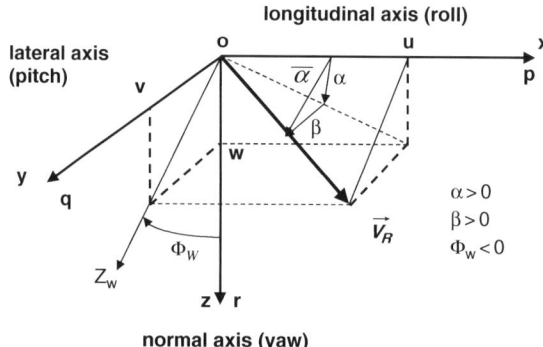

Fig. 4.2 Definition of the vehicle reference frame

$$u = V_R \cos \overline{\alpha}$$
$$v = -V_R \cdot \sin \overline{\alpha} \cdot \sin \phi_w$$
$$w = V_R \cdot \sin \overline{\alpha} \cdot \cos \phi_w$$

Total angle of attack is by definition $0 \leq \overline{\alpha} \leq 180°$ and windward meridian varies from $0° \leq \phi_w < 360°$

$$i = \overline{\alpha} \Leftrightarrow \cos \overline{\alpha} = \cos \beta \cdot \cos \alpha$$

For small angles, we have approximately:

$$i = \overline{\alpha} \approx \sqrt{\alpha^2 + \beta^2}$$

The air velocity relating to the vehicle is noted $\vec{V}_\infty$, it is such that $\vec{V}_\infty = -\vec{V}_R$. In the vehicle frame, components of the vector "angular rate" of the vehicle are $\vec{\Omega} = [p\ q\ r]$

2) Properties "at rest" of the upstream air:

  - In free flight the current altitude and a standard model of atmosphere, or a measured profile (case of an experimental flight)
  - In a wind tunnel experiment the thermodynamic state of the incidental air upstream of the model (not to be confused with "generating conditions")
  - In digital simulation one of the three possibilities according to studied case's

3) The situations described above apply to steady flow conditions such as a model at rest in a wind tunnel and lead to the measurement or the calculation of the "static" loads. In the case of a real free flight, the relative movement is an accelerated translation of the center of mass, combined with rotation of the vehicle. Instantaneous aerodynamic loads are "nonstationary" and thus depend on linear accelerations and angular velocities.

4.1 Aerodynamic Coefficients

The most natural reference frame is the vehicle frame. The aerodynamic loads are calculated or measured along "longitudinal," "transverse," and "normal" axes (Ox, Oy, and Oz) then converted into dimensionless aerodynamic coefficients by "reference" forces and moments.

The main reference parameters are:

- Dynamic pressure

$$\overline{q} = \frac{1}{2} \rho_\infty \cdot V_\infty^2$$

For a perfect gas (see Sect. 3.3):

$$\bar{q} = \frac{1}{2}\gamma \cdot p_\infty \cdot M_\infty^2$$

- Reference area, S_{ref}, generally defined as the area of maximum cross section for ballistic vehicles
- Reference length, L_{ref}, generally corresponding to overall length, or to maximum diameter (case of space probes)
- Thus the force and moment of references are $\bar{q} \cdot S_{ref}$ and $\bar{q} \cdot S_{ref} \cdot L_{ref}$.

4.1.1 Static Coefficients

Static aerodynamic coefficients are defined [NFX] starting from the components (X^A, Y^A, Z^A) of the aerodynamic force $\vec{R}^A$ and components (L, M, N) of the aerodynamic moment around origin O:

Relation of definition	Name	Symbol
$X^A = -A = -\bar{q} \cdot S_{ref} \cdot C_A$	Axial force coefficient	C_A
$Y^A = \bar{q} \cdot S_{ref} \cdot C_Y^A$	Lateral force coefficient	C_Y^A or C_Y
$Z^A = -\bar{q} \cdot S_{ref} \cdot C_N$	Normal force coefficient	C_N
$L = \bar{q} \cdot S_{ref} \cdot L_{ref} \cdot C_l$	Rolling moment coefficient	C_l
$M = \bar{q} \cdot S_{ref} \cdot L_{ref} \cdot C_m$	Pitching moment coefficient	C_m
$N = \bar{q} \cdot S_{ref} \cdot L_{ref} \cdot C_n$	Yawing moment coefficient	C_n

4.1.1.1 Changing the Origin of Moment Coefficients

The moments and associated coefficients are related to the origin O chosen for the vehicle frame. Coefficients relating to any different origin G are obtained starting from the preceding relations and rules to change origin of moments:

$$\vec{M}_G = \vec{M}_O - \vec{X}_G \wedge \vec{R}^A$$

which yields,

$$C'_l = C_l + \frac{y_G}{L_{ref}} \cdot C_N + \frac{z_G}{L_{ref}} \cdot C_Y$$

$$C'_m = C_m - \frac{x_G}{L_{ref}} \cdot C_N + \frac{z_G}{L_{ref}} \cdot C_A$$

$$C'_n = C_n - \frac{x_G}{L_{ref}} \cdot C_Y - \frac{y_G}{L_{ref}} \cdot C_A$$

4.1.1.2 Center of Pressure

The center of pressure corresponds to a location where the aerodynamic load is reduced to the resulting force $\vec{R}^A$, which implies $\vec{M}_{CP} = 0$, or:

$$\vec{X}_{CP} \wedge \vec{R}^A = \vec{M}_O$$

That is to say that one can define a unique center of pressure only if $\vec{M}_O$ is orthogonal with $\vec{R}^A$. This is generally not true, but on the other hand it is almost always possible to define a center of pressure for the loads in the normal plane (Ox, Oz) and another for the loads in the lateral plane. The normal force center of pressure is obtained from the pitching moment M and the components X^A and Z^A of the aerodynamic force in the pitch plane:

$$\frac{x_{CP}}{L_{ref}} = \frac{C_m}{C_N} + \frac{z_{CP}}{L_{ref}} \cdot \frac{C_A}{C_N}, \quad y_{CP} = 0, \quad z_{CP} \text{ arbitrary}$$

which gives an infinity of possible points located on the solution line according to z_{CP}.

Conventionally, one chooses $z_{CP} = 0$, which gives finally:

$$\frac{x_{CP}}{L_{ref}} = \frac{C_m}{C_N}$$

The lateral center of pressure is obtained from the lateral component Y^A of the aerodynamic force and the rolling and yawing moments (L, N):

$$\frac{x_{CPT}}{L_{ref}} = \frac{C_n}{C_Y} \quad \frac{y_{CPT}}{L_{ref}} \text{ arbitrary} \quad \frac{z_{CPT}}{L_{ref}} = -\frac{C_l}{C_Y}$$

The solution is also a line function of y_{CPT}, and the choice $y_{CPT} = 0$ determines the lateral center of pressure.

Finally, the static aerodynamic effects can be defined either by forces and moments coefficients (C_A, C_N, C_Y, C_l, C_m, C_n) or generally by the forces coefficient and centers of pressure $\left(C_A, C_N, C_Y, \frac{x_{CP}}{L_{ref}}, \frac{x_{CPT}}{L_{ref}}, \frac{z_{CPT}}{L_{ref}}\right)$. This is because the centers of pressure are sometimes undetermined, for example, when $C_N = 0$ and $C_m \neq 0$ or when $C_Y = 0$ and ($C_n \neq 0$ or $C_l \neq 0$).

It is important to note that all the results in this book correspond to the choice of the X axis directed forward along the vehicle.

4.1.1.3 Lift and Drag

These concepts originated with the birth of aeronautics. They use the aerodynamic frame (Ox_a, Oy_a, Oz_a) relating to the air flow. The aerodynamic frame is obtained from the vehicle frame (Ox, Oy, Oz) through a rotation $-\alpha$ around Oy, followed by a rotation β around Oz_a (Fig. 4.2). The axis Ox_a is along the relative velocity $\vec{V}_R$ of the vehicle, and axis Oz_a is normal to $\vec{V}_R$ in the pitch plane (symmetry plane of the vehicle). The drag and the lift are the components, with opposite signs, of the aerodynamic force along Ox_a and Oz_a. The sideslip force is along Oy_a.

The associated coefficients are:

Relation of definition	Name	Symbol
$X_a^A = -\bar{q} \cdot S_{ref} \cdot C_X$	Drag coefficient	C_D, C_x
$Y_a^A = \bar{q} \cdot S_{ref} \cdot C_Y$	Lateral force coefficient	C_y
$Z_a^A = -\bar{q} \cdot S_{ref} \cdot C_Z$	Lift coefficient	C_L, C_z

They are connected to the vehicle coefficients by:

$$C_D = (C_A \cdot \cos\alpha + C_N \cdot \sin\alpha) \cdot \cos\beta - C_Y \sin\beta$$
$$C_Y = C_Y^A \cdot \cos\beta + (C_A \cos\alpha + C_N \cdot \sin\alpha) \cdot \sin\beta$$
$$C_L = C_N \cdot \cos\alpha - C_A \cdot \sin\alpha$$

The lift concept is fundamental in two-dimensional incompressible, inviscid flows because in this case the drag is zero (D'Alembert paradox [RYH]), and the resulting $\vec{R}^A$ is equal to lift, normal to the relative velocity.

4.1.2 Dynamic Derivatives

Dynamic effects appear in presence of a rotational movement around the center of mass or a linear accelerated movement. In the most general case, they result in a modification of the static effects and can be expressed using dimensionless variable:

$p^* = \dfrac{p \cdot L_{ref}}{V_\infty}$	Reduced roll rate
$q^* = \dfrac{q \cdot L_{ref}}{V_\infty}$	Reduced pitch rate
$r^* = \dfrac{r \cdot L_{ref}}{V_\infty}$	Reduced yaw rate
$\dot{\alpha}^* = \dfrac{\dot{\alpha} \cdot L_{ref}}{V_\infty} = \dfrac{\dot{w} \cdot L_{ref}}{V_\infty^2}$	Reduced derivative of angle of incidence
$\dot{\beta}^* = \dfrac{\dot{\beta} \cdot L_{ref}}{V_\infty} = \dfrac{\dot{v} \cdot L_{ref}}{V_\infty^2}$	Reduced derivative of sideslip angle
$\dot{V}_R^* = \dfrac{\dot{V}_R \cdot L_{ref}}{V_R^2}$	Reduced tangential acceleration

However, in the majority of cases, and particularly those of interests to us, the reduced dynamic variable of state can be considered as first-order small terms. We can thus approximate the dynamic effects by a development limited to the first order

4.1 Aerodynamic Coefficients

of aerodynamic coefficients such as:

$$C'\left(x_i, x_j^*\right) = C(x_i) + \sum_j \frac{\partial C'}{\partial x_j^*} \cdot x_j^*$$

where $C(x_i)$ is a coefficient function of the static variables x_i, and x_j^* is a reduced variable representing the dynamic movement. Thus dynamic effects are characterized by the derivatives of forces and moment's coefficients relating to reduced dynamic variables.

In practice, in the case of hypersonic vehicles, theoretical and experimental evaluations showed that the only significant dynamic effects are the moments associated with angular rates. The corresponding coefficients are:

$\frac{\partial C_l}{\partial p^*}$	Roll damping coefficient	C_{lp}
$\frac{\partial C_m}{\partial q^*}$	Pitch damping coefficient	C_{mq}
$\frac{\partial C_n}{\partial r^*}$	Yaw damping coefficient	C_{nr}

Thus, hereafter we will neglect the dynamic effects on the aerodynamic force. Rules of transport of the static moment's coefficients do not apply to the dynamic derivatives. This will be developed later herein.

4.1.3 Axisymmetric Vehicles

In this case, we choose axis Ox along the symmetry axis. As the Oz axis can be arbitrarily selected in a plane orthogonal to Ox, we will choose it in the meridian half plane that includes velocity relative to air, thus:

$$Y^A = L = N = 0 \Leftrightarrow C_Y^A = C_l = C_n = 0$$

Moreover,

$$\beta = 0 \Rightarrow i = \overline{\alpha} = \alpha$$

In this "wind-fixed" frame (w), the number of coefficients necessary to define the static effects reduced from six to three, e.g., C_{Aw}, C_{Nw}, C_{mw}, functions of $\overline{\alpha}$. The damping moments exist along the three axes (x_w, y_w, z_w) and for small angles of attack we have $C_{nrw} \approx C_{mqw}$.

The reference frame (w) is not fixed relative to the vehicle. It rotates with the relative velocity vector; the corresponding meridian for an observer fixed with the vehicle is called "windward meridian" and is obtained from the meridian origin (Oz) of the vehicle frame through a rotation Φ_w around Ox.

$$\cos \Phi_w = \frac{\sin \alpha \cdot \cos \beta}{\sin \overline{\alpha}}$$
$$\sin \Phi_w = -\frac{\sin \beta}{\sin \overline{\alpha}}$$

Thus, one transforms the effects in the rotating frame (w) into the vehicle-fixed frame through a rotation $-\Phi_w$ around Ox.

$$C_A = C_{Aw}$$
$$C_N = C_{Nw} \cdot \cos \Phi_w$$
$$C_Y = C_{Nw} \cdot \sin \Phi_w$$
$$C_m = C_{mw} \cdot \cos \Phi_w$$
$$C_n = C_{mw} \cdot \sin \Phi_w$$
$$C_l = C_{lw}$$
$$C_{mq} = C_{mqw} \cdot \cos^2 \Phi_w + C_{nrw} \sin^2 \Phi_w$$
$$C_{nr} = C_{nrw} \cdot \cos^2 \Phi_w + C_{mqw} \sin^2 \Phi_w$$
$$C_{mr} = C_{nq} = (C_{mqw} - C_{nrw}) \cdot \sin \Phi_w \cdot \cos \Phi_w$$

The normal force center of pressure and the lateral force center of pressure are obviously the same:

$$\frac{x_{CP}}{L_{ref}} = \frac{x_{CPT}}{L_{ref}} = \frac{C_{mw}}{C_N}; \quad z_{CPT} = 0$$

Lift and drag coefficients for axisymmetric vehicles are:

$$C_{Dw} = C_{Aw} \cdot \cos \overline{\alpha} + C_{Nw} \cdot \sin \overline{\alpha}$$
$$C_{Yw} = C_{Yw}^A = 0$$
$$C_{Lw} = C_{Nw} \cdot \cos \overline{\alpha} - C_{Aw} \cdot \sin \overline{\alpha}$$

For small angle of attack, dynamic terms can be simplified and one obtains:

$$C_{mq} \approx C_{nr} \approx C_{mqw} \approx C_{nrw} \text{ and } C_{mr} \approx C_{nq} \approx 0.$$

4.2 Modes of Flow

4.2.1 Parameters of Similarity

In first analysis, the aerodynamic coefficients must be functions of variables representing the upstream flow listed in the preceding paragraph, and of geometry and dimensions of the vehicle. However, dimensional analysis [RYH] of the theoretical

4.2 Modes of Flow

flow equations shows that the number of parameters can be considerably reduced by the use of dimensionless "similarity parameters." In our case, the hypersonic aerodynamics coefficients' most commonly used parameters are:

Name	Definition	Physical significance
Mach number	$M_\infty = \dfrac{V_\infty}{a_\infty}$	$\dfrac{V_\infty^2}{a_\infty^2} \equiv \dfrac{kinetic \cdot energy}{internal \cdot energy}$
Reynolds number	$Re_\infty = \dfrac{\rho_\infty \cdot V_\infty \cdot L_{ref}}{\mu_\infty}$	$\dfrac{pressure\ forces}{viscous forces}$
Knudsen number	$Kn = \dfrac{\lambda_\infty}{L_{ref}}$	$\dfrac{mean\ free\ path}{vehicle\ length}$

For an axisymmetric vehicle, one often uses a functional aerodynamic model as follows:

$$C_{Aw}(M_\infty, Re_\infty, Kn_\infty, \overline{\alpha})$$
$$C_{Nw}(M_\infty, Re_\infty, Kn_\infty, \overline{\alpha})$$
$$C_{mw}(M_\infty, Re_\infty, Kn_\infty, \overline{\alpha})$$

However, this represents only one generally acceptable simplification. The aerothermal complexity of the phenomena may require, in some situations, the use of alternative variables. Thus, to model the influence of thermodynamic nonequilibrium phenomena [AND] (chemical or vibration nonequilibrium), it can be necessary to use dimensioned variables such as altitude, relative velocity, and density of the gas [PAT].

4.2.2 Characteristics of the Main Flow Modes

The Knudsen number characterizes the rarefaction status of the air with respect to dimensions of the vehicle:

1. At very high altitude, $Kn \gg 1$, there are practically no collisions between molecules in the vicinity of the vehicle. We are in "free molecular" mode, and each molecule interacts individually with the vehicle wall according to the statistical theory of gases.
2. At low altitudes, when $Kn \ll 1$, we are in "continuous flow" mode, governed by classical fluid mechanics theory (Euler equations for the inviscid flows, or Navier–Stokes equations to include viscous effects).
3. The intermediate section or "rarefied" mode can be approached by either using the Navier–Stokes equations with slip conditions at the wall (for flows close to continuous) or using Boltzman equations (valid everywhere else) solved by the direct methods of resolution or by direct Monte Carlo simulation (DSMC) methods.

In addition to these modes related to the number of molecules per unit volume, one can further classify the flows with respect to chemical reactions, and reactions of dissociation and their kinetics. There are primarily related to upstream enthalpy $h_\infty \approx \frac{\rho_\infty V_\infty^2}{2}$ and depend on transit time of the species through the shock layer. We will not treat this subject and invite the reader to refer to those books and papers specialized in aerothermal flow interactions [AND].

We will see that the dominant mode for reentry mechanics is the continuous mode where high aerodynamic loads occur (same order of magnitude as the weight or much higher).

The "rarefied" mode corresponds to altitude section where the aerodynamics forces are much lower than the weight, but the aerodynamic moments start to be significant (same order of magnitude as the gyroscopic moments), which are potentially critical in the case of some planetary probes.

The free molecular mode is important only for small objects or objects in slowly decaying orbits. For the needs of flight mechanics, we will be satisfied to determine the effects in free molecular mode by using a simplified model of the interactions to the wall.

Fast methods to estimate aerodynamic coefficients at the various flow modes will be treated in Sects. 4.3 and 4.4.

4.3 Continuous Mode

In continuous flow, we must examine the dependence on variable $(M_\infty, Re_\infty, \overline{\alpha})$. The problem is to model the aerodynamic coefficients in the "flight parameter's map" characteristic of the missions of the vehicle. Three methods may be employed.

4.3.1 Experimental

This primarily consists in measuring aerodynamic effects on a model in a wind tunnel. The Mach conditions are difficult to achieve along with viscous effects related to the Reynolds number (turbulent flow) and with surface conditions representative of flight. Indeed, Reynolds numbers are weak in wind tunnel conditions, because of size of the models and low gas density. Real gas effects, dependent on the dissociation of the air at high temperature, are also difficult to simulate because they require a very high level of upstream enthalpy. Instrumented experimental flights are often required to obtain these data.

4.3.2 Numerical [AND]

- Traditionally, computational methods solved "Euler" inviscid two- or three-dimensional equations (calculation of pressure and velocities valid only in

4.3 Continuous Mode

nonseparated zones) as inputs to viscous calculations of the boundary layer on the wall by exact or integral methods (wall friction). The turbulent flow computation, even on smooth walls, relies strongly on experimental results. Loads on the separated flow at the base of the vehicle (base drag) are determined empirically using correlations derived from experimental flight and wind tunnel data. The real gas effects can be taken into account in the computation by using the equation of state of air at thermodynamic equilibrium at the local conditions of temperature and pressure. Inviscid treatment of the subsonic zones (for example, around the stagnation point) requires the resolution the nonstationary Euler equations (time marching method). Treatment of the supersonic zones is generally optimized by using "space marching" methods. "Shock capturing" methods use fixed mesh in which the shock discontinuity is treated as the other points of the flow. The "shock fitting methods" suppose that the points of the upstream boundary of the mesh is the shock, and that location is determined iteratively from the Rankine–Hugoniot oblique-shock analytic solution. For the last method, the mesh evolves and is restricted to the layer between the body and the shock surface. It is quite obvious that the shock wave is neither a perfect discontinuity (case of shock fitting) nor the result of the numerical pseudo viscosity (case of shock capturing).

Many standard codes exist in aerospace industry, fast and accurate in their application field. Their application field is typically:

Vehicle geometry	2D or 3D on axisymmetrical shapes
Minimum Mach number	0.01 (Depending on mesh)
Maximum incidence	°–35° (Depending on geometry, Mach number, and on the method used to solve subsonic flow at stagnation point)
Limitations	Not applicable in the case of a separated flow

- "Exact" methods have been developed to solve Navier–Stokes equations or "parabolized" Navier–Stokes equations (case of steady state flows, in supersonic zone)[AND]. The laminar flow solution is theoretically well formulated; however, the turbulent flow solution is less satisfactory and relies on models based on experimental results. This kind of code requires long computational times. However, the progress in numerical methods is relentless, as the computers' capacities are enhanced, but the limits of the theory of turbulence could be the Achilles' heel of computation flow accuracy.

4.3.3 Approximate Analytical Method [TRU]

Many more or less sophisticated analytical methods were developed before the present era of digital codes. The only one, which still has a practical utility is "Newton impact theory," imagined by the father of mechanics to model the effects on a body moving through an incompressible fluid. According to this theory, pressure p undergone by an element of surface whose normal has a tilt angle θ relating to impinging flow direction is given by:

$$\frac{p - p_\infty}{\bar{q}} = 2 \cdot \cos^2 \theta$$

This theory, which was wrong in the assumed field of application, proved several centuries later to be valid near the stagnation point in hypersonic flow. Indeed, in this case the shock layer is very thin and approximately parallel to the wall. Thus, we can consider according to a naive vision that incident molecules individually impact the wall while yielding the normal component of their mV_N of their linear momentum and retaining the parallel component mV_P (Fig. 4.3).

Let us consider in this hypothesis a tube of incident flow at upstream relative velocity $\vec{V}_\infty$ crossing an element of wall area. Through the section of area normal to the tube $S \cdot \cos \theta$, flux of mass and linear momentum normal to the wall per unit of time are, respectively,

$$\frac{dm}{dt} = \rho_\infty \cdot V_\infty \cdot S \cdot \cos \theta$$

$$\frac{d(mV_N)}{dt} = V_N \cdot \frac{dm}{dt} = \rho_\infty \cdot V_\infty^2 \cdot \cos^2 \theta \cdot S$$

Linear momentum yielded to the wall per unit of time corresponds to the normal force F_N exerted on the element of area S and to an increment of pressure $p - p_\infty = \frac{F_N}{S}$ relating to the static pressure in the gas,

$$p - p_\infty = \rho_\infty \cdot V_\infty^2 \cdot \cos^2 \theta = 2 \cdot \bar{q} \cdot \cos^2 \theta$$

We finally obtain the pressure coefficient on the wall,

$$c_p = \frac{p - p_\infty}{\bar{q}} = 2 \cdot \cos^2 \theta$$

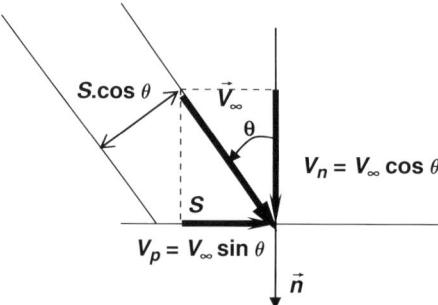

Fig. 4.3 Newtonian impact geometry

4.3 Continuous Mode

This expression can be empirically refined by observing that factor 2 corresponds to the maximum of pressure at the stagnation point of the fluid on the wall: ($\theta = 0$). We can generalize the expression in:

$$c_p = c_{p,stag} \cdot \cos^2\theta = c_{p,stag} \left(\frac{\vec{V}_\infty \cdot \vec{N}}{V_\infty}\right)^2$$

An exact value of $c_{p,stag}$ is obtained by calculating the stagnation pressure behind a normal shock using Hugoniot equations and isentropic compression:

$$p_{stag} = p_{shock}\left[1 + \frac{\gamma-1}{2} \cdot \frac{1 + \frac{\gamma-1}{2}M_\infty^2}{\gamma \cdot M_\infty^2 - \frac{\gamma-1}{2}}\right]^{\frac{\gamma}{\gamma-1}}$$

$$p_{shock} = p_\infty\left[1 + \frac{2\gamma}{\gamma+1}\left(M_\infty^2 - 1\right)\right]$$

Historically, some authors tried to refine the Newtonian value of pressure coefficient, for example, by taking into account a pressure correction associated with the centrifugal force induced by the wall curvature $\delta p = -\rho_\infty \frac{(V_\infty \sin\theta)^2}{R}$. All these improvements proved completely illusory, because the initial estimate has no true physical justification; therefore, it is preferred to use the simplest approximation $c_{p,stag} = 2$.

4.3.4 Continuous Coefficients by Newton's Method

4.3.4.1 Static Coefficients

Pressure Coefficients

We obtain the aerodynamic coefficients of a convex shape body by integrating the elementary surface effects in the areas exposed to incident flow, corresponding to $\vec{V}_\infty \cdot \vec{N} \geq 0$ conditions, where $\vec{N}$ is the normal to the wall:

$$d\vec{F} = \bar{q} \cdot c_{p,arret} \cdot \left(\frac{\vec{V}_\infty \cdot \vec{N}}{V_\infty}\right)^2 \cdot \vec{N} \cdot dS$$

$$d\vec{M} = \bar{q} \cdot c_{p,arret} \cdot \left(\frac{\vec{V}_\infty \cdot \vec{N}}{V_\infty}\right)^2 \cdot \vec{r} \wedge \vec{N} \cdot dS$$

$\vec{r} = \vec{OM}$ is the radius vector between the origin of vehicle frame and the center of the surface element.

For the areas sheltered from the flow such as ($\vec{V}_\infty \cdot \vec{N} \leq 0$), we have:

$$d\vec{F} = d\vec{M} \approx 0$$

We obtain finally the expression of the aerodynamic coefficients:

Name	Expression
Axial force coefficient	$C_{Aw} = -\dfrac{c_{p,arret}}{S_{ref}} \displaystyle\int\int_{\vec{V}_\infty \cdot \vec{N} \geq 0} \left(\dfrac{\vec{V}_\infty \cdot \vec{N}}{V_\infty}\right)^2 \cdot N_x \cdot dS$
Normal force coefficient	$C_{Nw} = -\dfrac{c_{p,arret}}{S_{ref}} \displaystyle\int\int_{\vec{V}_\infty \cdot \vec{N} \geq 0} \left(\dfrac{\vec{V}_\infty \cdot \vec{N}}{V_\infty}\right)^2 \cdot N_z \cdot dS$
Pitching moment coefficient	$C_{mw} = \dfrac{c_{p,arret}}{S_{ref} \cdot L_{ref}} \displaystyle\int\int_{\vec{V}_\infty \cdot \vec{N} \geq 0} \left(\dfrac{\vec{V}_\infty \cdot \vec{N}}{V_\infty}\right)^2 \cdot (z \cdot N_x - x \cdot N_z).dS$

In the case of a complicated geometry, it is preferable to carry out a numerical integration of these expressions. The old "hypersonic arbitrary body program (HABP)" code of NASA proposes a choice of theoretical analytical expressions or pressure correlations to numerically calculate the coefficients of any arbitrary geometry [HAB].

In the case of some simple geometrical forms, the integration of these relations leads to analytical expressions. For example, in the case of a *circular cone of half apex angle* θ_a and for angle of attack $\bar{\alpha} \leq \theta_a$:

$$C_{Aw} = 2 \cdot \sin^2 \theta_a + \left(1 - 3 \cdot \sin^2 \theta_a\right) \sin^2 \bar{\alpha}$$
$$C_{Nw} = \cos^2 \theta_a \cdot \sin 2\bar{\alpha}$$
$$C_{mw} = -\frac{2}{3} \cdot \sin 2\bar{\alpha}$$
$$\frac{x_{CP}}{L_{ref}} = -\frac{2}{3 \cdot \cos^2 \theta_a}$$

The origin of moment and center of pressure is located at the apex of the cone, the reference area is the maximum cross section and the reference length is the length of the cone.

In the case of a *segment of sphere* with radius R we obtain (for $\bar{\alpha} \leq \theta_a$ and an origin O located to the upstream pole of the sphere):

$$C_{Aw} = 1 - \sin^4 \theta_a - \frac{(1 + 3\sin^2 \theta_a)\cos^2 \theta_a}{2} \cdot \sin^2 \bar{\alpha}$$
$$C_{Nw} = \frac{2}{3}\cos^4 \theta_a \cdot \sin 2\bar{\alpha}$$
$$C_{mw} = -C_{mw} \Leftrightarrow x_{CP} = -R$$

4.3 Continuous Mode

with reference area $S_{ref} = \pi \cdot R^2$, reference length $L_{ref} = R$, and θ_a, half apex angle of the cone tangent to trailing edge of the segment.

Coefficients of a geometry made up of simple element truncated cones and spherical segments are calculated easily from the above expressions by adding the forces on elements, and moments brought back to a common origin O by applying the rule for changing origin (Sect. "Changing the Origin of Moment Coefficients"):

$$C_{Aw} = \sum_i \frac{S_i}{S_{ref}} \cdot C_{Awi}$$

$$C_{Nw} = \sum_i \frac{S_i}{S_{ref}} \cdot C_{Nwi}$$

$$C_{mw/O} = \sum_i \frac{S_i}{S_{ref}} \left[\frac{L_i}{L_{ref}} \cdot C_{mwi} + \frac{x_i}{L_{ref}} C_{Nwi} \right]$$

$$x_{CP} = \frac{\sum_i S_i \cdot C_{Nwi} \cdot x_{CPi}}{S_{ref} \cdot C_{Nw}}$$

where for each element "i": S_i and L_i are the reference values, x_i the coordinate of the reference point of moment relative to the common origin O, and x_{CPi} the coordinate of the center of pressure relative to the common origin O.

An interesting property is, as the dependence of coefficients with incidence $\bar{\alpha}$ is identical for the sphere segment and the cone, it is also identical for all the bodies made up of such elements:

$$C_{Aw} = C_{Aw0} + C_{Aw\alpha} \sin^2 \bar{\alpha}$$
$$C_{Nw} = C_{Nw\alpha 0} \sin \bar{\alpha} \cos \bar{\alpha}$$
$$C_{mw/O} = C_{mw\alpha 0/O} \sin \bar{\alpha} \cos \bar{\alpha}$$
$$\frac{x_{CP}}{L_{ref}} = \frac{C_{mw\alpha 0/O}}{C_{Nw\alpha 0}} = cste$$

Axial Force on Base of the Vehicle

An estimate of the base axial force coefficient can be obtained by assuming that the base pressure is a constant ratio of the static free stream pressure in the incident gas, $p_c = \eta \cdot p_\infty$. This results in :

$$C_{Ac} = -\frac{p_c - p_\infty}{\bar{q}} = (1 - \eta) \frac{p_\infty}{\bar{q}}$$

$$\bar{q} = \frac{1}{2} \gamma p_\infty M_\infty^2 \quad \Rightarrow \quad C_{Ac} = (1 - \eta) \frac{2}{\gamma M_\infty^2}$$

As we can see in this formulation, base drag becomes a significant contributor only at low Mach number.

To fit the parameter η relating to a given configuration, one may use experimental data. A reasonable value often used as first estimate is simply $C_{Ac} \approx \frac{1}{M_\infty^2}$, which corresponds to $\eta = 0.4$.

Axial Friction Force

Certain authors propose analytical formulae for the friction drag using expression of shear on a flat plane in laminar or turbulent flow. One finds in [GLO] such formulae, in the case of sharp small angle cone:

- For cold wall conditions, laminar flow:

$$C_{A,f} = \frac{1.9}{\tan \theta_c} \left[\frac{(1 + \gamma M^2 \sin^2 \theta_c) \cos \theta_c}{Re} \right]^{\frac{1}{2}}$$

where $Re = \frac{\rho V L}{\mu}$
- For smooth, hot wall turbulent flow:

$$C_{A,f} = \frac{0.1}{Re^{\frac{1}{5}}} \frac{(\cos \theta_c)^{0.8}}{\tan \theta_c} \frac{(1 + \gamma M^2 \sin^2 \theta_c)^{0.8}}{(1 + 0.17 \cdot M^2)^{0.7}}$$

Behavior of the Axial Force Coefficient for a Slender Vehicle

Approximate expressions of Sects. "Pressure Coefficients," "Axial Force on Base of the Vehicle," "Axial Friction Force," have been applied to calculate the axial force coefficient of a sharp cone of 8° half apex angle and D = 0.50 m diameter (length is L = 1.78 m), represented in Figs. 4.4–4.7. This behavior is typical of the zero angle of attack drag coefficient of a slender axisymmetric vehicle.

For a vehicle with high ballistic coefficient, $\frac{m}{SC_A}$, of order of magnitude 10^4 kg/m², aerodynamic deceleration is initially low and velocity is near constant above 10 km altitude.

Figures 4.4 and 4.5 give, respectively, approximate components of the drag at continuous laminar or turbulent flow and constant velocity V = 6000 m/s, between 60 km and 10 km altitude. Figure 4.6 gives the evolution of the Reynolds number $Re = \frac{\rho V L}{\mu}$ in atmosphere US66. By admitting that laminar to turbulent flow transition begins toward $Re = 10^6$ and becomes complete for $Re = 10^7$, the flow is completely laminar above 45 km and mostly turbulent under 32 km altitude.

At altitudes range from 60 to 10 km, the pressure drag coefficient is nearly constant. The base drag coefficient is almost negligible. Variation of the total drag coefficient is primarily related to the evolution of friction drag with Reynolds number or altitude. In laminar flow, the drag coefficient has high initial value followed by a

4.3 Continuous Mode   47

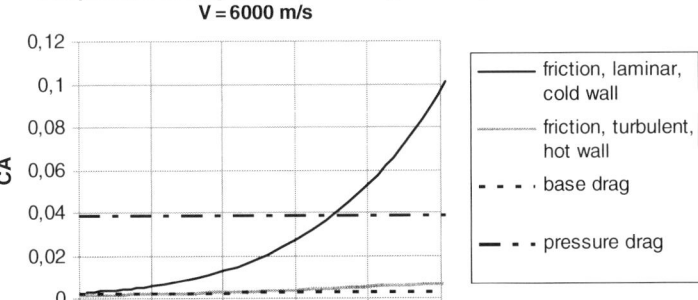

Fig. 4.4 Components of drag coefficient of 8° cone

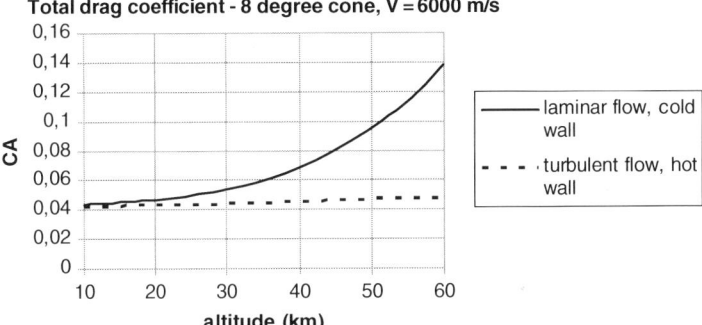

Fig. 4.5 Total drag coefficient of 8° cone

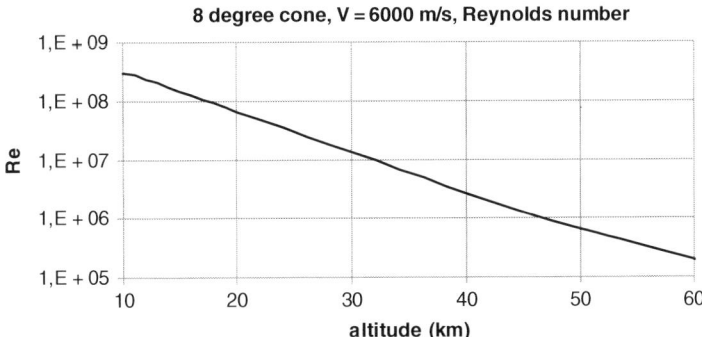

Fig. 4.6 Evolution of the Reynolds number $\mathrm{Re} = \frac{\rho V L}{\mu}$ in atmosphere US66

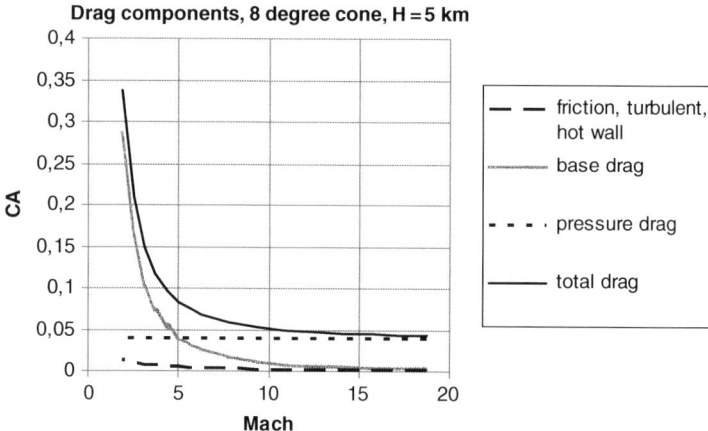

Fig. 4.7 Drag components of 8° cone

decrease with altitude. During the transition, one observes a gradual decrease until the full turbulence is reached where the drag coefficient remains nearly constant as long the velocity is sustained.

Under 10 km altitude, velocity and Mach number decrease sharply under drag effect, we observe the behavior in Fig. 4.7 dominated at Mach lower than 10 by the evolution of base drag with Mach number.

4.3.4.2 Dynamic Derivatives

The estimate of damping coefficients is possible theoretically using Newton's method or more elaborate methods, but quite as approximate (Ericsson [ERI] semiempirical method "Embedded Newtonian flow"). "Exact" dynamic numerical methods will likely be developed in future to solve this problem. Experimental approaches using wind tunnel measurements are possible but seldom used, as measurements are difficult and inaccurate. In fact, effort in this field remains very limited because we will see herein that the pitch damping moment coefficient, although having significant effect, is not considered in practice as a dimensioning parameter, except in the rare configurations of dynamic instability [KAR]. Consequently, the need for accuracy in this parameter is not critical and approximate estimate methods are adequate. Codes using Ericsson's method give damping coefficients of sphere cone configurations. This method takes into account the influence of the nosetip shock on downstream flow and allows the computation of Mach number effects. The Newtonian codes make it possible to calculate the dynamic derivatives on a convex arbitrary shape.

Indeed, Newtonian impact theory gives the pressure on a surface element from its motion relative to air. It can also be applied to dynamic effects in the flow.

4.3 Continuous Mode

Angular Damping Coefficients

Let us consider the case of a constant linear velocity $\vec{V}_R$ of the mass center and a constant angular rate $\vec{\omega} = \{p, q, r\}$ around it.

Velocity relative to air, for a surface element located at $\vec{r}$ relating to mass center, is:

$$\vec{V}_R(\vec{r}) = -\vec{V}_\infty(\vec{r}) = \vec{V}_R(0) + \vec{\omega} \wedge \vec{r}$$

That is to say,

$$\vec{V}_\infty(\vec{r}) = \vec{V}_\infty(0) - \vec{\omega} \wedge \vec{r}$$

The Newtonian pressure coefficient at $\vec{r}$ becomes:

$$c_p(\vec{r}) = c_{p,stag} \cdot \left(\frac{\vec{V}_\infty(\vec{r}) \cdot \vec{N}}{V_\infty(\vec{r})}\right)^2 \approx c_{p,stag} \cdot \left[\frac{\vec{V}_\infty(0) \cdot \vec{N}}{V_\infty(0)} - \frac{(\vec{\omega} \wedge \vec{r}) \cdot \vec{N}}{V_\infty(0)}\right]^2$$

This expression is simplified in the hypothesis $r\omega \ll V_\infty$ where one retains only the first-order terms in $\frac{r\omega}{V_\infty}$.

$$c_p(\vec{r}) \approx c_{p,stag} \cdot \left[\left(\frac{\vec{V}_\infty(0) \cdot \vec{N}}{V_\infty(0)}\right)^2 - 2\frac{\left(\vec{V}_\infty(0) \cdot \vec{N}\right)\left(\vec{r} \wedge \vec{N}\right) \cdot \vec{\omega}}{V_\infty^2(0)}\right]$$

$$c_p(\vec{r}) \approx c_p(\vec{\omega} = 0) + \delta c_p(\vec{\omega})$$

This involves a modification of the elementary static force,

$$\delta\left[d\vec{R}^A(\vec{\omega})\right] = \bar{q} \cdot \delta c_p(\vec{\omega}) \cdot \vec{N} \cdot dS$$

Let us choose the origin of the vehicle frame at its center of mass. The elementary moment of disturbance is written as:

$$d\delta\vec{M}(\vec{\omega}) = \bar{q} \cdot \delta c_p(\vec{\omega}) \cdot \vec{r} \wedge \vec{N} \cdot dS$$

We obtain the total disturbance of the force from integration of elementary forces on the impact area.

$$\delta\vec{R}^A(\vec{\omega}) = -2 \cdot \bar{q} \cdot c_{p,stag} \iint\limits_{\vec{V}_\infty \cdot \vec{N} \geq 0} \frac{\left(\vec{V}_\infty \cdot \vec{N}\right)\left(\vec{r} \wedge \vec{N}\right) \cdot \vec{\omega}}{V_\infty} \vec{N} \cdot dS$$

By projecting this expression along the three directions of the vehicle frame, we obtain the damping force coefficients:

$$C_{Nq} = -\frac{1}{\bar{q} \cdot S_{ref}} \frac{\partial \delta R_z^A (\vec{\omega})}{\partial \frac{q \cdot L_{ref}}{V_\infty}} = -\frac{2 \cdot c_{p,stag}}{S_{ref}} \iint_{\vec{V}_\infty \cdot \vec{N} \geq 0} \frac{(\vec{V}_\infty \cdot \vec{N})}{V_\infty} \frac{(\vec{r} \wedge \vec{N})_y \cdot N_z}{L_{ref}} dS$$

$$C_{Yr} = \frac{1}{\bar{q} \cdot S_{ref}} \frac{\partial \delta R_Y^A (\vec{\omega})}{\partial \frac{r \cdot L_{ref}}{V_\infty}} = -\frac{2 \cdot c_{p,stag}}{S_{ref}} \iint_{\vec{V}_\infty \cdot \vec{N} \geq 0} \frac{(\vec{V}_\infty \cdot \vec{N})}{V_\infty} \frac{(\vec{r} \wedge \vec{N})_z \cdot N_y}{L_{ref}} dS$$

$$C_{Nq} = -\frac{1}{\bar{q} \cdot S_{ref}} \frac{\partial \delta R_z^A (\vec{\omega})}{\partial \frac{q \cdot L_{ref}}{V_\infty}} = -\frac{2 \cdot c_{p,stag}}{S_{ref}} \iint_{\vec{V}_\infty \cdot \vec{N} \geq 0} \frac{(\vec{V}_\infty \cdot \vec{N})}{V_\infty} \frac{(\vec{r} \wedge \vec{N})_y \cdot N_z}{L_{ref}} dS$$

Dynamic forces at usual angular velocities have a negligible effect (Magnus lift effect included, as this book does not apply to projectile ammunition with high roll velocities) on the aerodynamic force. However, they are useful as they are used in the rules to change origin of damping moment coefficient.

Total disturbance moment is written:

$$\delta \vec{M}(\vec{\omega}) = -2 \cdot \bar{q} \cdot c_{p,stag} \iint_{\vec{V}_\infty \cdot \vec{N} \geq 0} \frac{(\vec{V}_\infty \cdot \vec{N})}{V_\infty} \frac{(\vec{r} \wedge \vec{N}) \cdot \vec{\omega}}{V_\infty} \vec{r} \wedge \vec{N} \cdot dS$$

The dynamic derivatives of moment are:

$$C_{lp} = \frac{1}{\bar{q} \cdot S_{ref} \cdot L_{ref}} \frac{\partial \delta \vec{M}(\vec{\omega})}{\partial \frac{p \cdot L_{ref}}{V_\infty}} = -\frac{2 \cdot c_{p,stag}}{S_{ref}} \iint_{\vec{V}_\infty \cdot \vec{N} \geq 0} \frac{(\vec{V}_\infty \cdot \vec{N})}{V_\infty} \frac{(\vec{r} \wedge \vec{N})_x^2}{L_{ref}^2} dS = 0$$

$$C_{mq} = \frac{1}{\bar{q} \cdot S_{ref} \cdot L_{ref}} \frac{\partial \delta \vec{M}(\vec{\omega})}{\partial \frac{q \cdot L_{ref}}{V_\infty}} = -\frac{2 \cdot c_{p,stag}}{S_{ref}} \iint_{\vec{V}_\infty \cdot \vec{N} \geq 0} \frac{(\vec{V}_\infty \cdot \vec{N})}{V_\infty} \frac{(\vec{r} \wedge \vec{N})_y^2}{L_{ref}^2} dS$$

$$C_{nr} = \frac{1}{\bar{q} \cdot S_{ref} \cdot L_{ref}} \frac{\partial \delta \vec{M}(\vec{\omega})}{\partial \frac{r \cdot L_{ref}}{V_\infty}} = -\frac{2 \cdot c_{p,stag}}{S_{ref}} \iint_{\vec{V}_\infty \cdot \vec{N} \geq 0} \frac{(\vec{V}_\infty \cdot \vec{N})}{V_\infty} \frac{(\vec{r} \wedge \vec{N})_z^2}{L_{ref}^2} dS$$

The application of these results to a *cone* with semiapex angle θ_a gives, for $\bar{\alpha} \leq \theta_a$ and origin at the apex:

4.3 Continuous Mode

$$C_{mqw} = C_{nrw} = -\left(1 + \tan^2 \theta_a\right) \cdot \cos \bar{\alpha}$$

$$C_{Nqw} = C_{Yrw} = \frac{4}{3} \cdot \cos \bar{\alpha}$$

Rules to Change Origin of Damping Moments Coefficients

When one examines the expression of damping moment coefficients, it appears a quadratic relationship with radius vector from the origin to the current surface element. The rules of change are obtained while replacing $\frac{\vec{r}}{L_{ref}}$ with $\frac{\vec{r}-\vec{r}_G}{L_{ref}}$ in the original expressions. The only case of interest for an axisymmetric vehicle is that of a displacement of the center of rotation along Ox.

The rigorous development of calculations led to the following relations, valid for $\bar{\alpha} \leq \theta_{min}$ (maximum incidence, such that the complete area upstream the maximum cross section, is impacted by the flow), in the wind frame "W":

$$C_{mqw/G} = C_{mqw/O} - 2 \cdot \frac{x_G}{L_{ref}} \cdot C_{Nqw/O} - \left(\frac{x_G}{L_{ref}}\right)^2 \cdot \frac{\cos \bar{\alpha}}{\cos 2\bar{\alpha}} \cdot C_{N\alpha w}$$

$$C_{nrw/G} = C_{nrw/O} - 2 \cdot \frac{x_G}{L_{ref}} \cdot C_{Yrw/O} + \left(\frac{x_G}{L_{ref}}\right)^2 \cdot C_{Y\beta w}$$

with,

$$C_{N\alpha w} = \frac{\partial C_{Nw}}{\partial \bar{\alpha}} = -2 \frac{c_{p,arret}}{S_{ref}} \cdot \cos 2\bar{\alpha} \cdot \iint_{\vec{V}_\infty \cdot \vec{N} \geq 0} N_x \cdot N_z^2 \cdot ds$$

$$C_{Y\beta w} = \frac{\partial C_Y}{\partial \beta} (\bar{\alpha}, \beta = 0) = 2 \frac{c_{p,arret}}{S_{ref}} \cdot \cos \bar{\alpha} \cdot \iint_{\vec{V}_\infty \cdot \vec{N} \geq 0} N_x \cdot N_Y^2 \cdot ds$$

The application to a sharp conical body shape gives:

$$C_{N\alpha w} = 2 \cdot \cos^2 \theta_a \cdot \cos 2\bar{\alpha}$$

$$C_{Y\beta w} = -2 \cdot \cos^2 \theta_a \cdot \cos \bar{\alpha}$$

It is useful to note that in the Newtonian approximation, dynamic derivatives C_{Nq} and C_{Yr} are dependent on the gradients of static moment coefficients C_m and C_n, respectively, which are easier to calculate. One can show indeed:

$$C_{m\alpha w} = \frac{\partial C_{mw}}{\partial \bar{\alpha}} = \sin \bar{\alpha} \cdot C_{Aqw} - \cos \bar{\alpha} \cdot C_{Nqw}$$

$$C_{n\beta w} = \frac{\partial C_n}{\partial \beta} (\bar{\alpha}, \beta = 0) = -C_{Yrw}$$

For low angle of attack, $\bar{\alpha}$, $\cos\bar{\alpha} \approx 1$ and $\sin\bar{\alpha} \approx 0$, we obtain the following rules of change:

$$C_{mqw/G} = C_{mqw/O} + 2 \cdot \frac{x_G}{L_{ref}} \cdot C_{m\alpha w/O} - \left(\frac{x_G}{L_{ref}}\right)^2 \cdot C_{N\alpha w}$$

$$= C_{mqw/O} + \left[2\frac{x_G}{L_{ref}}\frac{x_{CP}}{L_{ref}} - \left(\frac{x_G}{L_{ref}}\right)^2\right] C_{N\alpha w}$$

$$C_{nrw/G} = C_{mqw/G}$$

$$C_{m\alpha w/O} = \frac{x_{cp}}{L_{ref}} \cdot C_{N\alpha w}$$

This gives for the sharp conical body of length L_{ref}:

$$C_{mqw/G} = -(1 + \tan^2\theta_a) - \frac{8}{3}\cos\theta_a \cdot \frac{x_G}{L_{ref}} - 2 \cdot \cos^2\theta_a \cdot \left(\frac{x_G}{L_{ref}}\right)^2$$

Remarks:

- Do not forget that with our convention of direction for OX axis forwards vehicle, we have $\frac{x_G}{L_{ref}} < 0$ for an origin O at stagnation point.
- This often used method of change is empirical, because it is derived only from the Newtonian model. To our knowledge, there is no exact method at this time.

4.3.4.3 Influences of Linear Accelerations

The expression of the pressure coefficients in Newtonian flow depends only on the instantaneous speed of the surface element relating to the gaseous medium. From this fact, dynamic effects associated with the acceleration of the center of mass are null in the field of this approximation.

4.4 Rarefied Mode

4.4.1 Free Molecular Flow

In free molecular flow, effects on the surface element are calculated from analytical expressions for pressure coefficient and shear coefficient resulting from the kinetic theory of gases.

The more current model assumes that speed of incident molecules relating to the wall is the sum of two terms:

$$\vec{v}_R = \vec{V}_R + \vec{v}\,(T_\infty)$$

4.4 Rarefied Mode

- A term $\vec{v}\,(T_\infty)$ of random thermal motion in the gas at rest corresponding to a Maxwell distribution of velocities at equilibrium at the atmosphere temperature T_∞
- A term $\vec{V}_R$ representing the velocity of the vehicle relating to the reference frame of the atmosphere at rest

Other hypotheses are:

- All incidental molecules are reemitted after interaction with the wall with a velocity distribution relating to the wall such as:
  - In the case of a total accommodation with the wall, these molecules have a relative velocity distribution at the temperature of the wall T_w (completely diffuse reflection, with a Maxwell distribution over 2π steradians).
  - In the case of partial accommodation with the wall, a fraction of the molecules are reemitted diffusively at the wall temperature T_w, the remainder is specularly reflected (elastic reflection).

Pressure and shear stresses result from the assessments of linear momentum normal and parallel to the wall by unit of area and unit of time.

$$\Sigma(m\text{v})_{\text{wall}} = \Sigma(m\text{v})_{\text{inc}} - \Sigma(m\text{v})_{\text{ref}}$$

In practice, to take into account the experimental results, one is brought to use a different ratio of molecules reemitted diffusively (coefficients of accommodation) to calculate pressure and shear stresses:

σ' = accommodation coefficient normal to the wall,
σ = accommodation coefficient tangent to the wall.

Rigorously, we should have

$$\sigma = \sigma' = \frac{\text{number of diffuse reflections}}{\text{total number}}.$$

We will use a simplified assumption, valid at high Mach number, which neglects the random thermal velocity of the molecules in front of relative velocity due to the movement of the wall. In the case of the completely diffuse reflection $\sigma = \sigma' = 1$, the molecules yield their momentum to the wall, i.e., the normal component and the tangential component corresponds to a completely inelastic shock. In the case of specular reflection $\sigma = \sigma' = 0$, which corresponds to an elastic shock, the molecules yield twice their normal momentum to the wall, while retaining tangential momentum. Note that in Newtonian theory of impact, the molecules yield their normal momentum to the wall, but keep their parallel momentum as in specular reflection. Newtonian effects are bound to a pressure coefficient

$c_p = 2\cos^2\theta$ corresponding to $\sigma' = 1; \sigma = 0$. In the case of the specular reflection ($\sigma = \sigma' = 0$), the wall receives twice the incidental momentum normal, which corresponds to twice the Newtonian value, that is to say a pressure coefficient $c_p = 4\cos^2\theta$.

In the case of diffuse reflection we have, in addition to the Newtonian pressure, a shear stress, directed along the wall component of the incidental velocity. Momentum yielded to the wall in this direction per unit area and unit volume of incidental gas is $\rho V_\infty \sin\theta$, and the volume of incidental gas by unit of area and time is $V_\infty \cos\theta$. We obtain the shear stress per unit of area to the wall, equal to the product of the two terms:

$$\tau = \rho V_\infty^2 \sin\theta \cos\theta \Rightarrow c_\tau = \frac{\tau}{\bar{q}} = 2\sin\theta\cos\theta$$

We thus can summarize:

$\sigma = \sigma' = 1 \rightarrow$ Entirely diffuse reflection $\rightarrow c_p \approx 2\cdot\cos^2\theta, c_\tau \approx 2\cdot\sin\theta\cdot\cos\theta$
$\sigma = \sigma' = 0 \rightarrow$ Completely specular reflection $\rightarrow c_p \approx 4\cdot\cos^2\theta, c_\tau = 0.$

The general simplified expression for shearing and pressure coefficients for any accommodation coefficients is thus:

$$c_p = \sigma' \cdot 2 \cdot \cos^2\theta + (1-\sigma') \cdot 4 \cdot \cos^2\theta = (2-\sigma') \cdot 2 \cdot \cos^2\theta$$
$$c_\tau = \sigma \cdot 2 \cdot \sin\theta\cos\theta + (1-\sigma)\cdot 0 = \sigma \cdot 2 \cdot \sin\theta \cdot \cos\theta$$

In practice, a realistic assumption for the majority of materials, which is commonly used, corresponds to completely diffuse reflection, $\sigma = \sigma' = 1$, and:

$$c_p = 2 \cdot \cos^2\theta$$
$$c_\tau = 2 \cdot \sin\theta \cdot \cos\theta$$

This assumption results in a simple conclusion: *the total effect is limited to the drag component.*

$$c_d = \sqrt{c_p^2 + c_\tau^2} = 2\cos\theta.$$
$$c_p = c_d \cdot \cos\theta$$
$$c_\tau = c_d \cdot \sin\theta$$

The integration of these elementary coefficients on a convex body shape and on the zones directly impinged by the flow is by definition the *drag coefficient*,

$$C_D = \frac{2}{S_{ref}} \iint \cos\theta\, ds$$

4.4 Rarefied Mode

This coefficient is also written as:

$$C_D = 2\frac{S}{S_{ref}}$$

where S represents the surface of the body projected on a plane normal to the incidental direction.

We deduce immediately the drag coefficient of a sphere section, like those of disk and of cone at zero angle of attack, relating to area of their respective maximum cross sections, which are equal to 2. Within the field of this approximation, the lift coefficient C_L and the fineness coefficient $f = \frac{C_L}{C_D}$ of an arbitrarily shaped vehicle are zero, which is not very far from physical reality.

This does not mean normal force and pitching moment on a convex axisymmetric body are negligible. Indeed, normal and axial forces are obtained directly by projection of the drag force on the axes of the vehicle, but the moment must be computed either numerically by integrating the moments elements, or by geometric construction using the fact that *the resulting drag forces cross the former projected surface of the vehicle (on a plane normal to the direction of the incident flow) through the barycenter.*

We can now deduce easily with the help of the three preceding geometrical properties, force coefficients of an axisymmetric sphere cone with diameter D and semiapex angle θ_a (when all the surface upstream of the maximum cross section is impinged by the flow, i.e., for an incidence $\bar{\alpha} \leq \theta_a$):

$$C_D = 2\cos\bar{\alpha}.$$
$$C_N = C_X \sin\bar{\alpha} = 2\sin\bar{\alpha}\cos\bar{\alpha}$$
$$C_A = C_X \cos\bar{\alpha} = 2\cos^2\bar{\alpha}$$

By observing that the barycenter of the maximum cross section (which corresponds to the base plane) is the center of pressure, because it is aligned with that of projected surface normal to the upstream flow direction, we obtain the center of pressure location and the pitching moment coefficient relating to the cone apex and for $\bar{\alpha} \leq \theta_a$,

$$\frac{x_{cp}}{D} = -\frac{1}{2tg\theta_a}$$
$$C_{m/A} = C_N \frac{x_{cp}}{D} = -\frac{1}{2tg\theta_a} C_D \sin\bar{\alpha} = -\frac{1}{tg\theta_a}\sin\bar{\alpha}\cos\bar{\alpha}$$

We can observe that these expressions are applicable to a segment of sphere by using an angle θ_a corresponding to half apex angle of cone tangent at the trailing edge.

In order to assess the error associated with the assumption neglecting the random thermal velocity of gas in front of the mean macroscopic velocity, we compare hereafter the result of the simplified method with the exact analytical expression using the kinetic theory of gases for the drag coefficient of a sphere, with the assumption of completely diffuse reflection [HBK]:

$$C_D = \frac{2}{3}\frac{\sqrt{\pi}}{S}\sqrt{\frac{T_b}{T_\infty}} + \frac{2S^2+1}{\sqrt{\pi}S^3}e^{-\frac{S^2}{2}} + \frac{4S^4+4S^2-1}{2S^4}erf(S)$$

$$S = \frac{\sqrt{\gamma}}{2}Mach$$

This exact analytical estimate, with the conditions Mach 20, $T_\infty = 180.65$ K; $T_b = 300$ K; $\gamma = 1.4$, *corresponds to* $C_D = 2.098137$.

The error of the simplified estimate $C_D = 2$. relating to kinetic theory of gases is about 5%.

4.4.2 Intermediate Flow

This flow represents the transition between free molecular flow and continuous flow. Its exact theoretical calculation requires DSMC codes. They are very expensive with respect to memory and in computing times. NASA [PAR] leads efforts in this area. Certain organizations successfully use Navier–Stokes codes in the field corresponding to the beginning of rarefied flow, others develop methods called Bathnagar, Gross, and Krook [BGK] in the rarefied field. Historically, and often now, the estimates used correlation of experimental data in rarefied flow together with theoretical results in adjacent flow modes (free molecular and continuous). Models of correlation rely upon "bridging function" between free molecular and continuous flow. These bridging functions generally use Knudsen number as correlation parameter. Standard expression used for modeling the aerodynamic coefficients is:

$$C = C_c + \phi(Kn_\infty) \cdot (C_m - C_c)$$

where C_m and C_c are the respective boundary values of the coefficient in free molecular and continuous flow. The bridging functions $\phi(Kn)$ vary from zero to unity between a low value of Knudsen $Kn_c \approx 10^{-3}$ limit at the end of the continuous flow and a high value $Kn_m \approx 10$ to 100 corresponding to the beginning the free molecular flow. A kind of bridging function often used for planetary probes applications and particularly for the US space shuttle [RAT] is:

$$\phi(Kn_\infty) = \sin^2\left[\pi\left(a_1 + a_2 \log_{10} Kn_\infty\right)\right]$$

where a_1 and a_2 are constants related to Kn_m and Kn_c, which define the limits of free molecular and continuous flow.

4.4 Rarefied Mode

Another function was used for Soyuz capsule and corresponds to [RAT]:

$$\phi(Kn_\infty) = \frac{1}{2}\left[1 + \text{erf}\left(\frac{\sqrt{\pi}}{\Delta Kn}\ln\left\{\frac{Kn_\infty}{Kn_{mi}}\right\}\right)\right]$$

where Kn_{mi} corresponds to the middle of the intermediate flow, such as $\phi(Kn_{mi}) = 0.5$ and ΔKn is the logarithmic width of the intermediate zone $\Delta Kn = \ln(Kn_m) - \ln(Kn_c)$.

Figure 4.8, from [RAT] shows the comparison between the two bridging functions, NASA DSMC computations and the flight data for the Martian probe Viking 1.

We can observe that the present value of the parameter $\frac{C_N}{C_A} \approx 0.184$ in free molecular flow, determined for Viking at the angle of attack $\bar{\alpha} = 11°$, is in good agreement with the value $\frac{C_N}{C_A} = tg\bar{\alpha} - 0.194$ predicted by the simplified theory of Sect. 4.4.1. We will note that $\frac{C_N}{C_A}$ does not represent the fineness of the body, which is equal to

$$f = \frac{C_L}{C_D} = \frac{\frac{C_N}{C_A} - tg\bar{\alpha}}{1 + \frac{C_N}{C_A}tg\bar{\alpha}}.$$

Fineness of Viking 1 thus determined is equal to -0.16 in continuous flow and -0.01 in free molecular, which is in good agreement with the fineness zero predicted in free molecular with the simplified method of this book.

From the same origin, one may propose for the "Erf-Log" correlation a common value for the parameter $\Delta Kn = \ln(500) = 6.2146$ well suited for highly blunted probe shapes and a methodology, based on a single DSMC calculation in intermediate flow for each aerodynamic coefficient to fit the value of Kn_{mi}.

Results of this method relating to different probes and various aerodynamic coefficients are presented in Tables 4.1 and 4.2 whose data are from [RAT].

This method, validated here for highly blunted planet probes, must, a priori, be applicable to other kind of vehicles.

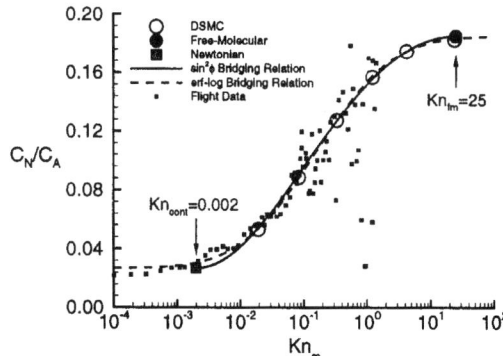

Fig. 4.8 Viking 1, comparison of bridging functions' estimates with exact computations and experimental data

Table 4.1 Parameters of bridging functions for aerodynamic coefficients of Mars'01

Coefficient	Square sine correlation			Erf-Log correlation		
	a_1	a_2	Error RMS (%)	ΔKn	Kn_{mi}	Error RMS (%)
C_A	0.2836	0.1	3.9	6.5	0.3576	4.1
C_N	0.3107	.0926	17.3	6.3	0.1804	7.0
C_m	0.3367	.0921	19.8	5.9	.0825	12.1

Table 4.2 Parameters of Erf-Log bridging function for C_A and for different vehicles obtained with the fitting method with a single DSMC computation

Vehicle	ΔKn	Kn_{mi}	Error RMS (%)
Viking	6.2146	0.1805	4.3
Pathfinder	6.2146	0.1804	4.6
Mars'01	6.2146	0.4894	4.8
Stardust	6.2146	0.0756	4.8
Microprobe	6.2146	0.0475	5.8

4.5 Qualities of Flight

Flight quality standards are specific to the particular type of aerodynamic vehicle. They are different for a sailplane, a delta plane, a transport aircraft, a stunt-flying plane or a fighter, a ballistic reentry vehicle, or a shuttle.

For a nonpowered vehicle, they are associated with the mass and inertial properties and with the aerodynamic properties. There are three important criteria:

- The velocity criterion, associated with weight and drag
- The maneuverability criterion, associated with weight and lift
- The controllability criterion, associated with rolling, yawing, pitching moments, and inertia tensor

Velocity criterion: We limit the analysis case to an axisymmetric ballistic reentry vehicle. The simplified evolution of the module of velocity during aerodynamic reentry is governed by

$$\frac{dV}{dt} = -\frac{1}{2}\rho V^2 \frac{S_{ref}C_D}{m} = -\frac{\bar{q}}{\beta}.$$

The "ballistic coefficient" parameter $\beta = \frac{m}{S_{ref}C_D}$, expressed in kg/m², characterizes the response of the vehicle to aerodynamic braking.

- For a planetary entry probe, one seeks effective braking at high altitude and low atmosphere density (to lessen heat stresses, and to allow soft landing), the parameter β must be low. The typical values are 50–100 kg/m²

4.5 Qualities of Flight

- For a military vehicle, we wish to maintain a high velocity; the parameter β must be high. The typical values are 5000–10000 kg/m².

Criterion of maneuverability: This criterion adapted to actively controlled vehicles is not very useful for a ballistic vehicle. It is characterized by the parameter "fineness ratio" $f = \frac{C_L}{C_D}$.

A high fineness ratio achieves large trajectory deviations while minimizing the velocity loss associated with the angle of attack (lift induced drag) and with the increased range of trajectory. This makes it possible either to obtain high lateral load factors for maneuvering or to increase the ground range compared to purely ballistic range.

One example corresponds to space reentry vehicles, like Apollo capsule and space shuttle. Among military vehicles, one distinguishes the maneuvers intended to avoid possible interceptions and those intended to correct trajectory to the target.

Controllability criterion: This function consists to control actively or passively the rotational movement around the center of mass in order to follow a trajectory intended for the mission.

In the case of purely ballistic vehicles two passive means of stabilization are used:

- Aerodynamic stability around the ideal angle of attack, along the pitching and yaw axes
- Gyroscopic stability, obtained using a constant rotation rate along rolling axis (spin)

4.5.1 Static Stability

4.5.1.1 Vehicle with a Symmetry Axis

In the case of a vehicle with an aerodynamic axis of symmetry, centered in $G = [x_G \ 0 \ 0]$ on the axis of symmetry, there are at least two positions of static trim such that

$$C_{mw/G}(\bar{\alpha}) = 0.$$

They correspond to the angles of attack $\bar{\alpha} = 0°$ and $\bar{\alpha} = 180°$, respectively.

The ideal incidence of flight, for which the vehicles are designed and, which minimizes the transverse effects and deviations of trajectory, is zero. The vehicle has in general an initial angle of attack at the beginning of reentry, resulting from errors at separation with the missile. This incidence becomes increasingly harmful because of growth of the dynamic pressure. It is thus necessary, in order to control this angle of attack, that there is a static restoring moment toward the flight path vector relating to air during all the reentry. In the wind-fixed frame (w) centered at G, this restoring moment is equivalent to a strictly negative coefficient of moment around pitching axis Y'_w, Fig. 4.3,

$$C_{mw/G} (\bar{\alpha}, \text{Mach}, \text{Re}, \quad \text{Kn}) = C_{mw/O} - \frac{x_G}{L_{ref}} C_{Nw} < 0.$$

Using the definition of center of pressure for a body of revolution (Sect. 4.1.3), we can write an equivalent condition,

$$C_{mw/O} - \frac{x_G}{L_{ref}} C_{Nw} < 0 \quad \Leftrightarrow \quad -\frac{x_G - x_{CP}}{L_{ref}} C_{Nw} < 0.$$

With our convention, axis OX is directed forward the vehicle. Provided $C_{Nw} > 0$ (which is the usual case, Fig. 4.9, but there are some exceptions), this is equivalent to a condition of centering the vehicle along OX forward of the center of pressure.

The static margin parameter must be strictly positive in all the flight conditions:

$$\frac{x_G - x_{CP}}{L_{ref}} > 0$$

This static stability condition for convergence toward the zero trim angle of attack is necessary, but not sufficient. We will see that there is also a dynamic stability condition.

Figure 4.10 represents, for a sphere–cone body of half apex angle 8° with a flat base and a bluntness ratio $\frac{R_N}{R_B} = \mathbf{0.1}$, the influence of center of mass location on the pitching moment from a Newtonian calculation. The center of pressure of this shape exists for the incidences $0 < \bar{\alpha} < 172°$; it is such that $\frac{x_{CP}}{L_{ref}} \approx -0.65$. For $\bar{\alpha} \geq 172°$, the pitching moment is zero and stability is indifferent. We can verify that the object

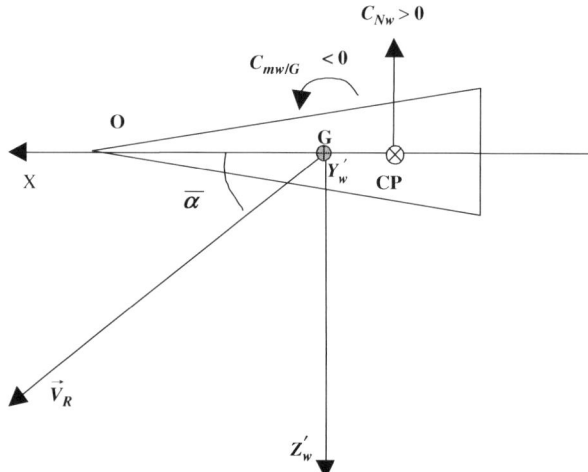

Fig. 4.9 Static stability of an axisymmetric body

4.5 Qualities of Flight

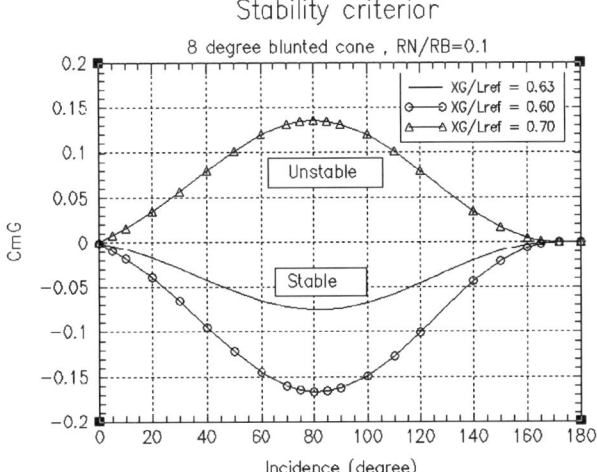

Fig. 4.10 Sensitivity of axial CG location on static stability for an axisymmetrical body

becomes unstable when static margin becomes negative and the moment increases in absolute value when static margin increases.

Except for resonance phenomena, which will be studied in Chaps. 11 and 12, two conditions, static and dynamic, are sufficient to ensure converging toward low incidence. However, in order to prevent the risks of resonance, which appear primarily at two altitudes range, the control of trim eventually require

- A more severe condition on the static margin, such as $\frac{x_G - x_{CP}}{L_{ref}} >$ strictly positive minimum value to be defined.
- Conditions on the maximum acceptable level of aerodynamic and inertial asymmetries.

We will observe that, in practice, the strict static stability condition can be relaxed in some circumstances.

It is indeed possible to fly with some level of static instability in rarefied flow at low dynamic pressure, which can be compensated by gyroscopic stability (case of planetary entry probes). When the vehicle has nonlinear behavior in incidence (e.g., an aftward motion of the center of pressure with angle of attack), it is possible to accept temporarily, in continuous flow and in a range of altitude where the dynamic pressure is still moderate, some level of instability at low angle of attack. This requires good aerodynamic tools and a good control of the design.

Note: We repeat our warning with respect to center of pressure and static margin concept. The only physical criterion, which allows pertain stability is the behavior of the moment around the center of mass. Indeed, the condition of positive static margin is generally valid, but not in all circumstances.

4.5.1.2 Nonaxisymmetric Vehicle

For a more general vehicle, including plane symmetric (like sailplane), the aerodynamic model corresponds to no separable functions of incidence and sideslip angle. In addition, the trim state does not necessarily correspond to zero incidence and/or sideslip angles,

$$C_{m/G}\left(\alpha_{eq}, \beta_{eq}\right) = 0$$
$$C_{n/G}\left(\alpha_{eq}, \beta_{eq}\right) = 0$$

The differential stability condition for small deviations $\delta\alpha, \delta\beta$ from the trim state $\lfloor \alpha_{eq} \quad \beta_{eq} \rfloor$ corresponds to a static restoring moment. Taking into account the sign conventions (Fig. 4.1) and definitions of moment coefficients (Sect. 4.1.1), this condition is:

$$\left.\frac{\partial C_{m/G}}{\partial \alpha}\right|_{\alpha_{eq},\beta_{eq}} < 0$$

$$\left.\frac{\partial C_{n/G}}{\partial \beta}\right|_{\alpha_{eq},\beta_{eq}} > 0$$

In the same way, the general condition of stability in a finished field $D(\alpha, \beta)$ around $\lfloor \alpha_{eq}\beta_{eq} \rfloor$ corresponds to a criterion associated with the sign of the moment coefficients (in order to have a restoring moment). With the previous sign and moments coefficient conventions, the criterion is

$$Sign\left\{C_{m/G}(\alpha, \beta)\right\} = -sign\left\{\alpha - \alpha_{eq}\right\}$$
$$Sign\left\{C_{n/G}(\alpha, \beta)\right\} = +sign\left\{\beta - \beta_{eq}\right\}$$

As we pointed out previously, static margin criteria, associated to center of pressure and focus concepts, are not universal and are to be handled with caution because they sometimes lead to false conclusions.

We will say that in general, differential stability around any axis (pitching or yawing) is obtained under the condition the center of mass is before "aerodynamic focus" associated with corresponding pitching/yawing moment and normal/lateral force.

Aerodynamic focus are the correspondents of the centers of pressure with respect to small variation of aerodynamic static force and moment around a given equilibrium state.

$$\frac{x_{FN}}{L_{ref}} = \frac{\partial C_m}{\partial C_N}; \quad \frac{x_{FT}}{L_{ref}} = \frac{\partial C_n}{\partial C_Y}; \quad \frac{z_{CPT}}{L_{ref}} = -\frac{\partial C_l}{\partial C_Y}$$

A body having a plane of symmetry generally possesses a different focus (or center of pressure) for normal force (C_N) and lateral force (C_Y).

4.5 Qualities of Flight

For an axisymmetric body, we can define a single focus (and also a single center of pressure) such as $\frac{x_{FN}}{L_{ref}} = \frac{x_{FT}}{L_{ref}} = \frac{x_F}{L_{ref}}; \frac{z_{FT}}{L_{ref}} = 0$.

Analysis of stability around trim generally needs to use the criterion on sign of the moment coefficients.

The simplest example of nonsymmetric behavior is a body with axisymmetric aeroshell, whose mass center is not on the symmetry axis. We assume that the normal force coefficient has a "usual" behavior, i.e., $C_{Nw}(\bar{\alpha}) \geq 0$ for all $\bar{\alpha}$. We choose the symmetry plane like the pitching plan, i.e., the meridian plane around the axis of symmetry $O'X$ that includes the center of mass, Fig. 4.11.

Pitching and yawing moments are given by:

$$C_{m/G} = C_{m/O} - \frac{x_G}{L_{ref}} \cdot C_N + \frac{z_G}{L_{ref}} \cdot C_A = -\frac{x_G - x_{CP}}{L_{ref}} \cdot C_N + \frac{z_G}{L_{ref}} \cdot C_A$$

$$C_{n/G} = C_{n/O} - \frac{x_G}{L_{ref}} \cdot C_Y = -\frac{x_G - x_{CP}}{L_{ref}} \cdot C_Y.$$

Taking into account the relations of Sect. 4.1.3, these equations are rewritten using wind-fixed (w) coefficients:

$$C_{m/G} = -\frac{x_G - x_{CP}(\bar{\alpha})}{L_{ref}} \cdot \frac{C_{Nw}(\bar{\alpha})}{\sin\bar{\alpha}} \cos\beta\sin\alpha + \frac{z_G}{L_{ref}} \cdot C_{Aw}(\bar{\alpha})$$

$$C_{n/G} = -\frac{x_G - x_{CP}}{L_{ref}} \cdot C_Y = \frac{x_G - x_{CP}(\bar{\alpha})}{L_{ref}} \cdot \frac{C_{Nw}(\bar{\alpha})}{\sin\bar{\alpha}} \sin\beta$$

The zero value of the yawing moment for $\beta = 0$ shows zero sideslip is a possible trim condition. Variations of moment around $\beta = 0$ show this trim is stable in the interval around trim where static margin is positive (function $\frac{C_{Nw}(\bar{\alpha})}{\sin\bar{\alpha}}$ is by hypothesis strictly positive).

Analysis of the pitching moment for $\beta = 0$ results in a trim angle of incidence such that,

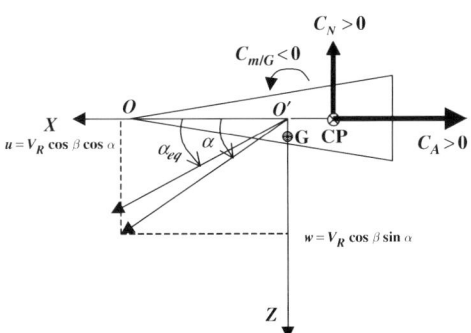

Fig. 4.11 Stability of a nonsymmetrical object

$$C_{Nw}\left(|\alpha_{eq}|\right) \cdot \text{sign}\left(\alpha_{eq}\right) \doteq \frac{z_G}{x_G - x_{CP}\left(|\alpha_{eq}|\right)} \cdot C_{Aw}\left(|\alpha_{eq}|\right)$$

We observe that the absolute value of the trim incidence is inversely proportional to the static margin.

From analysis of pitching moment variations around α_{eq}, assuming the axial force coefficient is near constant, we conclude that the static stability conditions is equivalent to a positive static margin. This behavior is represented in Fig. 4.12 for the same body as in Fig. 4.9, computed in the Newtonian approximation and for center of mass locations $\frac{x_G}{L_{ref}} = 0.63; 0.62$ & $\frac{z_G}{L_{ref}} = 0.0; 0.1$.

4.5.2 Gyroscopic Stability

A given rotation rate "p" of the vehicle around its longitudinal axis (roll) insures some gyroscopic stiffness along this axis. This method is effective during the ballistic phase and during high-altitude reentry where external moments are low compared to gyroscopic moment. We will see also that for a ballistic vehicle a minimum level of constant sign roll rate is needed during the whole reentry to insure the stability of the flight path angle.

This topic will be developed later during the study of dynamic reentry behavior.

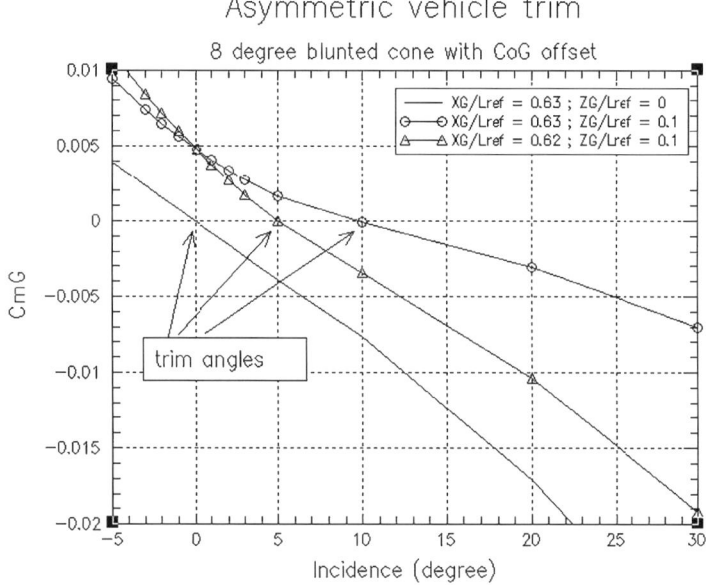

Fig. 4.12 Static stability of nonsymmetrical body

4.6 Characteristics of a Family of Sphere Cones

To illustrate the previous chapter and because we will need some of these results later, we present the aerodynamic coefficients of a series of sphere–cone reentry vehicles (Fig. 4.13) whose characteristics are:

- Common origin at apex of the sharp cone
- Half apex angle, $\theta_c = 8°$
- Base radius, $R_b = 0.25$ m
- Common reference length is sharp cone length $L_{apex} = \frac{R_b}{\tan \theta_c} = 1.788$ m
- Common reference area is that of maximum cross section $S_{ref} = \pi R_b^2 = 0.19635 \, \text{m}^2$

Nosetip radius R_n, bluntness ratio $e = \frac{R_n}{R_b}$, and overall length L_v of each vehicle are given in Table 4.3.

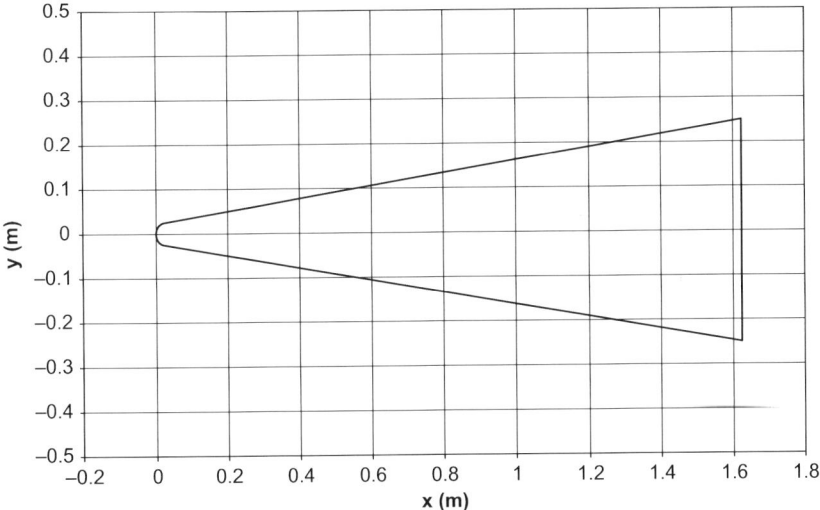

Fig. 4.13 Blunted cone, 8 degree, e = 0.1

Table 4.3 Nosetip radius R_n, bluntness ratio $e = \frac{R_n}{R_b}$, and overall length L_v of each vehicle

R_n/R_b	0	0.05	0.1	0.15	0.2
R_n(mm)	0	12.5	25	37.5	50
L_v(m)	1.7788	1.7015	1.6242	1.5469	1.4695

Overall length, $L_v = L_{apex} - R_n \left(\frac{1}{\sin \theta_c} - 1 \right)$.

4.6.1 Mach Number Influence

Center of pressure evolution at 1° angle of attack (Euler code results) is shown in Fig. 4.14, evolution of static margin for center of mass location $\frac{x_G}{L_{ref}} = -65.3\%$ is shown in Fig. 4.15, and normal force coefficient at 1° angle of attack is shown in Fig. 4.16.

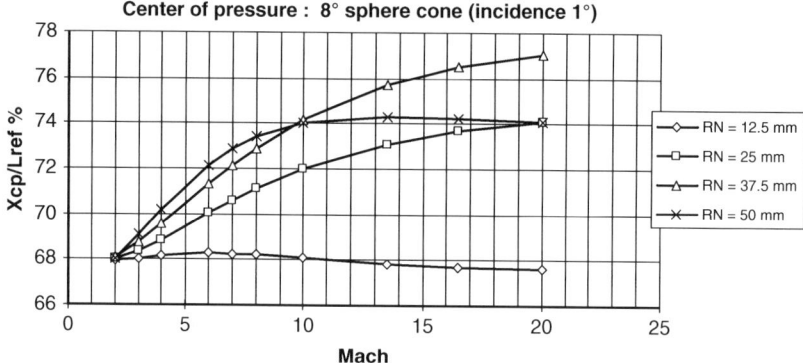

Fig. 4.14 Center of pressure location

The normal forces' coefficient at 1° incidence is often used to approximate the gradient of normal force coefficient:

$$C_{N\alpha}(\text{degree})^{-1} \approx C_N(1 \text{ degree})$$

Evolution of axial force coefficients at zero angle of attack is shown in Figs. 4.17–4.19.

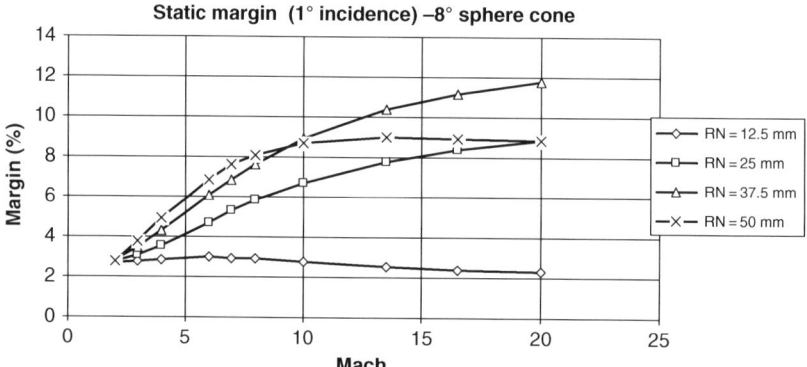

Fig. 4.15 Static margin

4.6 Characteristics of a Family of Sphere Cones

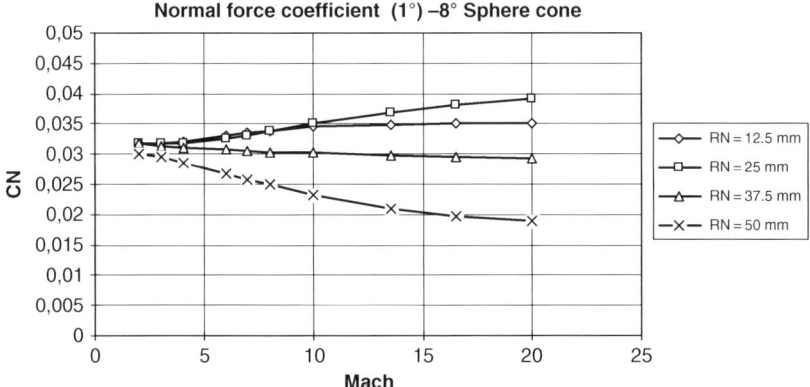

Fig. 4.16 Normal force coefficient

We can observe that pressure and friction components are significantly dependent on bluntness ratio, unlike the base pressure component. However, when we compare total axial force coefficients at 5 km altitude, we can see that influence of bluntness ratio is weaker, thanks to a compensation effect, because pressure drag grows with nose radius while friction drag decreases (Fig. 4.20). Moreover, we can see that at high Mach number, there is an optimum bluntness ratio $e \approx 0.06$ corresponding to $R_N \approx 30$ mm for which total drag coefficient is minimum.

The friction component of axial coefficient depends strongly on altitude as well as flow turbulence and wall roughness as illustrated in Figs. 4.21 and 4.22 for $R_N = 25$ mm (Euler code + integral boundary-layer solution).

Base pressure component of axial force coefficient is estimated from experimental correlations of the base pressure and upstream conditions computed at the boundary layer limit at the end of the body (using entropy swallowing method [AND], which consists in determining in the inviscid flow conditions at the boundary layer

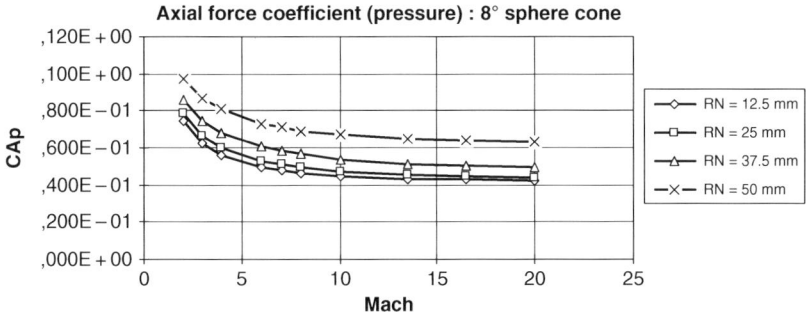

Fig. 4.17 Pressure component of axial force coefficient

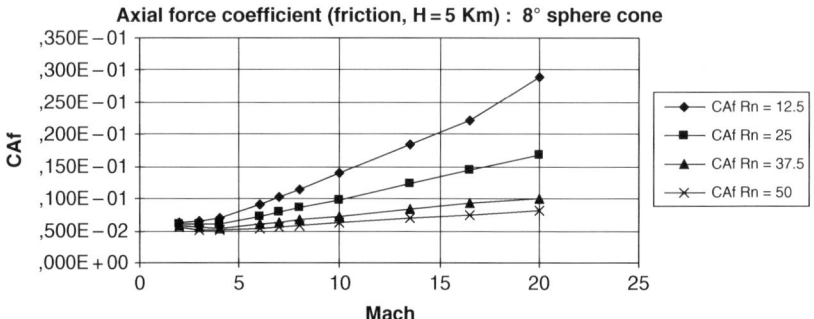

Fig. 4.18 Friction component of axial force coefficient

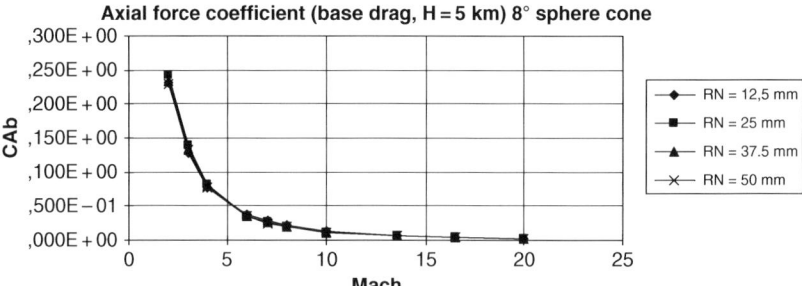

Fig. 4.19 Base pressure component of axial froce coefficient

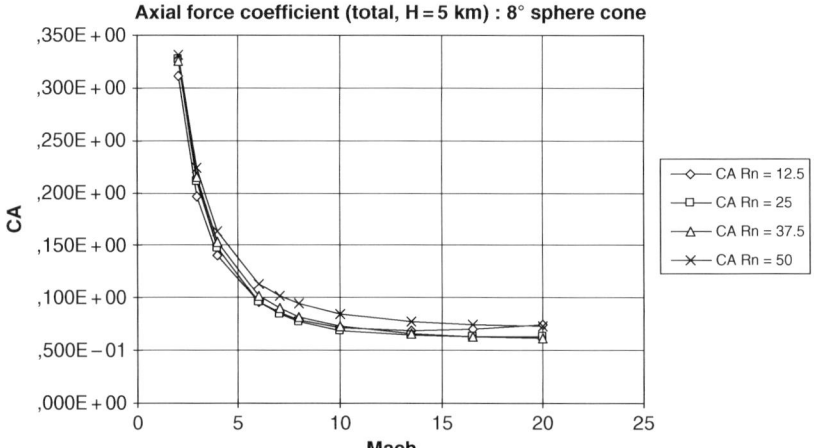

Fig. 4.20 Total axial force coefficient

4.6 Characteristics of a Family of Sphere Cones

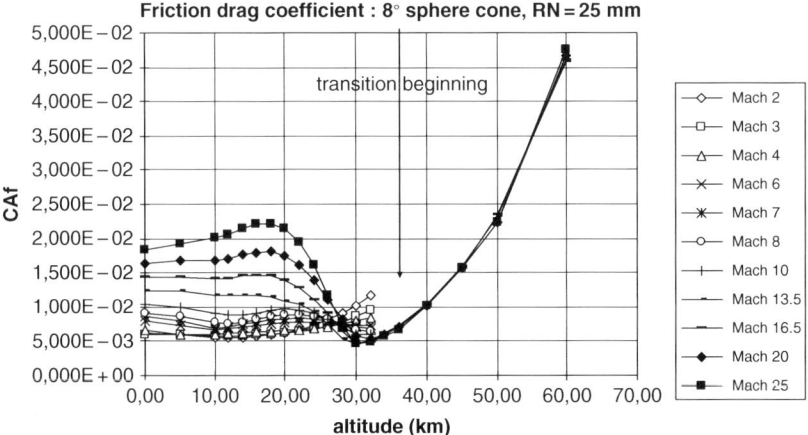

Fig. 4.21 Friction drag coefficient versus altitude, smooth wall

edge). As we can notice in Fig. 4.23, for $R_N = 25$ mm, this coefficient is weakly depending on altitude.

The previous results justify modeling the total axial force coefficient at zero angle of attack as $C_A = C_{Ap}(Mach) + C_{Af}(Mach, altitude) + C_{Ab}(Mach)$.

Evolution of pitch damping moment coefficient versus Mach number for a center of mass location $\frac{x_G}{L_{ref}} = -65.3\%$ is represented in Fig. 4.24.

We can notice in Figs. 4.15 and 4.24 that the increase in the nose radius increases the static stability (static margin) at high Mach number but decreases dynamic stability (damping coefficient).

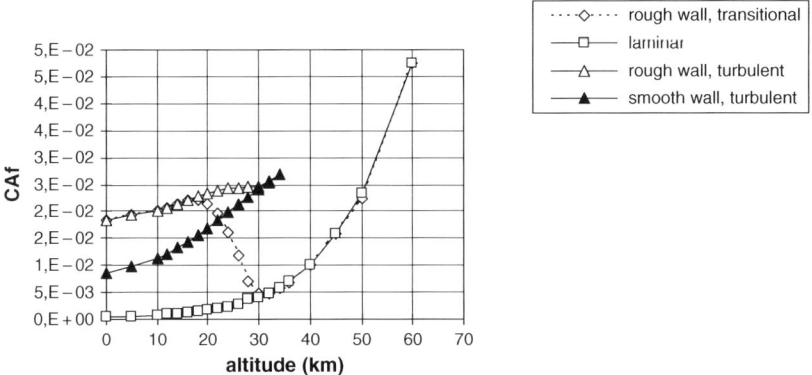

Fig. 4.22 Friction drag coefficient versus altitude, influence of transition and roughness

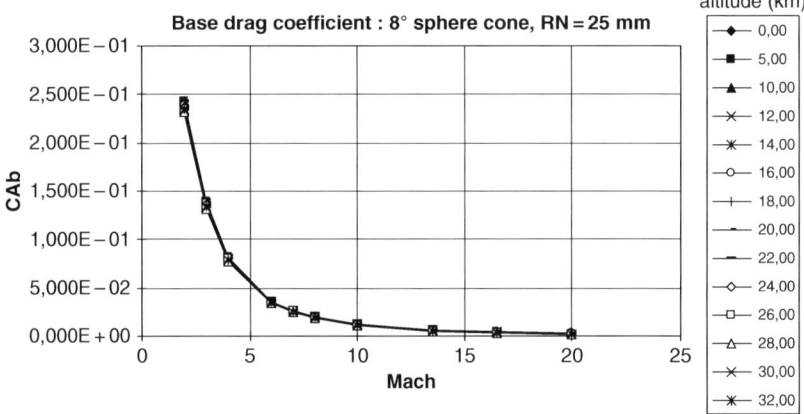

Fig. 4.23 Base axial force component versus Mach and altitude

Influence of center of mass location on pitch damping moment coefficient for the sharp cone (Newtonian estimate) is given in Fig. 4.25.

The roll damping coefficient assessment for $R_N = 25$ mm (Euler code + integral boundary layer solution), is represented in Fig. 4.26.

4.6.2 Influence of Angle of Attack

The estimates of inviscid aerodynamic coefficients up to $8°$ incidence (Euler code) are represented in Figs. 4.27–4.30.

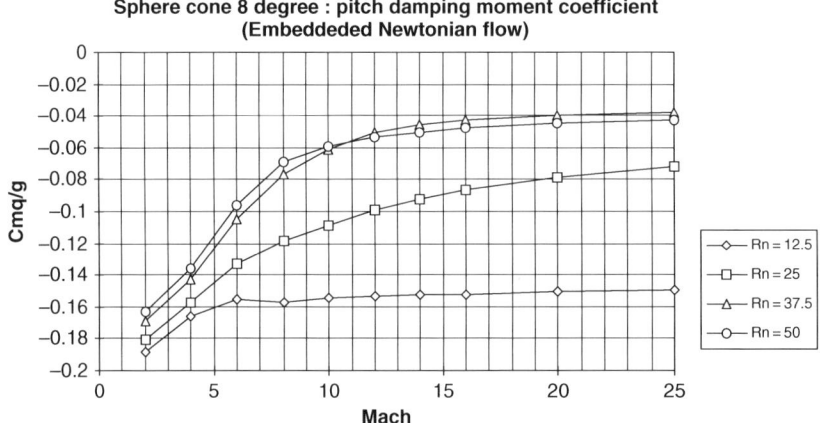

Fig. 4.24 Pitch damping moment coefficient

4.6 Characteristics of a Family of Sphere Cones

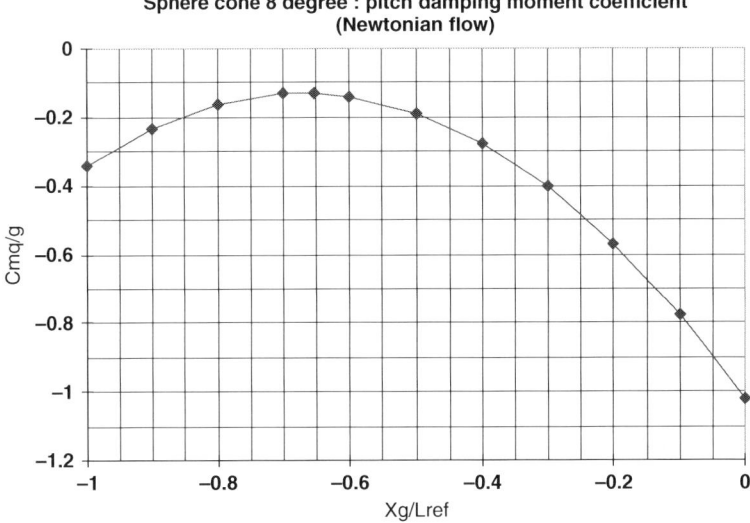

Fig. 4.25 Pitch damping moment coefficient

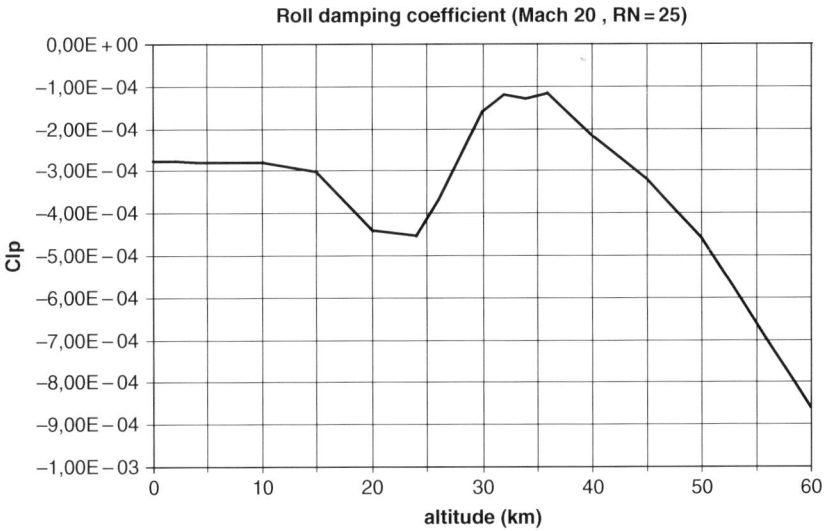

Fig. 4.26 Roll damping coefficient

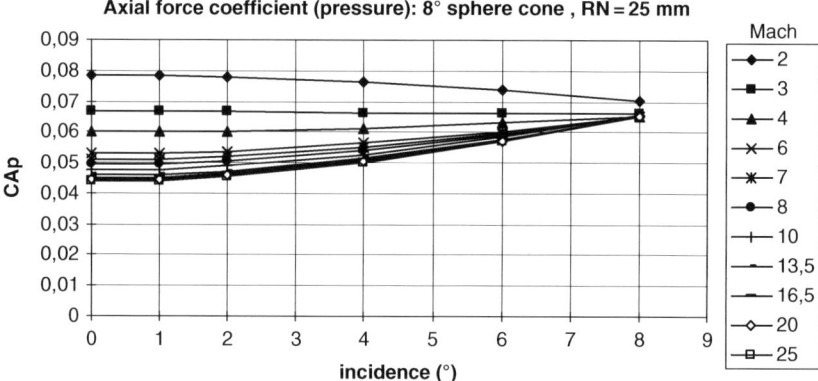

Fig. 4.27 Evolution of axial force pressure component

- Digital code solutions considered to be exact (Euler, Navier–Stokes) are currently applicable to most configurations. However, tractable computations [Euler + boundary-layer codes, parabolized Navier–Stokes (PNS) codes] are available up to 30°–35° incidence for most robust codes, 10°–15° incidence depending on geometries for more accurate but less robust codes. Estimates available for higher incidences are generally derived from experimental results (ground or flight data) or from codes using empirical pressure correlations like Newton or HABP (NASA code). There is good agreement between the various numerical methods around 30°–35° incidence and a correct agreement between Newtonian and experimental pressures at higher angle of attack. Thus, Newtonian estimates represent a reasonable first approximation of the inviscid aerodynamic coefficients at high angle of attack in the conceptual design phase (experimentation is still essential during the subsequent phases). Considering now the viscous drag

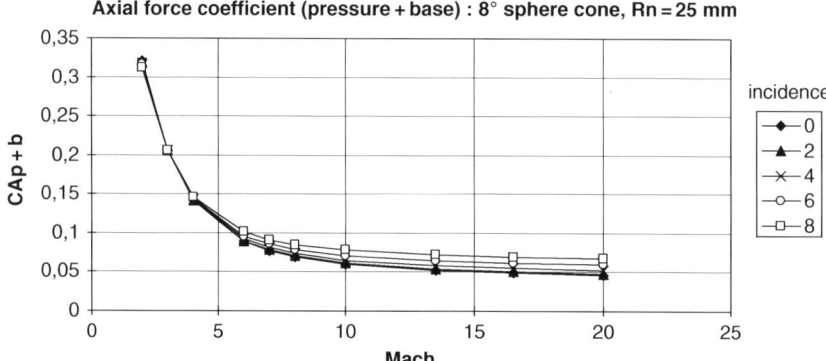

Fig. 4.28 Evolution of pressure plus base component of axial force coefficient

4.6 Characteristics of a Family of Sphere Cones 73

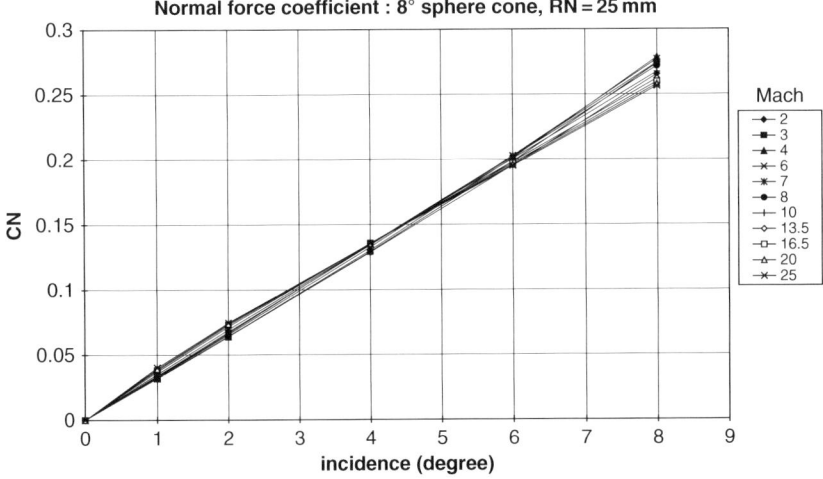

Fig. 4.29 Evolution of normal force coefficient versus angle of attack

components at high angle of attack (friction and base pressure components) in continuous flow:
- High angles of attack are generally limited to high altitudes. In this altitude range, ballistic reentry vehicles have high Mach number and base drag is negligible;
- At higher angles of attack, friction drag also becomes negligible compared with pressure drag.

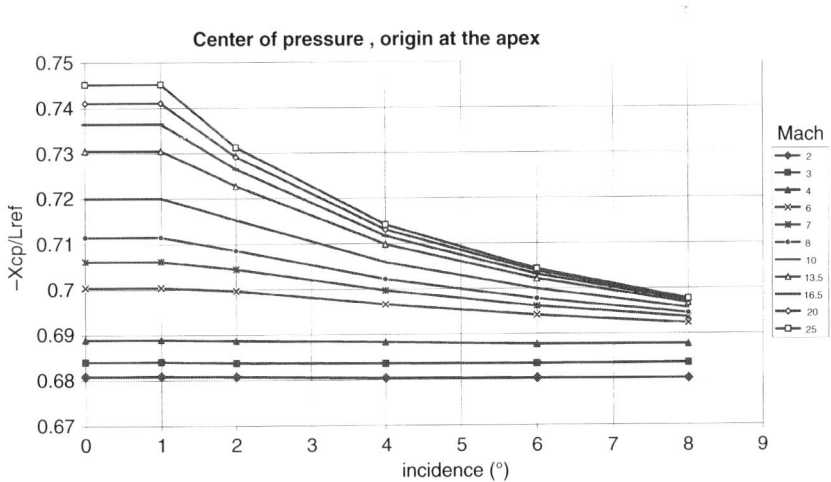

Fig. 4.30 Evolution of center of pressure location versus angle of attack

The Newtonian estimate ($C_{p,stag} = 2$) of pressure drag for a 8° cone with 25 mm nosetip radius is compared with more exact codes in Figs. 4.31–4.33.

At Mach 10, we have agreement at incidences lower than 8° with axial and normal force estimates from Euler codes; at decreasing Mach number, agreement is very poor. For the center of pressure, the best agreement with Euler results occurs at Mach 2 and 3 and the agreement improves at higher angle of attack for all Mach numbers.

4.6.3 Aerodynamic Modeling for Trajectory Codes

Taking into account the state of the art, the ideal model for up to 30° angles of attack would use coefficients as a function of Mach number, altitude and incidence including pressure and friction from Navier–Stokes or PNS codes. An approximation commonly used models the normal forces and the pitching moment (or center of pressure) with inviscid Euler code results. The axial force is then modeled as the sum of a pressure component and a viscous drag component including the base, using integral boundary layer solutions and appropriate base pressure correlations.

$$C_N = C_N (\text{Mach}, \bar{\alpha})$$

$$\frac{x_{CP}}{L_{rel}} = \frac{x_{CP}}{L_{rel}} (\text{Mach}, \bar{\alpha})$$

$$C_A = C_{Ap} (Mach, \bar{\alpha}) + C_{Af} (Mach, H or \text{Re}) + C_{Ab} (Mach, H or \text{Re})$$

At incidences higher than the capability of Euler or Navier–Stokes numerical codes, the model is derived either from pressure correlation (Newton or HABP) or

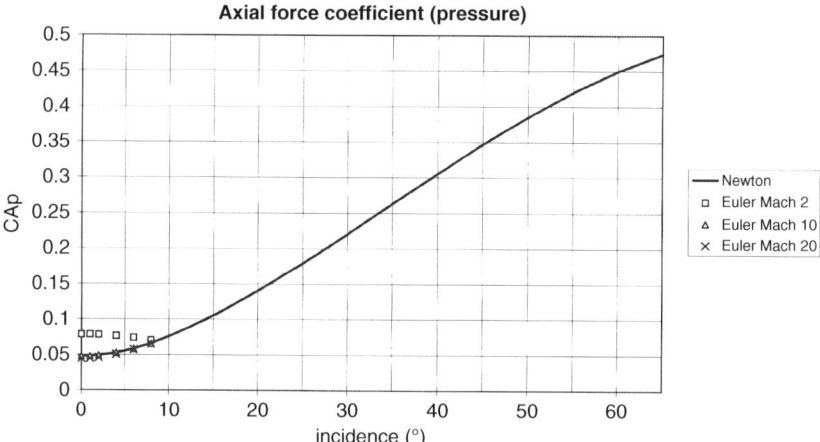

Fig. 4.31 Euler–Newton comparison

4.6 Characteristics of a Family of Sphere Cones

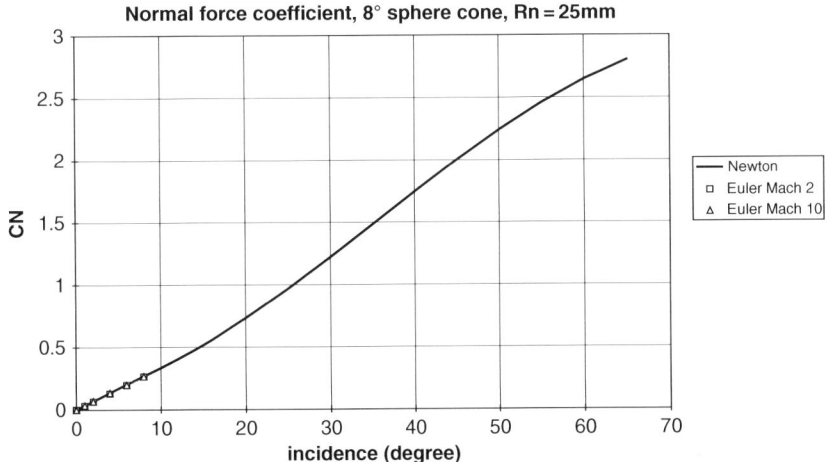

Fig. 4.32 Normal force coefficient Euler–Newton comparison

from experimental data. For practical reasons, the need for accuracy in this range is less critical and usually related to stability of the vehicle (static margin).

Note 1: Model based on altitude uses a standard atmosphere; use of Reynolds number is valid for any atmosphere (of same chemical mixture).

Note 2: This viscous axial force model is independent of incidence. It was the most realistic model for a reasonable cost at the time of the first edition. Computer and numerical algorithm progress make it possible now to use Navier–Stokes or PNS calculations at angle of attack.

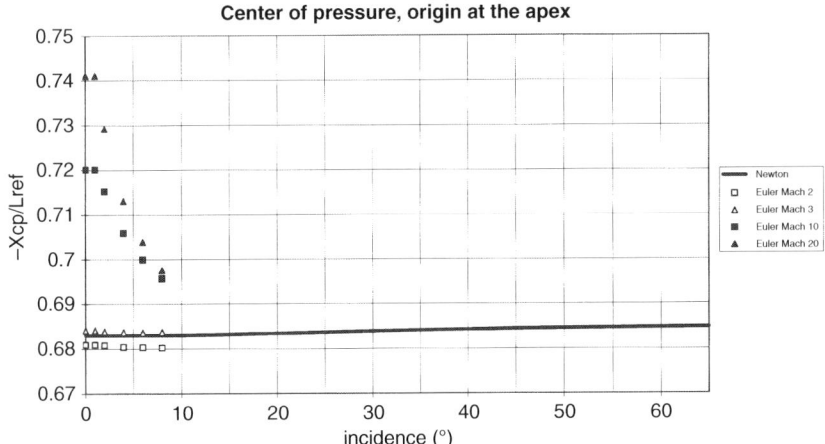

Fig. 4.33 Center of pressure Euler–Newton comparison

Note 3: Euler inviscid calculation is not valid when separation of the flow from the wall occurs. This occurs when angular surface angles are abruptly increased. It also occurs on the leeward side at high angle of attack and on the base in most conditions.

4.7 Planetary Entry Capsule

For planetary entry capsules, the aerodynamic entry requirements are quite different from military reentry vehicles. The objective is to decelerate a scientific payload or a human crew from an elliptic or hyperbolic outer atmosphere trajectory to a low supersonic velocity at a suitable altitude in order to initiate a soft landing. To alleviate thermal and mechanical stresses, deceleration must occur at the highest possible altitude. This requires for (see Sect. 8.2) a shallow trajectory (small path angle) and a blunt aeroshell, which offers a low ballistic coefficient $\beta = \frac{m}{S_{ref}C_D}$. Hence, the aeroshell shape must have high value of drag parameter $S_{ref}C_D$ (with dimensions compatible with launcher diameter) and stability compatible with the payload constraints (CG location).

The most usual shapes are:

- Sphere or spherical cap manned capsule (NASA Mercury, Gemini and Apollo, Russian Vostok, Soyuz, Chinese Shenzhou, ESA Aerodynamic Reentry Demonstrator (ARD), and German Mirka)
- Sphere–cone with high semiapex angle (NASA Viking and Mars Pathfinder, ESA Huygens probes, and Japan Orex)
- Sphere–cone with medium semiapex angle (NASA Pioneer Venus, Galileo, and Stardust probes) (Fig. 4.35)

Table 4.4 gives the front shield geometrical parameters' values of some of these vehicles.

Apollo [APO], Mars Pathfinder [PAT], Pioneer Venus, and Galileo aeroshell are shown in Fig. 4.35.

Table 4.4 The front shield geometrical parameters' values

Vehicle	Geometry	Bluntness ratio $\frac{R_N}{D}$	Semiapex angle θ	Shoulder radius at maximum diameter $\frac{R_A}{D}$
Mercury	Spherical cap	4.106	no	
Gemini	Spherical cap	1.6	no	
Apollo, ARD	Spherical cap	1.2	no	0.05
Vostock, Soyuz, Mirka, Shenzhou	Sphere	0.5	no	
Galileo	Sphere–Cone	0.176	44.85°	No
Huygens	Sphere–Cone	0.463	60°	0.0185
Pathfinder	Sphere–Cone	0.25	70°	0.025

4.7 Planetary Entry Capsule

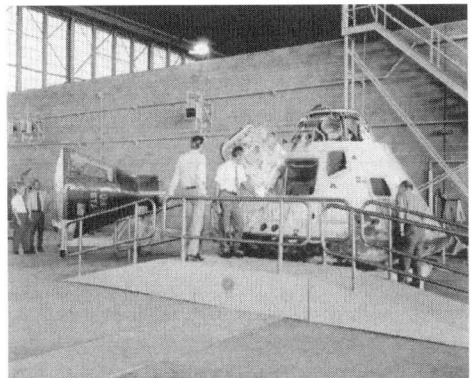

Fig. 4.34 Mercury and Apollo capsules (pictures from NASA)

Near zero angle of attack, these shapes have a high axial force coefficient, $C_A \approx$ 1 to 1.7 and a small positive normal force coefficient C_N above Mach 1.5. Their lift coefficients, mainly generated by the axial force, are negative up to moderate angle of attack,

$$C_L = C_N \cos\overline{\alpha} - C_A \sin\overline{\alpha} \approx -C_A \sin\overline{\alpha}.$$

This lift coefficient behavior is quite different from slender shapes of Sect. 4.5, which are generated mainly by normal force, which is positive up to moderate angle of attack,

$$C_L = C_N \cos\overline{\alpha} - C_A \sin\overline{\alpha} \approx C_N \cos\overline{\alpha}.$$

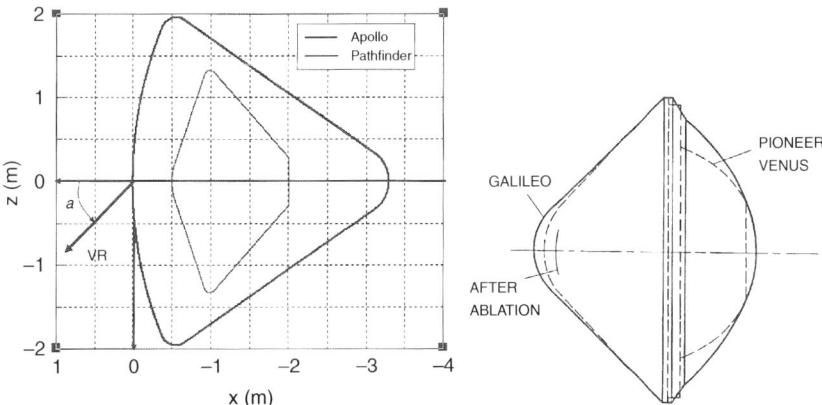

Fig. 4.35 Apollo, Mars Pathfinder, Pioneer Venus and Galileo aeroshell

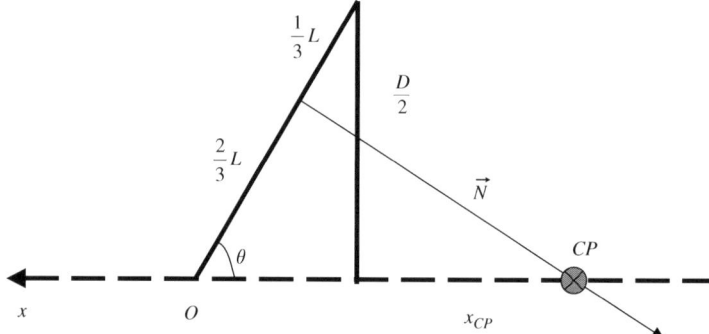

Fig. 4.36 Construction for the center pressure of a cone

The center of pressure of Apollo and Soyuz shapes is naturally close to the center of the spherical cap or the sphere. The center of pressure of sharp conical shapes can be derived using the simple geometrical method shown in Fig. 4.35. Indeed, we can approximate the cone surface using small isosceles triangles with origins at the apex; assuming the pressure is uniform, the center of pressure is the common intersection of normal (forces) from those triangles with the cone axis.

This approximate construction gives the same result as the Newtonian approximation:

$$\frac{x_{CP}}{D} \approx -\frac{1}{3\sin\theta\cos\theta}$$

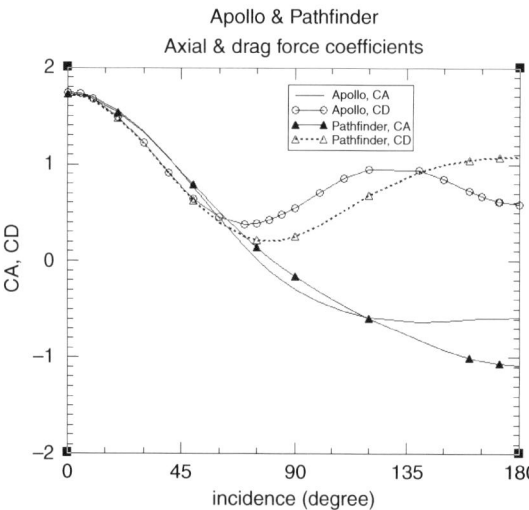

Fig. 4.37 Newtonian aerodynamic axial and drag force coefficients for Apollo and Mars Pathfinder

4.7 Planetary Entry Capsule

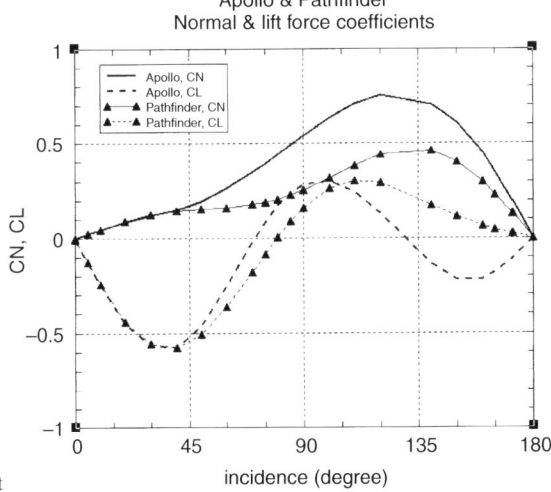

Fig. 4.38 Newtonian aerodynamic normal and lift force coefficients for Apollo and Mars Pathfinder

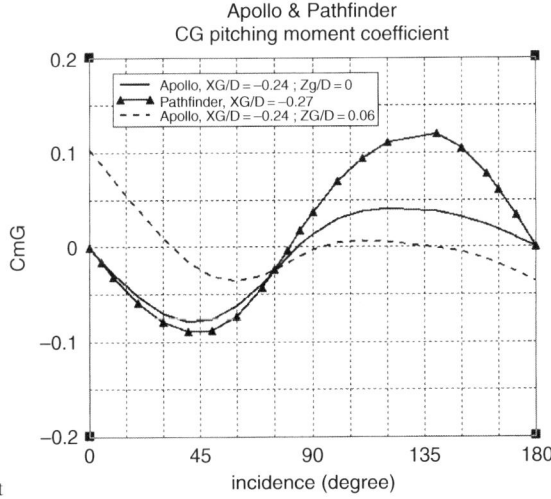

Fig. 4.39 Newtonian aerodynamic CG pitching moment coefficients for Apollo and Mars Pathfinder

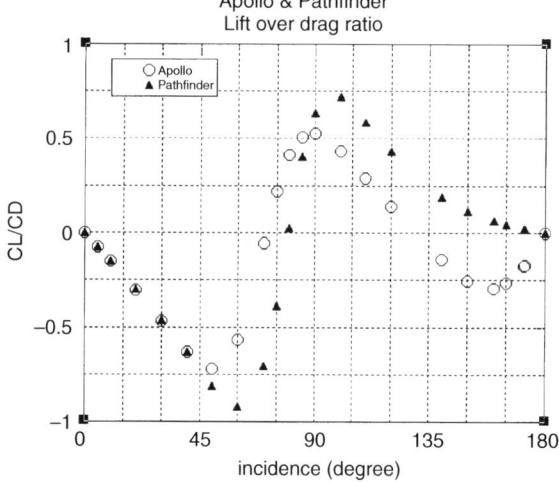

Fig. 4.40 Newtonian aerodynamic lift over drag ratio coefficients for Apollo and Mars Pathfinder

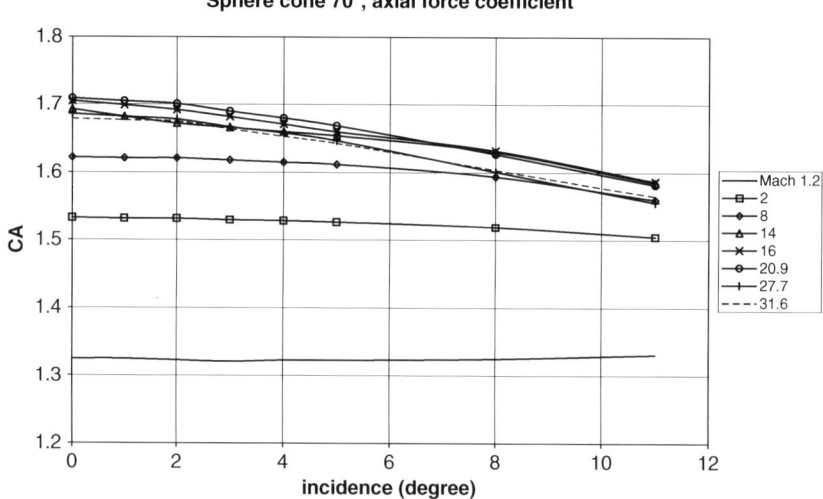

Fig. 4.41 Axial force coefficients in the continuous regime of a 70° sphere cone configuration similar to Pathfinder

It is thus easy to understand why the center of pressure of a high-angle cone is so far aft, which is vital for centering the payload of the vehicle.

Figures 4.37–4.40 compare Newtonian aerodynamic coefficients ($C_{p,stag} = 2$) for Apollo and Mars Pathfinder.

4.7 Planetary Entry Capsule

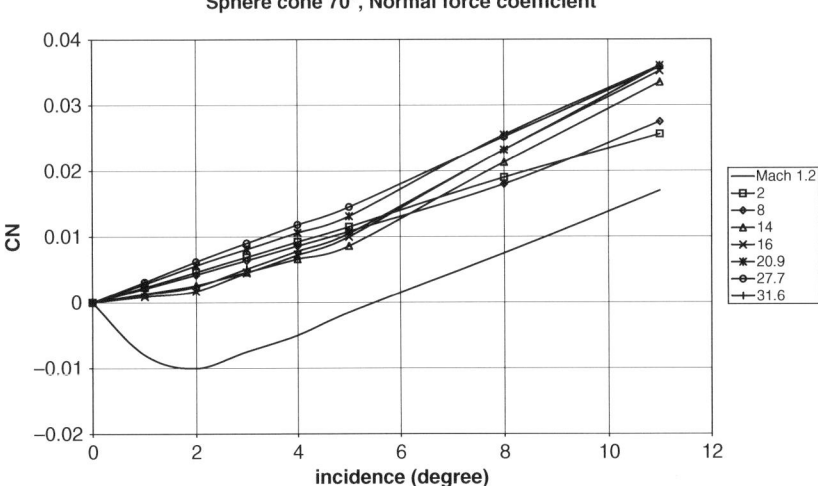

Fig. 4.42 Normal force coefficients in the continuous regime of a 70° sphere cone configuration similar to Pathfinder

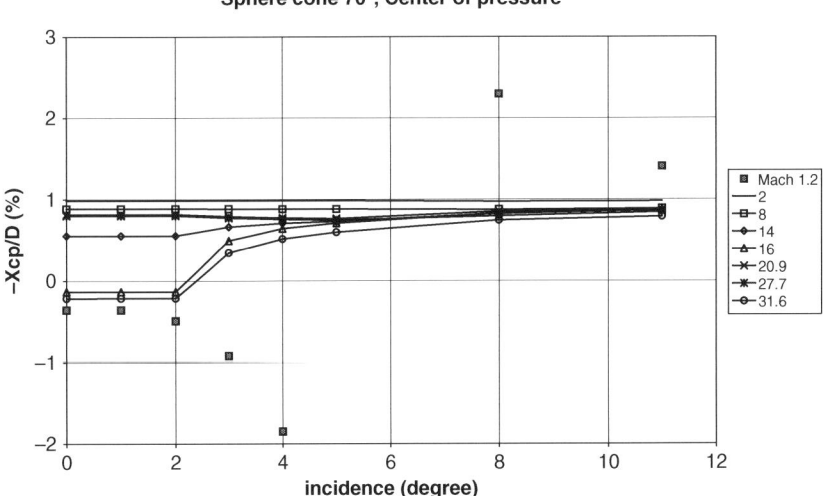

Fig. 4.43 Center of pressure coefficients in the continuous regime of a 70° sphere cone configuration similar to Pathfinder

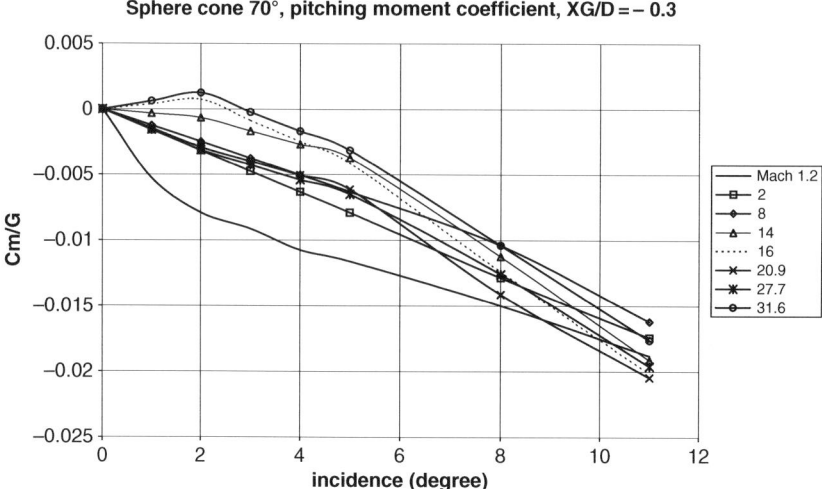

Fig. 4.44 Pitching moment coefficients in the continuous regime of a 70° sphere cone configuration similar to Pathfinder

We notice that aerodynamic coefficients of the two shapes are similar up to 40° or 50° angle of attack. Pathfinder has a slightly better static stability (calculations correspond to a center of mass location $x_G/D = -0.24$ for Apollo and $x_G/D = -0.27$ for Pathfinder). The two vehicles are stable from 0° to 75° angle of attack

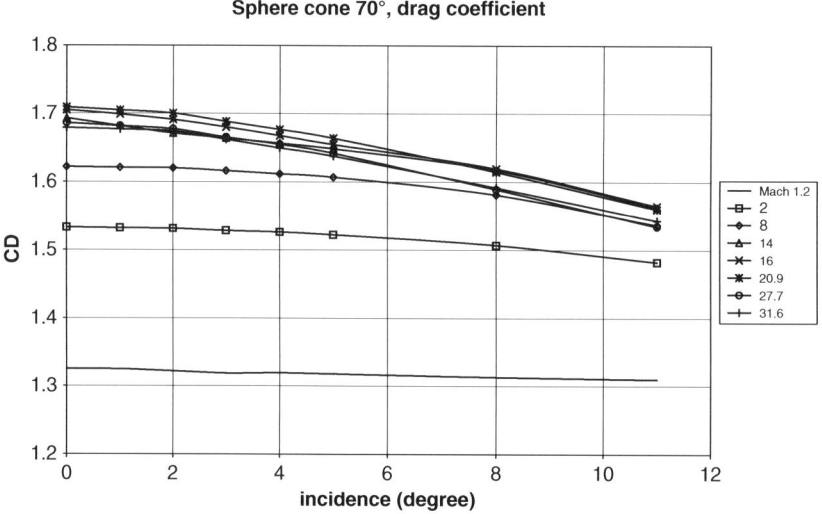

Fig. 4.45 Drag coefficients in the continuous regime of a 70° sphere cone configuration similar to Pathfinder

4.7 Planetary Entry Capsule

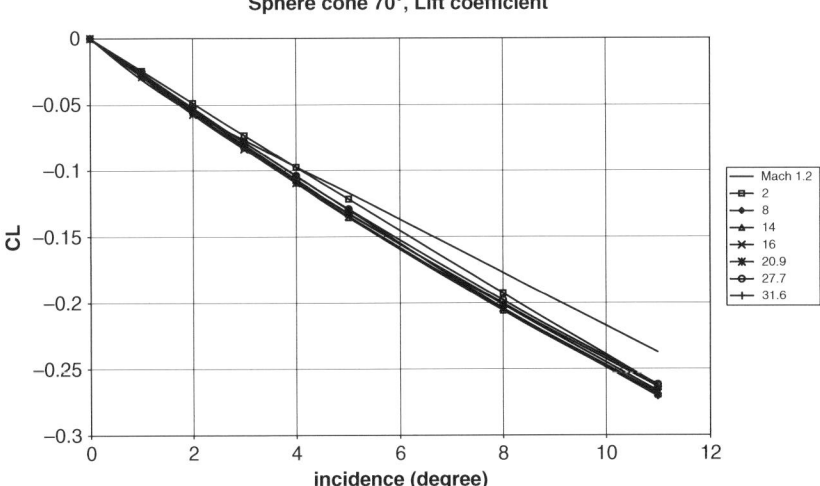

Fig. 4.46 Lift coefficients in the continuous regime of a 70° sphere cone configuration similar to Pathfinder

(Fig. 4.39), above this range they become stable with aft cover forward (around 180° trim angle of attack).

A model for static coefficients in the continuous regime [PAT],[VIK],[GAL] of a 70° sphere cone configuration similar to Pathfinder is given in Figs. 4.41–4.46. This model is valid in a CO_2 atmosphere, for a Mach range from 1.2 to 31.6 and incidence angles from 0° to 11°.

It is worth noting the abnormal behavior of normal force coefficient and center of pressure [VIK] versus angle of attack at 1.2 Mach number. The C_N coefficient is negative, and center of pressures is well ahead of the stagnation point at less than 5° AoA. This phenomenon relates to windward flow separation aft of the maximum cross section. The pitch moment coefficient curve (Fig. 4.44) shows that this behavior does not involve a static instability of the vehicle. This illustrates the caveat of Sect. 4.5.1.1 about use of center of pressure and static margin as stability criterion. In the present circumstances, the vehicle remains stable although static margin is negative. The only universal stability criterion is based on the pitching moment coefficient.

Chapter 5
Inertial Models

For a homogeneous and compact body such as a ballistic reentry vehicle or a planetary entry capsule, the offset between center of mass and center of gravity is very small, and its effect on flight mechanics, even out of the atmosphere, is negligible. In this book we will not distinguish between them.

By experience, the most appropriate choice of the Eulerian frame associated with vehicle is such that longitudinal axis Gx is parallel to the aerodynamic axis of symmetry (roll axis).

Indeed, we could choose alternatively Gx along the corresponding main direction of inertia, but the previous choice is far better as it corresponds to the simplest aerodynamic model in the equations of motion.

5.1 Moments of Inertia

In ideal circumstances, the CG is along the aerodynamic axis of symmetry, and the longitudinal axis of inertia has the same direction. The whole configuration is close to axisymmetric, and the other Eulerian axes may be chosen arbitrarily such that the frame (Gx_E, Gy_E, Gz_E) is orthogonal and direct. To define the orientation of the main inertia frame, we can use the rotation angle φ along Gx_E, which transforms Gy_E and Gz_E into the lateral main directions of inertia. For configurations of interest, the transverse inertia are usually close, $I_y \approx I_z$. Longitudinal-to-lateral inertia ratio $\mu = I_x/I_y$ is typically around 1/10 for slender high-β vehicles and more than one for blunted low-β planetary vehicles.

With the previous hypothesis, the angle φ is a second-order parameter with negligible influence on flight mechanics, thus we will use $\varphi = 0$ for this book. Hence, in these ideal circumstances Gy_E and Gz_E are principal axes of inertia and the inertia matrix in the E frame is diagonal. For the exact calculation of the moments of inertia, the reader is referred to the definitions of Chap. 1 and mechanics textbooks.

To obtain orders of magnitude, examples are derived for moments of inertia of a homogeneous solid right circular cone of mass M, height L, and maximum radius R:

- Center of mass "G" (origin to the cone apex "A"):

$$\frac{x_G}{L} = \frac{3}{4} \tag{5.1}$$

- Roll moment of inertia:

$$I_x = \frac{3}{10} M R^2 \tag{5.2}$$

- Transverse moments of inertia (origin to the cone apex "A"):

$$I_{y/A} = I_{z/A} = \frac{3}{5} M L^2 + \frac{I_x}{2} \tag{5.3}$$

- Transverse moments of inertia relating to the center of mass "G":

$$I_{y/G} = I_{z/G} = I_{y/A} - M x_G^2 \tag{5.4}$$

For a pointed cone of semiapex angle $8°$, radius $R = 0.25$ m, length $L = 1.788$ m, and M for $\rho = 1000$ kg/m^3, we obtain:

$$M = \frac{\pi}{3} \rho L R^2 \approx 117 \text{ kg}$$

$$I_x \approx 2.19 \text{ m}^2\text{kg}$$

$$I_{y/G} \approx 15.2 \text{ m}^2\text{kg}$$

5.2 CG Offset and Principal Axis Misalignment

In general, small inertial asymmetries always exist, resulting from the practical constraints and integration. We will see herein that levels of acceptable asymmetries are often very constraining.

Significant inertial asymmetries are of two types:

- CG offset (Fig. 5.1) characterizes the lateral shift (y_G, z_G) of center of mass from the geometrical axis of symmetry OX_A (aerodynamic axis)
- Principal axis misalignment (Fig. 5.2) characterizes the misalignment θ_I from Gx_E axe. The meridian line around Gx_E containing the principal axis of inertia is characterized by its angle Φ_I with the origin of meridian lines Gy_E.
- A third angle φ_I (not shown) is necessary to define the inertial frame; it corresponds to the location of the transverse principal axes of inertia in the plane normal to Gx_I and does not play a significant role during reentry.

5.2 CG Offset and Principal Axis Misalignment

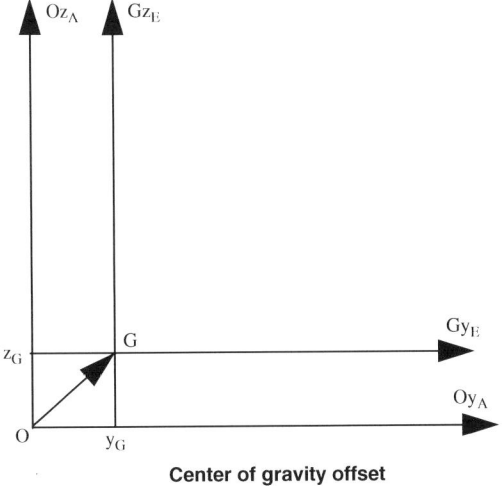

Center of gravity offset

Fig. 5.1 Definition of CG offset

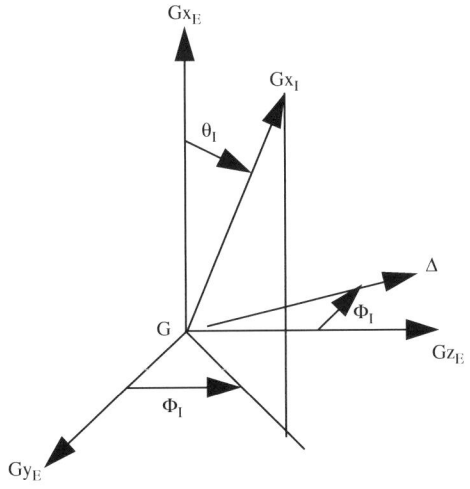

Principal axis misaligment

Fig. 5.2 Definition of principal axe misalignment

Thus, in the general case, the geometrical reference frame (aerodynamic) and the Eulerian frame have parallel axes. One transfers the aerodynamic axes into Eulerian by a translation $\vec{OG}$ (Fig. 5.1) of components in the transverse plane y_G along Gy_E and z_G along Gz_E. One then transfers from the Eulerian frame to the principal inertial axis frame (Gx_I, Gy_I, Gz_I) by a rotation θ_I around axis Δ, which is derived from Gz_E by rotation of Φ_I around Gx_E, followed by a rotation φ_I around Gx_I. The rotation matrix R that gives the components of $\vec{\omega}$ in E from its components in the principal inertial axis frame may be derived using (6.60). This expression is:

$$\begin{bmatrix} 1 & 0 & 0 \\ 0 & \cos\varphi_I & \sin\varphi_I \\ 0 & -\sin\varphi_I & \cos\varphi_I \end{bmatrix} \begin{bmatrix} \cos\theta_I & \sin\theta_I \cos\Phi_I & \sin\theta_I \sin\Phi_I \\ -\sin\theta_I \cos\Phi_I & \cos\theta_I + (1-\cos\theta_I)\sin^2\Phi_I & \sin\Phi_I \cos\Phi_I(\cos\theta_I - 1) \\ -\sin\theta_I \sin\Phi_I & \sin\Phi_I \cos\Phi_I(\cos\theta_I - 1) & \cos\theta_I + (1-\cos\theta_I)\cos^2\Phi_I \end{bmatrix}$$

(5.5)

The inertia tensor in E is obtained from (1.43):

$$[I'] = [R]^T [I] [R]$$

The detailed calculations are cumbersome and exact results rarely have practical interest. The expression becomes more interesting when we use a first-order approximation suited to a low principal axis misalignment θ_I, which is sufficiently accurate for most practical applications:

$$[R] \approx \begin{bmatrix} I_x & 0 & 0 \\ 0 & \cos\varphi_I & \sin\varphi_I \\ 0 & -\sin\varphi_I & \cos\varphi_{Iz} \end{bmatrix} \begin{bmatrix} 1 & \theta_I \cos\Phi_I & \theta_I \sin\Phi_I \\ -\theta_I \cos\Phi_I & 1 & 0 \\ -\theta_I \sin\Phi_I & 0 & 1 \end{bmatrix}$$

$$[I'] \approx \begin{bmatrix} I_x & -(I_y - I_x)\theta_I \cos\Phi_I & -(I_z - I_x)\theta_I \sin\Phi_I \\ -(I_y - I_x)\theta_I \cos\Phi_I & I_y + (I_z - I_y)\sin^2\varphi_I & -(I_z - I_y)\sin\varphi_I \cos\varphi_I \\ -(I_z - I_x)\theta_I \sin\Phi_I & -(I_z - I_y)\sin\varphi_I \cos\varphi_I & I_z - (I_z - I_y)\sin^2\varphi_I \end{bmatrix}$$

(5.6)

We observe that the angle φ_I appears only in the components H_x and H_y of angular momentum and only modifies the moments of inertia along Gy and Gz axes and coupling terms I_{yz} between them.

This expression shows that the principal axis misalignment θ_I is connected to the products of inertia I_{xy} and I_{xz} by the relation:

$$\theta_I \approx \frac{\sqrt{I_{xy}^2 + I_{xz}^2}}{I_y - I_x} \qquad (5.7)$$

Levels of acceptable asymmetries depend primarily on the size of the vehicle, on the reentry velocity and flight path angle, and on the levels of aerodynamic asymmetries. Corresponding studies represent a significant design effort for a reentry vehicle. They are developed in the Chap. 12, which treats the roll resonance phenomenon.

Chapter 6
Changing of Reference Frame

Our essential need is to determine the evolution of center of mass location and instantaneous vehicle orientation relative to a terrestrial frame, rotating or not. For this purpose, we must use linear momentum equations for the CG location and angular momentum equations for the vehicle orientation. Expressions of the fundamental principles in a rotating reference frame were developed in Chap. 1 and are well adapted to the calculation of the trajectory of the center of mass relating to a terrestrial observer. We also showed that the equation of evolution of the angular momentum is greatly simplified when one uses an Eulerian frame rigidly linked to the vehicle. These two systems of equations being coupled, we can then transform components of vectors in the axes of the vehicle to earth related components (for example, the aerodynamic resulting force), or transform earth related vector components to vehicle related components (for example, linear relative velocity).

Let us consider initially an observer in linear translation (accelerated or not) relating to inertial frames. This observer is rigidly linked to a reference frame (K) whose axes have a fixed orientation (parallel to those of an inertial frame). We consider a solid with a fixed point located at origin O of (K), involved in an arbitrary rotation movement. We associate the solid in frame (E) rigidly linked, centered out of O. Thus the reference frame (E) is in no uniform rotation at angular velocity $\vec{\omega}_{E/K}(t)$ relating to (K). We must characterize the orientation of $\vec{e}_i$ (E) axes relating to (K) axes and in addition express the components of a vector relative to (E) from its components in (K) and reciprocally.

Notice: K has a pure translation movement relating to inertial frames and the rotational movements of E relative to K and to any inertial frame are identical.

6.1 Direction Cosine Matrices

Each frame K and E is associated with a set of three mutually orthogonal axes crossing at the common origin O, with unit vectors of same modulus. The set of axes i and unit vectors $\vec{e}_i$ are numbered with index i selected such that their relative orientations satisfy the rule: "an observer along axis 1, looking at axis 2 direction, has axis 3 on his left (classical right-handed rule)." By definition, the frames and associated set of axes E and K result one from another by rotation, so the relative orientations of their axes is maintained through the movement. Let us name $\vec{e}_i$ and $\vec{e}'_i$, $i \in \{1, 2, 3\}$ the respective unit vectors of (K) and (E). We can write the general relations between unit vectors by using the rule of summation on the repeated indices (this convention is only one convenience to simplify the notations in the event of summation):

$$\vec{e}'_i = a_{ik}\,\vec{e}_k$$
$$\vec{e}_k = b_{kj}\,\vec{e}'_j \tag{6.1}$$

From the preceding assumptions, we can deduce the general properties of these relations.

While replacing $\vec{e}_k$ with its value drawn from the second expression we obtain:

$$\vec{e}'_i = a_{ik}\,b_{kj}\vec{e}'_j \Rightarrow a_{ik}\,b_{kj} = \delta_{ij}$$
$$i \neq j \rightarrow \delta_{ij} = 0$$
$$i = j \rightarrow \delta_{ii} = 1 \tag{6.2}$$

where one recognizes the symbol of Kronecker δ_{ij}. This can be expressed in matrix form:

$$AB = I \Leftrightarrow A = B^{-1}, B = A^{-1}$$
$$A = [a_{ij}]\,;\, B = [b_{ij}]\,;\, I = \begin{bmatrix} 1 & 0 & 0 \\ 0 & 1 & 0 \\ 0 & 0 & 1 \end{bmatrix} \tag{6.3}$$

In addition, the axes of each set being mutually orthogonal with unit vectors of equal length, we obtain:

$$a_{ik} = \vec{e}'_i \cdot \vec{e}_j = \cos(\Phi_{ik})\,;\, b_{ij} = \vec{e}_i \cdot \vec{e}'_j = \cos(\Phi_{ji}) \tag{6.4}$$

where A and B are obviously named direction cosine matrices.

Then, we directly obtain transformation rules for vector components:

$$\vec{X} = x_i\,\vec{e}_i = x'_j\,\vec{e}'_j \Rightarrow x_i\,b_{ij}\vec{e}'_j = x'_j\,\vec{e}'_j$$
$$\Rightarrow x'_j = x_i b_{ij} \Leftrightarrow [x']^T = [x]^T\,[B]$$
$$\Rightarrow x_j = x'_i a_{ij} \Leftrightarrow [x]^T = [x']^T\,[A] \tag{6.5}$$

$$[x]^T = \begin{bmatrix} x_1 & x_2 & x_3 \end{bmatrix} = \begin{bmatrix} x_1 \\ x_2 \\ x_3 \end{bmatrix}^T,\quad [x']^T = \begin{bmatrix} x'_1 & x'_2 & x'_3 \end{bmatrix} = \begin{bmatrix} x'_1 \\ x'_2 \\ x'_3 \end{bmatrix}^T$$

with,
where index T indicates the transpose matrix.

6.1 Direction Cosine Matrices

While transposing these matrix relations, let us obtain a more common expression to represent the transformations of the vectors components:

$$[x']^T = [x]^T [B] \Rightarrow [x'] = [B]^T [x]$$
$$[x]^T = [x']^T [A] \Rightarrow [x] = [A]^T [x']$$
$$a_i^{Tj} = a_j^i, \quad b_i^{Tj} = b_j^i \tag{6.6}$$

A very useful fundamental property of these matrix results from conservation of vector length (or scalar product) through a rotation transform:

$$\left\| \vec{X} \right\|^2 = [x]^T [x] = [x']^T [x'] = [x]^T [B][B]^T [x] = [x']^T [A][A]^T [x']$$
$$\Rightarrow [A][A]^T = [B][B]^T = [I] \Leftrightarrow [A]^T = [A]^{-1} = [B]; \quad [B]^T = [B]^{-1} = [A] \tag{6.7}$$

Thus, the inverse matrixes are equal to the transpose matrixes (they are orthogonal), which makes their computation very simple!

6.1.1 Angular Velocity

In order to deal with the problem of evolution of the transformation matrices, we need to mathematically define the instantaneous angular velocity or the rotating frame.

For this purpose, let us seek the expression of the linear velocity relating to (K) of an arbitrary point P of the solid. By definition, this point is at rest relating to (E). The velocity of P relating to (K) is entirely associated with the rotation movement of (E) axes:

$$\vec{X} = x_i \vec{e}_i = x'_i \vec{e}'_i$$
$$\Rightarrow \dot{\vec{X}} = \dot{x}_i \vec{e}_i = x'_i \dot{\vec{e}}'_i$$

Let us name Ω'_{ij} the components expressed in (E) of vector $\dot{\vec{e}}'_i$, time derivative of $\vec{e}'_i$:

$$\dot{\vec{e}}'_i = \Omega'_{ij} \vec{e}'_j \tag{6.8}$$

Components of $\dot{\vec{X}}$ in the rotating frame (E) are then written as:

$$\dot{\vec{X}} = v'_j \vec{e}'_j = x'_i \dot{\vec{e}}'_i = x'_i \Omega'_{ij} \vec{e}'_j \Rightarrow v'_j = x'_i \Omega'_{ij} \tag{6.9}$$

$\dot{\vec{X}}$ and $\vec{X}$ being vectors, it results from the above equality Ω'_{ij} is a second order tensor: it is named "angular velocity." This tensor is skew symmetric. Indeed, base vector set $\{\vec{e}'_i\}$ being orthogonal and unit:

$$\vec{e}'_i \cdot \vec{e}'_j = \delta_{ij} \Rightarrow \dot{\vec{e}}'_i \cdot \vec{e}'_j + \vec{e}'_i \cdot \dot{\vec{e}}'_j = 0$$

which gives:

$$\left(\Omega'_{ik}\vec{e}'_k\right) \cdot \vec{e}'_j + \vec{e}'_i \cdot \left(\Omega'_{jk}\vec{e}'_k\right) = 0 \Rightarrow \delta_{jk}\Omega'_{ik} + \delta_{ik}\Omega'_{jk} = 0$$

and finally:

$$\Omega'_{ij} + \Omega'_{ji} = 0 \Rightarrow \Omega'_{ji} = -\Omega'_{ij} \tag{6.10}$$

The transpose tensor is equal to the opposite.

This skew symmetric tensor $\vec{\vec{\Omega}}$ can be represented by a vector $\vec{\omega}$ while posing:

$$\omega'_1 = p' = \Omega'_{23}, \quad \omega'_2 = q' = \Omega'_{31}, \quad \omega'_3 = r' = \Omega'_{12}$$

$$\vec{\vec{\Omega}} = [\Omega'] = \begin{bmatrix} 0 & -\omega'_3 & \omega'_2 \\ \omega'_3 & 0 & -\omega'_1 \\ -\omega'_2 & \omega'_1 & 0 \end{bmatrix} \tag{6.11}$$

Thus, we obtain according to (6.9):

$$\begin{aligned} v'_1 &= x'_i \, \Omega'_{i\,1} = x'_2 \, \omega'_3 - x'_3 \omega'_2 \\ v'_2 &= x'_i \, \Omega'_{i\,2} = -x'_1 \, \omega'_3 + x'_3 \omega'_1 \\ v'_3 &= x'_i \, \Omega'_{i\,3} = x'_1 \, \omega'_2 - x'_2 \omega'_1 \end{aligned} \tag{6.12}$$

Velocity $\dot{\vec{X}} = \vec{v}$ relative to (K) of any point P of the solid is thus the vector product of angular rotation vector $\vec{\omega}$ by the location vector from the origin $\vec{OP}$. The rotational movement of E being the same relative to K and to any inertial frame $\vec{\vec{\Omega}}$, and $\vec{\omega}$ have a physical meaning independent of the observation frame (they are respectively a tensor and a vector). Thus, we can just as easily define their components in (K):

$$\dot{\vec{e}}_i = \Omega_{ij}\,\vec{e}_j \quad (\Omega_{ij} \text{ angular velocity tensor}) \tag{6.13}$$

from where we obtain components in (K) of the velocity of the point P relative to (K),

$$v_j = x_i \Omega_{ij} \tag{6.14}$$

The transformation of the components of $\vec{\vec{\Omega}}$ results from that of ordinary vectors $\vec{v}$ and $\vec{x}$:

$$[v]^T = [x]^T [\Omega] \Rightarrow [v']^T [A] = [x']^T [A] [\Omega]$$

6.1 Direction Cosine Matrices

Then,

$$[v']^T = [x']^T [A][\Omega][A]^{-1} \Rightarrow [\Omega'] = [A][\Omega][A]^{-1}, [\Omega] = [B][\Omega'][B]^{-1} \quad (6.15)$$

It is easy to check that the skew symmetric character of the tensor is maintained through the change of frame. Thus we can write in the frame (K):

$$\vec{\vec{\Omega}} = [\Omega] = \begin{bmatrix} 0 & -\omega_3 & \omega_2 \\ \omega_3 & 0 & -\omega_1 \\ -\omega_2 & \omega_1 & 0 \end{bmatrix} \rightarrow \vec{\omega} = \begin{bmatrix} \omega_1 \\ \omega_2 \\ \omega_3 \end{bmatrix}_{\text{(angular rate vector)}} \quad (6.16)$$

$$\begin{aligned} v_1 &= x_2\omega_3 - x_3\omega_2 \\ v_2 &= -x_1\omega_3 + x_3\omega_1 \\ v_3 &= x_1\omega_2 - x_2\omega_1 \end{aligned} \quad (6.17)$$

It is easy, but somewhat tedious, to demonstrate that for rotation transform, the angular velocity vector has the same behavior as that of ordinary vectors.

Notice: The relation (6.17) expresses velocity $\vec{v} = \vec{\omega} \wedge \vec{r}$ of a point of the solid E relating to K. K has a single linear nonuniform movement relating to inertial frames. The velocity of point P relating to an arbitrary inertial frame is:

$$\dot{\vec{V}} = \dot{\vec{V}}_e + \vec{\omega} \wedge \vec{r} \quad (6.18)$$

where $\vec{V}_e(t)$ is the instantaneous linear velocity of origin O of K relating to this inertial frame.

The result of this is

$$\ddot{\vec{V}} = \dot{\vec{V}}_e + \dot{\vec{\omega}} \wedge \vec{r} + \vec{\omega} \wedge \dot{\vec{r}} \quad (6.19)$$

With $\dot{\vec{r}} = \vec{v} = \vec{\omega} \wedge \vec{r}$, we obtain.

$$\dot{\vec{V}} = \dot{\vec{V}}_e + \dot{\vec{\omega}} \wedge \vec{r} + \vec{\omega} \wedge (\vec{\omega} \wedge \vec{r}) \quad (6.20)$$

Let us observe that the Coriolis acceleration term does not appear here because positions of points $\vec{r}$ of solid are fixed in E and thus have a pure rotational movement relating to K. In the more general case of a point P having a linear translation and a rotation movement, we obtain the results (1.2)–(1.4) of Chap. 1, whose demonstration is proposed as an exercise.

6.1.2 Composition of Angular Velocities

Let us consider three observation frames with the same origin O:

- K_0 whose axes remain parallel to some inertial frame ones
- E_1 in rotation at the absolute velocity $\Omega\,(1,0)$ (relating to K_0)
- E_2 in rotation at the relative velocity $\Omega\,(2,1)$ relating to E_1

Let us seek the absolute rotation velocity $\Omega\,(2,2)$ of E_2 relating to K_0. According to the (6.8) and (6.13) already established, we have:

$$\dot{\vec{e}}_i\,(1) = \Omega_{ij}\,(1,0)\,\vec{e}_j\,(0) = \Omega_{i\,j}\,(1,1)\,\vec{e}_j\,(1) \tag{6.21}$$

$$\dot{\vec{e}}_i\,(2) = \Omega_{ij}\,(2,0)\,\vec{e}_j\,(0) = \Omega_{i\,j}\,(2,2)\,\vec{e}_j\,(2) \tag{6.22}$$

In addition, according to (6.1):

$$\vec{e}_i\,(2) = a_{ij}\,(2/1)\,\vec{e}_j\,(1) \Rightarrow \dot{\vec{e}}_i\,(2) = \dot{a}_{ij}\,(2/1)\,\vec{e}_j\,(1) + a_{ij}\,(2/1)\,\dot{\vec{e}}_j\,(1) \tag{6.23}$$

from where:

$$\dot{\vec{e}}_i\,(2) = a_{ik}\,(2/1)\,\Omega_{kj}(2/1,1)\,\vec{e}_j\,(1) + a_{ij}\,(2/1)\,\Omega_{j\,k}(1,1)\vec{e}_k\,(1)$$
$$\Rightarrow \dot{\vec{e}}_i\,(2) = \left[a_{ik}\,(2/1)\,\Omega_{kj}(2/1,1) + a_{ik}\,(2/1)\,\Omega_{kj}(1,1)\right]\vec{e}_j\,(1)$$
$$\Rightarrow \dot{\vec{e}}_i\,(2) = \left[a_{ik}\,(2/1)\,\Omega_{kj}(2/1,1) + a_{ik}\,(2/1)\,\Omega_{kj}(1,1)\right]b_{j\,l}\,(2/1)\,\vec{e}_l\,(1) \tag{6.24}$$

Thus we obtain,

$$\dot{\vec{e}}_i\,(2) = \Omega_i^l\,(2,2)\,\vec{e}_l\,(2)$$
$$= \left[a_{ik}\,(2/1)\,\Omega_{kj}(2/1,1)b_{j\,l}\,(2/1) + a_{ik}\,(2/1)\,\Omega_{kj}(1,1)b_{j\,l}\,(2/1)\right]\vec{e}_l\,(1)$$
$$\Rightarrow \Omega_i^l\,(2,2) = \Omega_i^l(2/1,2) + \Omega_i^l(1,2) \tag{6.25}$$

The property established in E_2 frame can be immediately applied to the K_0 and E_1 frames using transformation rule for angular velocity tensor.

$$\Rightarrow \Omega_i^l\,(2,0) = \Omega_i^l(2/1,0) + \Omega_i^l(1,0)$$
$$\Rightarrow \Omega_i^l\,(2,1) = \Omega_i^l(2/1,1) + \Omega_i^l(1,1) \tag{6.26}$$

Thus, the tensor "absolute rotation velocity of E_2" is equal to the sum of tensors "angular rotation velocity E_2 relating to E_1" and "absolute rotation velocity of E_1." This is also clearly valid for the angular velocity vectors.

Thus, the Galilean principle of addition of linear velocities applies also to the angular velocities. This result will be applied, in particular, to the case where E_1 is a frame linked with the rotating earth and E_2 is a frame linked with the vehicle in rotation relative to earth.

6.1.3 Evolution of the Direction Cosine Matrices

We answered part of the question up to now, which relates to the form and the properties of transformation matrices for the components of a vector. Now let us seek the time law of evolution of these matrices. In order to obtain this evolution, let us examine the movement relative to rotating frame E of an arbitrary vector $\vec{X}_0 = x_{0i}\vec{e}_i = x'_{0i}\vec{e}'_i$ constant in the K frame:

$$\dot{\vec{X}}_0 = 0 = \dot{x}'_{0i}\,\vec{e}'_i + x'_{0i}\,\dot{\vec{e}}'_i = \dot{x}'_{0i}\,\vec{e}'_i + x'_{0i}\Omega'_{ij}\vec{e}'_j \qquad (6.27)$$

Again, this expresses the fact that the movement of $\vec{X}_0$ relating to (E) is entirely related to the movement of observation axes.

By expressing the components of the vector $\vec{X}_0$ using the matrix B, we obtain:

$$\dot{b}_{ji}x_{0j}\,\vec{e}'_i + b_{ji}x_{0j}\Omega'_{ik}\,\vec{e}'_k = 0 \qquad (6.28)$$

The vector $\vec{X}_0$ being arbitrary, we choose it successively equal to the unit vectors $\vec{e}_j$ of system K:

$$\Rightarrow \forall j,\, \dot{b}_{ji}\,\vec{e}'_i = -b_{ji}\Omega'_{ik}\,\vec{e}'_k = -b_{jk}\Omega'_{ki}\,\vec{e}'_i \qquad (6.29)$$

By identifying the coefficients of the base vectors $\vec{e}'_i$, we finally obtain:

$$\dot{b}_{ji} = -b_{jk}\,\Omega'_{ki} \Leftrightarrow [\dot{B}] = -[B][\Omega'] \qquad (6.30)$$

The evolution of the inverse matrix A is determined by transposing the two members of the preceding equation:

$$[\dot{B}]^T = [\dot{A}] = -[\Omega']^T[B]^T = [\Omega'][A] \Leftrightarrow \dot{a}_{ij} = \Omega'_{ik}a_{kj} \qquad (6.31)$$

We can summarize in symbolic form these two equations:

$$\begin{aligned}{}[\dot{B}] &= -[B][\Omega'] \\ [\dot{A}] &= [\Omega'][A]\end{aligned} \qquad (6.32)$$

They can be used to determine the evolution of the attitude of a rotating vehicle, either starting from angular velocities provided by the moment equation or from the angular velocity measured inboard the vehicle. These equations can be transformed using the angular velocity solved in the fixed frame, by using the rule of transformation for angular velocity tensor (in both cases the velocity of E relating to K is expressed either in the rotating or in the fixed frame):

$$[\dot{B}] = -[\Omega][B] \quad \Leftrightarrow \quad \dot{b}_{ij} = -\Omega_{ik}b_{kj}$$
$$[\dot{A}] = [A][\Omega] \quad \Leftrightarrow \quad \dot{a}_{ij} = a_{ik}\Omega_{kj} \tag{6.33}$$

6.2 Euler Angles

The preceding tools are strictly sufficient to determine the movement of the solid around the center of gravity, when one knows the angular acceleration. However, this description is neither simplest nor most convenient. Indeed, we must determine the evolution of nine elements of the transformation matrix A or B. In fact, these nine quantities are not independent since the matrices have to verify (6.7):

$$AA^T = I = BB^T$$

These equalities are equivalent to six independent relations between elements of A or B matrix (and not nine, because matrices AA^T and BB^T are symmetrical). We are left with only three independent quantities. Another drawback, of practical nature, is that it is difficult to quickly assess the relative position of the frames E and K starting from the matrix elements. On the other hand, the method invented by Leonard Euler is optimum from two points of view. It uses the minimum number of parameters, and thus makes it possible to directly visualize the relative position of the two frames.

The transformation process from the "fixed" starting frame (K) to the "moving" final frame (E) is divided according to an ordered sequence of three independent rotations (Fig. 6.1). The first axis of rotation is arbitrarily selected along any of the three axes $\vec{e}_i(0)$ of the initial reference frame and the corresponding amplitude of rotation is θ_1. We arrive at a first intermediate frame E_1 with axes $\vec{e}_j(1)$ such as $\vec{e}_i(1) = \vec{e}_i(0)$. From this intermediate frame, we perform a second rotation of angle θ_2 around an arbitrary axis $\vec{e}_j(1)$ different from $\vec{e}_i(1)$. Thus we arrive at a second intermediate reference frame E_2, whose axes $\vec{e}_k(2)$ are such that $\vec{e}_j(2) = \vec{e}_j(1)$. Finally, the frame E is completed by rotation of angle θ_3 around an arbitrary axis $\vec{e}_k(2) = \vec{e}_k(3)$ different from $\vec{e}_j(2)$. Thus we have $3 \times 2 \times 2 = 12$ available choices for the rotation axes $\vec{e}_i(1), \vec{e}_j(2)$, and $\vec{e}_k(3)$. The choice ($i = 1, j = 2, k = 3$) is represented in Fig. 6.1.

6.2.1 Euler Rotation Matrix

For the preceding choice, matrices of passage from coordinates in the old axes to coordinates in the new axes are:

6.2 Euler Angles

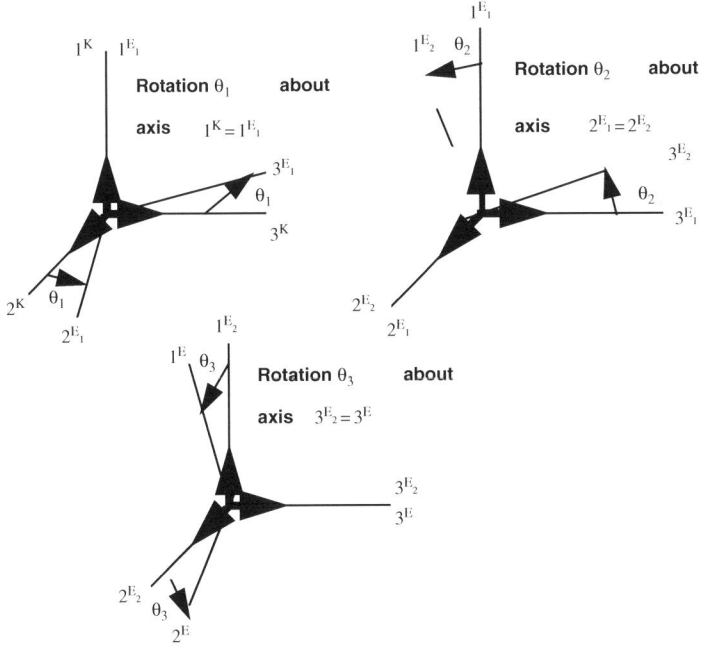

Fig. 6.1 Definition of Euler's rotation axes

$$A_1(\theta_1) = \begin{bmatrix} 1 & 0 & 0 \\ 0 & \cos\theta_1 & \sin\theta_1 \\ 0 & -\sin\theta_1 & \cos\theta_1 \end{bmatrix} \tag{6.34}$$

$$A_2(\theta_2) = \begin{bmatrix} \cos\theta_2 & 0 & -\sin\theta_2 \\ 0 & 1 & 0 \\ \sin\theta_2 & 0 & \cos\theta_2 \end{bmatrix} \tag{6.35}$$

$$A_3(\theta_3) = \begin{bmatrix} \cos\theta_3 & \sin\theta_3 & 0 \\ -\sin\theta_3 & \cos\theta_3 & 0 \\ 0 & 0 & 1 \end{bmatrix} \tag{6.36}$$

The lower index in the left-hand matrix corresponds to the choice of the rotation axes; lower index of the angle corresponds to the rank of the intermediate rotation.

The transformation matrix from coordinates in the "fixed" frame K to coordinates in the moving frame E is the ordered product of the three elementary matrixes:

$$A(\theta_1, \theta_2, \theta_3) = A_3(\theta_3) A_2(\theta_2) A_1(\theta_1) \tag{6.37}$$

We have seen previously that the inverse rotation matrix is equal to the transposed matrix. We thus obtain:

$$B(\theta_1, \theta_2, \theta_3) = A_1^T(\theta_1) A_2^T(\theta_2) A_3^T(\theta_3) = A_1(-\theta_1) A_2(-\theta_2) A_3(-\theta_3) \quad (6.38)$$

In the general case, we will have:

$$A(\theta_1, \theta_2, \theta_3) = A_{k(3)}(\theta_3) A_{j(2)}(\theta_2) A_{i(1)}(\theta_1) \quad (6.39)$$

Indices i, j, and k $\in \{1, 2, 3\}$ are such that $j \neq i$ and $k \neq j$, and elementary matrices of the form:

$$A_1(\theta_n) = \begin{bmatrix} 1 & 0 & 0 \\ 0 & \cos\theta_n & \sin\theta_n \\ 0 & -\sin\theta_n & \cos\theta_n \end{bmatrix} \quad (6.40)$$

$$A_2(\theta_n) = \begin{bmatrix} \cos\theta_n & 0 & -\sin\theta_n \\ 0 & 1 & 0 \\ \sin\theta_n & 0 & \cos\theta_n \end{bmatrix} \quad (6.41)$$

$$A_3(\theta_n) = \begin{bmatrix} \cos\theta_n & \sin\theta_n & 0 \\ -\sin\theta_n & \cos\theta_n & 0 \\ 0 & 0 & 1 \end{bmatrix} \quad (6.42)$$

The combination of axes $\{i(1) = 1, j(2) = 2, k(3) = 1\}$, is frequently used, associated with the notation $\theta_1 = \psi$, $\theta_2 = \theta$, $\theta_3 = \varphi$ (for example in Sect. 7.2 for the coning motion study). We have:

$$A_1(\theta_1 = \psi) = \begin{bmatrix} 1 & 0 & 0 \\ 0 & \cos\theta_1 & \sin\theta_1 \\ 0 & -\sin\theta_1 & \cos\theta_1 \end{bmatrix} = \begin{bmatrix} 1 & 0 & 0 \\ 0 & \cos\psi & \sin\psi \\ 0 & -\sin\psi & \cos\psi \end{bmatrix} \quad (6.43)$$

$$A_2(\theta_2 = \theta) = \begin{bmatrix} \cos\theta_2 & 0 & -\sin\theta_2 \\ 0 & 1 & 0 \\ \sin\theta_2 & 0 & \cos\theta_2 \end{bmatrix} = \begin{bmatrix} \cos\theta & 0 & -\sin\theta \\ 0 & 1 & 0 \\ \sin\theta & 0 & \cos\theta \end{bmatrix} \quad (6.44)$$

$$A_1(\theta_3 = \varphi) = \begin{bmatrix} 1 & 0 & 0 \\ 0 & \cos\theta_3 & \sin\theta_3 \\ 0 & -\sin\theta_3 & \cos\theta_3 \end{bmatrix} = \begin{bmatrix} 1 & 0 & 0 \\ 0 & \cos\varphi & \sin\varphi \\ 0 & -\sin\varphi & \cos\varphi \end{bmatrix} \quad (6.45)$$

$$A(\psi, \theta, \varphi) = A_1(\varphi) A_2(\theta) A_1(\psi) \quad (6.46)$$

$$A(\psi, \theta, \varphi) = \begin{bmatrix} \cos\theta & \sin\theta\sin\psi & -\sin\theta\cos\psi \\ \sin\varphi\sin\theta & \cos\varphi\cos\psi - \sin\varphi\cos\theta\sin\psi & \cos\varphi\sin\psi + \sin\varphi\cos\theta\cos\psi \\ \cos\varphi\sin\theta & -\sin\varphi\cos\psi - \cos\varphi\cos\theta\sin\psi & -\sin\varphi\sin\psi + \cos\varphi\cos\theta\cos\psi \end{bmatrix} \quad (6.47)$$

Combinations used in this book for the reentry studies correspond to $\{i(1) = 3, j(2) = 2, k(3) = 1\}$, where initial axis 3 is along the downward vertical and axis 1 is along the geographic north direction of the observer. The initial axis 2 is thus directed toward the east of the observer, and the initial plane 1, 2 is horizontal. The angle $\theta_1 = \psi$ is the azimuth of axis 1 of the moving frame, the angle $\theta_2 = \theta$ is the longitudinal inclination, and the angle $\theta_3 = \varphi$ is the roll angle around longitudinal axis 1 of the moving frame, relating to the horizontal plane.

Thus, the elementary matrixes of rotation are written as:

$$A_3(\theta_1 = \psi) = \begin{bmatrix} \cos\psi & \sin\psi & 0 \\ -\sin\psi & \cos\psi & 0 \\ 0 & 0 & 1 \end{bmatrix} \quad (6.48)$$

$$A_2(\theta_2 = \theta) = \begin{bmatrix} \cos\theta & 0 & -\sin\theta \\ 0 & 1 & 0 \\ \sin\theta & 0 & \cos\theta \end{bmatrix} \quad (6.49)$$

$$A_1(\theta_3 = \varphi) = \begin{bmatrix} 1 & 0 & 0 \\ 0 & \cos\varphi & \sin\varphi \\ 0 & -\sin\varphi & \cos\varphi \end{bmatrix} \quad (6.50)$$

$$A(\psi, \theta, \varphi) = A_1(\varphi) A_2(\theta) A_3(\psi) \quad (6.51)$$

$$A(\psi, \theta, \varphi) = \begin{bmatrix} \cos\theta\cos\psi & \cos\theta\sin\psi & -\sin\theta \\ -\cos\varphi\sin\psi + \sin\varphi\sin\theta\cos\psi & \cos\varphi\cos\psi + \sin\varphi\sin\theta\sin\psi & \sin\varphi\cos\theta \\ \sin\varphi\sin\psi + \cos\varphi\sin\theta\cos\psi & -\sin\varphi\cos\psi + \cos\varphi\sin\theta\sin\psi & \cos\varphi\cos\theta \end{bmatrix} \quad (6.52)$$

6.2.2 Evolution of Euler Angles

The instantaneous angular velocity vector of the moving frame can be represented either by its components on an arbitrary frame,

$$\vec{\omega} = \omega_1 \vec{e}_1(0) + \omega_2 \vec{e}_2(0) + \omega_3 \vec{e}_3(0) = \omega'_1 \vec{e}_1(3) + \omega'_2 \vec{e}_2(3) + \omega'_3 \vec{e}_3(3) \quad (6.53)$$

or by its components according to Euler intermediate axes of rotation:

$$\vec{\omega} = \dot{\theta}_1 \, \vec{e}_{i(1)}(1) + \dot{\theta}_2 \, \vec{e}_{j(2)}(2) + \dot{\theta}_3 \, \vec{e}_{k(3)}(3) \tag{6.54}$$

As these axes are not orthogonal, the most straightforward method is to express them as a function of mobile frame unit vectors. According to Sect. 6.1, we know that the matrices of transformation of the unit vectors are identical to those of co-ordinates, which are products of the elementary matrices of rotation detailed in the preceding paragraph. We obtain finally a matrix relation between $\vec{\omega}$ components in the moving frame and Euler angles time derivatives.

$$\begin{bmatrix} \omega'_1 \\ \omega'_2 \\ \omega'_3 \end{bmatrix} = [E(\theta_1, \theta_2, \theta_3)] \begin{bmatrix} \dot{\theta}_1 \\ \dot{\theta}_2 \\ \dot{\theta}_3 \end{bmatrix} \tag{6.55}$$

The inverse matrix generally exists (except for some particular relative orientations, depending on the choice of Euler angles) and one obtains the equation of evolution of the Euler angles:

$$\begin{bmatrix} \dot{\theta}_1 \\ \dot{\theta}_2 \\ \dot{\theta}_3 \end{bmatrix} = [E(\theta_1, \theta_2, \theta_3)]^{-1} \begin{bmatrix} \omega'_1 \\ \omega'_2 \\ \omega'_3 \end{bmatrix} \tag{6.56}$$

Knowing the angular velocity and initial values of Euler angles, we are able to determine their evolution and the transformation matrix A. We thus brought back the problem of evolution from nine to three parameters.

Any medal having its reverse introduces a new problem, which is the appearance of singular configurations for which the angles cannot be defined. This corresponds to a zero value of the matrix E determinant and an indeterminant solution of the system (6.55).

Let us develop the expression of E in the case of the previous sequence 1 2 1:

$$\begin{bmatrix} \vec{e}_1(1) \\ \vec{e}_2(1) \\ \vec{e}_3(1) \end{bmatrix} = A_2^{-1}(\theta) A_1^{-1}(\varphi) \begin{bmatrix} \vec{e}_1(3) \\ \vec{e}_2(3) \\ \vec{e}_3(3) \end{bmatrix}$$

$$= \begin{bmatrix} \cos\theta & 0 & \sin\theta \\ 0 & 1 & 0 \\ -\sin\theta & 0 & \cos\theta \end{bmatrix} \begin{bmatrix} 1 & 0 & 0 \\ 0 & \cos\varphi & -\sin\varphi \\ 0 & \sin\varphi & \cos\varphi \end{bmatrix} \begin{bmatrix} \vec{e}_1(3) \\ \vec{e}_2(3) \\ \vec{e}_3(3) \end{bmatrix}$$

6.2 Euler Angles

$$\begin{bmatrix} \vec{e}_1(2) \\ \vec{e}_2(2) \\ \vec{e}_3(2) \end{bmatrix} = A_1^{-1}(\varphi) \begin{bmatrix} \vec{e}_1(3) \\ \vec{e}_2(3) \\ \vec{e}_3(3) \end{bmatrix} = \begin{bmatrix} 1 & 0 & 0 \\ 0 & \cos\varphi & -\sin\varphi \\ 0 & \sin\varphi & \cos\varphi \end{bmatrix} \begin{bmatrix} \vec{e}_1(3) \\ \vec{e}_2(3) \\ \vec{e}_3(3) \end{bmatrix}$$

$$\Rightarrow \vec{e}_1(1) = \cos\theta \vec{e}_1(3) + \sin\theta \sin\varphi \vec{e}_2(3) + \sin\theta \cos\varphi \vec{e}_3(3) \tag{6.57}$$

$$\Rightarrow \vec{e}_2(2) = \cos\varphi \vec{e}_2(3) - \sin\varphi \vec{e}_3(3) \tag{6.58}$$

By transforming the expression of (6.54) using (6.57) and (6.58), we obtain:

$$\vec{\omega} = \dot{\theta}_1 \vec{e}_1(1) + \dot{\theta}_2 \vec{e}_2(2) + \dot{\theta}_3 \vec{e}_1(3) = \dot{\psi}\vec{e}_1(1) + \dot{\theta}\vec{e}_2(2) + \dot{\varphi}\vec{e}_1(3)$$
$$\vec{\omega} = \dot{\psi}\left[\cos\theta \vec{e}_1(3) + \sin\theta \sin\varphi \, \vec{e}_2(3) + \sin\theta \cos\varphi \, \vec{e}_3(3)\right]$$
$$\quad + \dot{\theta}\left[\cos\varphi \, \vec{e}_2(3) - \sin\varphi \, \vec{e}_3(3)\right] + \dot{\varphi} \vec{e}_1(3)$$
$$\vec{\omega} = \left[\dot{\psi}\cos\theta + \dot{\varphi}\right]\vec{e}_1(3) + \left[\dot{\psi}\sin\theta \sin\varphi + \dot{\theta}\cos\varphi\right]\vec{e}_2(3)$$
$$\quad + \left[\dot{\psi}\sin\theta \cos\varphi - \dot{\theta}\sin\varphi\right]\vec{e}_3(3) \tag{6.59}$$

From which,

$$\omega'_1 = p = \dot{\psi}\cos\theta + \dot{\varphi}$$
$$\omega'_2 = q = \dot{\psi}\sin\theta \sin\varphi + \dot{\theta}\cos\varphi$$
$$\omega'_3 = r = \dot{\psi}\sin\theta \cos\varphi - \dot{\theta}\sin\varphi \tag{6.60}$$

Finally, we obtain:

$$\dot{\psi}\cos\theta + \dot{\varphi} = p$$
$$\dot{\psi}\sin\theta = q\sin\varphi + r\cos\varphi$$
$$\dot{\theta} = q\cos\varphi - r\sin\varphi \tag{6.61}$$

This system admits a single solution in $\dot{\psi}, \dot{\theta}, \dot{\varphi}$, except when $\theta = 0 \text{ or } \pi$ where it is indeterminant in $\dot{\psi}, \dot{\varphi}$:

$$\dot{\psi} = \frac{q\sin\varphi + r\cos\varphi}{\sin\theta}$$
$$\dot{\theta} = q\cos\varphi - r\sin\varphi$$
$$\dot{\varphi} = p - \cos\theta \frac{q\sin\varphi + r\cos\varphi}{\sin\theta} \tag{6.62}$$

In the case of the choice of Euler axes of rotation more appropriate to flight mechanics needs (3, 2, 1), we note that the last two axes of the sequence are identical with the preceding case. According to this fact, intermediate matrices of rotations in θ and φ have identical expressions (the angle values are obviously different). The first axis of rotation is written as:

$$\Rightarrow \vec{e}_3(1) = -\sin\theta\, \vec{e}_1(3) + \cos\theta \sin\varphi\, \vec{e}_2(3) + \cos\theta \cos\varphi\, \vec{e}_3(3) \quad (6.63)$$

$$\vec{\omega} = \dot{\psi}\, \vec{e}_3(1) + \dot{\theta}\, \vec{e}_2(2) + \dot{\varphi}\, \vec{e}_3(3)$$

$$\vec{\omega} = \dot{\psi}\left[-\sin\theta\, \vec{e}_1(3) + \cos\theta \sin\varphi\, \vec{e}_2(3) + \cos\theta \cos\varphi\, \vec{e}_3(3)\right]$$

$$+ \dot{\theta}\left[\cos\varphi\, \vec{e}_2(3) - \sin\varphi\, \vec{e}_3(3)\right] + \dot{\varphi}\, \vec{e}_1(3)$$

$$\vec{\omega} = \left[-\dot{\psi}\sin\theta + \dot{\varphi}\right]\vec{e}_1(3) + \left[\dot{\psi}\cos\theta \sin\varphi + \dot{\theta}\cos\varphi\right]\vec{e}_2(3)$$

$$+ \left[+\dot{\psi}\cos\theta \cos\varphi - \dot{\theta}\sin\varphi\right]\vec{e}_3(3) \quad (6.64)$$

This involves,

$$p = -\dot{\psi}\sin\theta + \dot{\varphi}$$
$$q = \dot{\psi}\cos\theta \sin\varphi + \dot{\theta}\cos\varphi$$
$$r = \dot{\psi}\cos\theta \cos\varphi - \dot{\theta}\sin\varphi \quad (6.65)$$

that is to say,

$$-\dot{\psi}\sin\theta + \dot{\varphi} = p$$
$$\dot{\psi}\cos\theta = q\sin\varphi + r\cos\varphi$$
$$\dot{\theta} = q\cos\varphi - r\sin\varphi \quad (6.66)$$

This sequence thus admits the case of indeterminant $\theta = \pm\frac{\pi}{2}$, excepted for which the solution is,

$$\dot{\psi} = \frac{q\sin\varphi + r\cos\varphi}{\cos\theta}$$
$$\dot{\theta} = q\cos\varphi - r\sin\varphi$$
$$\dot{\varphi} = p + \sin\theta\frac{q\sin\varphi + r\cos\varphi}{\cos\theta} \quad (6.67)$$

Happily, the indeterminant case corresponds to a very improbable orientation for airplanes and reentry vehicles and does not preclude the use of these Euler angles.

6.3 Representations with Four Parameters

The apparent simplicity of the Euler representation presents two drawbacks. First, the existence of singular configurations constitutes a risk in digital codes; in addition, the use of the trigonometric functions is expensive in computing times. Representations with four parameters may represent the best compromise.

6.3.1 Vectorial Representation

Let us consider a rotation transform of angle φ around an axis with unit vector $\vec{\Delta}$ (Fig. 6.2):

$$\vec{X} = \vec{OM} \xrightarrow{\text{rotation } \vec{\Delta},\varphi} \vec{X}' = \vec{OM}'$$

The component of vectors along $\vec{\Delta}$ remains constant through the rotation, and the normal component rotates an angle φ:

$$\vec{X} = \left(\vec{X} \cdot \vec{\Delta}\right)\vec{\Delta} + \vec{X}_\perp \Rightarrow \vec{X}_\perp = \vec{X} - \left(\vec{X} \cdot \vec{\Delta}\right)\vec{\Delta}$$

$$\vec{X}' = \left(\vec{X} \cdot \vec{\Delta}\right)\vec{\Delta} + \vec{X}'_\perp = \left(\vec{X} \cdot \vec{\Delta}\right)\vec{\Delta} + \vec{X}_\perp \cos\varphi + \left(\vec{\Delta} \wedge \vec{X}_\perp\right)\sin\varphi$$

From which,

$$\vec{X}' = \left(\vec{X} \cdot \vec{\Delta}\right)\vec{\Delta} + \left[\vec{X} - \left(\vec{X} \cdot \vec{\Delta}\right)\vec{\Delta}\right]\cos\varphi + \left(\vec{\Delta} \wedge \vec{X}\right)\sin\varphi \qquad (6.68)$$

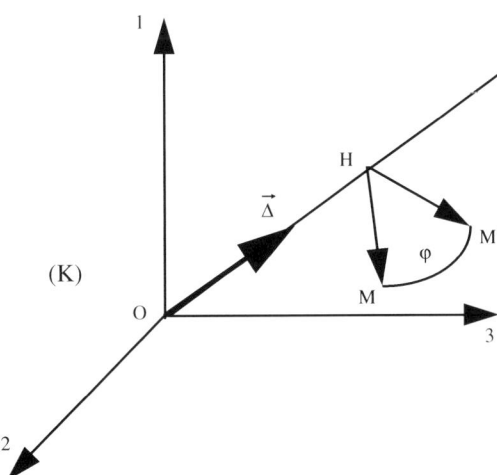

Fig. 6.2 Vectorial representation for rotations

Such is the vectorial representation of an arbitrary rotation of angle φ around an axis $\vec{\Delta}$. We verify that it is a representation with four parameters, composed of a scalar φ and the three components of the vector $\vec{\Delta}$ (not independent since it is a unit vector).

6.3.2 Quaternion

6.3.2.1 Definition

A quaternion $\underline{Q}$ is a quadruplet of real numbers composed of a scalar q_0 and of the three components of a vector $\vec{q} = (q_1 q_2 q_3)$.

It is formally represented by $\underline{Q} = q_0 + \vec{q}$ and the conjugate quaternion is by definition $\underline{Q}^* = q_0 - \vec{q}$. The ensemble obeys the equality rule, the addition operations, and multiplication by a scalar similar to those of R^4 vectors.

Moreover, the product obeys the following definition:

$$\underline{AB} = (a_0 + \vec{a})(b_0 + \vec{b}) = a_0 b_0 - \vec{a} \cdot \vec{b} + a_0 \vec{b} + b_0 \vec{a} + \vec{a} \wedge \vec{b} \quad (6.69)$$

It clearly appears that this product is distributive, but noncommutative (presence of a vector product). It is left to the reader to check that this product is associative, i.e., $(\underline{AB})\underline{C} = \underline{A}(\underline{BC})$. A useful immediate consequence is the expression of the norm of the quaternion:

$$\underline{Q}^* \underline{Q} = \underline{Q} \underline{Q}^* = q_0^2 + \vec{q} \cdot \vec{q} = \|\underline{Q}\|^2 \quad (6.70)$$

Another useful property is:

$$\underline{A}^* \underline{B}^* = (a_0 - \vec{a})(b_0 - \vec{b}) = a_0 b_0 - \vec{a} \cdot \vec{b} - a_0 \vec{b} - b_0 \vec{a} - \vec{a} \wedge \vec{b} = [\underline{BA}]^* \quad (6.71)$$

Let us close these definitions noting, like vectors, $\vec{\underline{Q}}$, pure vectorial quaternion of the type $\underline{Q} = (0 + \vec{q})$.

6.3.2.2 Quaternion and Rotations

Let us consider the quaternion:

$$\underline{R} = \cos\frac{\varphi}{2} + \sin\frac{\varphi}{2}\vec{\Delta} \quad (6.72)$$

It is built using the parameters of the previous vectorial representation of rotations. The norm is that this quaternion is equal to 1:

$$\|\underline{R}\|^2 = \cos^2\frac{\varphi}{2} + \sin^2\frac{\varphi}{2} = 1 \quad (6.73)$$

6.3 Representations with Four Parameters

Consider vector $\vec{X}$ defined in K, it is easy to show by using the preceding rules that the product $\underline{R}\vec{X}\underline{R}^*$ represents the vector $\vec{X}'$ transformed from $\vec{X}$ by rotation φ around axis $\vec{\Delta}$:

$$\underline{R}\vec{X}\underline{R}^* = \left(\vec{X}\cdot\vec{\Delta}\right)\vec{\Delta} + \left[\vec{X} - \left(\vec{X}\cdot\vec{\Delta}\right)\vec{\Delta}\right]\cos\varphi + \left(\vec{\Delta}\wedge\vec{X}\right)\sin\varphi = \vec{X}' \quad (6.74)$$

A rotation φ of observed vector being equivalent to a rotation $-\varphi$ of observation axes, the preceding operator allows us to determine the new components of $\vec{X}$ in a frame E transformed from K by rotation $-\varphi$ around $\vec{\Delta}$. In the case of a rotation $+\varphi$ of axes, we must use the quaternion $\underline{R}' = \cos\frac{\varphi}{2} - \sin\frac{\varphi}{2}\vec{\Delta} = \underline{R}^*$. The expression for the changing of axes thus becomes $\vec{X}' = R^*\vec{X}R$.

6.3.2.3 Evolution of the Quaternion

Let us indicate formally by $\vec{X}_K$ and $\vec{X}_E$ representations in K and in E of a single vector $\vec{X}$. We have just established the relations:

$$\vec{X}_K = \underline{R}\vec{X}_E\underline{R}^*$$
$$\vec{X}_E = \underline{R}^*\vec{X}_K\underline{R} \quad (6.75)$$

According to (6.70) and (6.73), they are equivalent to:

$$R^*\vec{X}_K = \vec{X}_E R^*$$
$$\underline{R}\vec{X}_E = \vec{X}_K\underline{R} \quad (6.76)$$

Let us assume that the vector $\vec{X}$ is constant in the E frame itself in rotation with the instantaneous angular velocity $\vec{\omega} = \dot{\varphi}\vec{\Delta}$ relative to K. We obtain according to (6.75) and (6.76):

$$\dot{\vec{X}}_E = 0 \Rightarrow \dot{\vec{X}}_K = \underline{\dot{R}}\vec{X}_E\underline{R}^* + \underline{R}\vec{X}_E\underline{\dot{R}}^*$$
$$\Rightarrow \dot{\vec{X}}_K = \underline{\dot{R}R^*}\vec{X}_K + \vec{X}_K\underline{R\dot{R}^*} \quad (6.77)$$

While using (6.71) and (6.77), we obtain:

$$\vec{X}_K\underline{R\dot{R}^*} = \vec{X}_K\left[\underline{\dot{R}R^*}\right]^* = \left[-\left[\underline{\dot{R}R^*}\right]\vec{X}_K\right]^* = -\left[\left[\underline{\dot{R}R^*}\right]\vec{X}_K\right]^*$$
$$\Rightarrow \dot{\vec{X}}_K = \underline{\dot{R}R^*}\vec{X}_K + -\left[\left[\dot{R}R^*\right]\vec{X}_K\right]^* = 2\times \text{vectorial part}\left\{\underline{\dot{R}R^*}\vec{X}_K\right\} \quad (6.78)$$

Moreover, $\underline{\dot{R}R^*}$ is pure vectorial, indeed,

$$\underline{RR^*} = \|\underline{R}\|^2 = 1 \Rightarrow \frac{d}{dt}\underline{RR^*} = \underline{\dot{R}R^*} + \underline{R\dot{R}^*} = \underline{\dot{R}R^*} + \left[\underline{\dot{R}R^*}\right]^*$$
$$= 2 \times real\ part\ \{\underline{\dot{R}R^*}\} = 0$$

Let us denote $\vec{W}$ the pure vector $\underline{\dot{R}R^*}$, we obtain:

$$\dot{\vec{X}}_K = 2 \times vectorial\ part\ \{\underline{\dot{R}R^*}\vec{X}_K\} = 2\vec{W} \wedge \vec{X}_K \Rightarrow \vec{W} = \frac{\vec{\omega}_K}{2} \qquad (6.79)$$

where $\vec{\omega}_K$ denotes the expression of $\vec{\omega}$ in K, that is to say,

$$\underline{\dot{R}R^*} = \frac{1}{2}\vec{\omega}_K \Leftrightarrow \underline{\dot{R}} = \frac{1}{2}\vec{\omega}_K\,\underline{R} = \frac{1}{2}\underline{R}\vec{\omega}_E \qquad (6.80)$$

The details of this equality between quaternion, expressed in E gives:

$$\left(\dot{r}_0 + \dot{\vec{r}}\right) = \frac{1}{2}(r_0 + \vec{r})(0 + \vec{\omega}_E)$$
$$\Leftrightarrow \dot{r}_0 = -\frac{1}{2}\vec{r} \cdot \vec{\omega}_E$$
$$\dot{\vec{r}} = \frac{1}{2}\{r_0\vec{\omega}_E + \vec{r} \wedge \vec{\omega}_E\} \qquad (6.81)$$

The equation of evolution for the components of the quaternion is thus:

$$\dot{r}_0 = -\frac{1}{2}(r_1\omega_{1E} + r_2\omega_{2E} + r_3\omega_{3E})$$
$$\dot{r}_1 = \frac{1}{2}\{r_0\omega_{1E} + r_2\omega_{3E} - r_3\omega_{2E}\}$$
$$\dot{r}_2 = \frac{1}{2}\{r_0\omega_{2E} + r_3\omega_{1E} - r_1\omega_{3E}\}$$
$$\dot{r}_3 = \frac{1}{2}\{r_0\omega_{3E} + r_1\omega_{2E} - r_2\omega_{1E}\} \qquad (6.82)$$

The corresponding transformation matrix for coordinates result from the expression of the rotation operator:

$$\vec{X}_E = (r_0 - \vec{r})(0 + \vec{X}_K)(r_0 + \vec{r}) \Leftrightarrow \vec{X}_E = A\vec{X}_K \qquad (6.83)$$

$$[A] = \begin{bmatrix} r_0^2 + r_1^2 - r_2^2 - r_3^2 & 2(r_0r_3 + r_1r_2) & 2(r_1r_3 - r_0r_2) \\ 2(r_1r_2 - r_0r_3) & r_0^2 + r_2^2 - r_1^2 - r_3^2 & 2(r_0r_1 + r_2r_3) \\ 2(r_0r_2 + r_1r_3) & 2(r_2r_3 - r_0r_1) & r_0^2 + r_3^2 - r_1^2 - r_2^2 \end{bmatrix} \qquad (6.84)$$

6.3.2.4 Quaternions and Euler Angles

Let us denote $R_{i(n)}(\theta_n)$ the quaternion associated with nth elementary rotation around the Euler axis $\vec{i}(n)$. The total operator of rotation is written as:

$$\vec{X}_E = R^*_{i(3)}(\theta_3)\, R^*_{i(2)}(\theta_2)\, R^*_{i(1)}(\theta_1)\, \vec{X}_K\, R_{i(1)}(\theta_1)\, R_{i(2)}(\theta_2)\, R_{i(3)}(\theta_3) \quad (6.85)$$

The quaternion corresponding to the product of the three elementary rotations is thus written as:

$$R_{i(1),i(2),i(3)}(\theta_1, \theta_2, \theta_3) = R_{i(1)}(\theta_1)\, R_{i(2)}(\theta_2)\, R_{i(3)}(\theta_3) \quad (6.86)$$

with:

$$R_{i(n)}(\theta_n) = \cos\frac{\theta_n}{2} + \sin\frac{\theta_n}{2}\vec{e}_i(n)(n)$$

Let us evaluate this expression in the case of Euler angles corresponding to sequence 3, 2, 1:

$$R_3(\psi)\, R_2(\theta)\, R_1(\varphi) = \left\{\cos\frac{\psi}{2} + \sin\frac{\psi}{2}\begin{bmatrix}0\\0\\1\end{bmatrix}\right\} \left\{\cos\frac{\theta}{2} + \sin\frac{\theta}{2}\begin{bmatrix}0\\1\\0\end{bmatrix}\right\}$$

$$\left\{\cos\frac{\varphi}{2} + \sin\frac{\varphi}{2}\begin{bmatrix}1\\0\\0\end{bmatrix}\right\}$$

$$R_3(\psi)\, R_2(\theta)\, R_1(\varphi) = \left\{\cos\frac{\psi}{2} + \sin\frac{\psi}{2}\begin{bmatrix}0\\0\\1\end{bmatrix}\right\} \left\{\cos\frac{\theta}{2}\cos\frac{\varphi}{2} + \begin{bmatrix}\cos\frac{\theta}{2}\sin\frac{\varphi}{2}\\ \sin\frac{\theta}{2}\cos\frac{\varphi}{2}\\ -\sin\frac{\theta}{2}\sin\frac{\varphi}{2}\end{bmatrix}\right\}$$

$$R_3(\psi)\, R_2(\theta)\, R_1(\varphi) = \left\{\cos\frac{\psi}{2}\cos\frac{\theta}{2}\cos\frac{\varphi}{2} + \sin\frac{\psi}{2}\sin\frac{\theta}{2}\sin\frac{\varphi}{2}\right.$$

$$\left. + \begin{bmatrix}\cos\frac{\psi}{2}\cos\frac{\theta}{2}\sin\frac{\varphi}{2} - \sin\frac{\psi}{2}\sin\frac{\theta}{2}\cos\frac{\varphi}{2}\\ \cos\frac{\psi}{2}\sin\frac{\theta}{2}\cos\frac{\varphi}{2} + \sin\frac{\psi}{2}\cos\frac{\theta}{2}\sin\frac{\varphi}{2}\\ -\cos\frac{\psi}{2}\sin\frac{\theta}{2}\sin\frac{\varphi}{2} + \sin\frac{\psi}{2}\cos\frac{\theta}{2}\cos\frac{\varphi}{2}\end{bmatrix}\right\}$$

From which we obtain components of the quaternion $R_{3\,2\,1}(\psi, \theta, \varphi)$ equivalent with the three rotations:

$$r_0 = \cos\frac{\psi}{2}\cos\frac{\theta}{2}\cos\frac{\varphi}{2} + \sin\frac{\psi}{2}\sin\frac{\theta}{2}\sin\frac{\varphi}{2}$$

$$r_1 = \cos\frac{\psi}{2}\cos\frac{\theta}{2}\sin\frac{\varphi}{2} - \sin\frac{\psi}{2}\sin\frac{\theta}{2}\cos\frac{\varphi}{2}$$

$$r_2 = \cos\frac{\psi}{2}\sin\frac{\theta}{2}\cos\frac{\varphi}{2} + \sin\frac{\psi}{2}\cos\frac{\theta}{2}\sin\frac{\varphi}{2}$$

$$r_3 = -\cos\frac{\psi}{2}\sin\frac{\theta}{2}\sin\frac{\varphi}{2} + \sin\frac{\psi}{2}\cos\frac{\theta}{2}\cos\frac{\varphi}{2}. \tag{6.87}$$

Chapter 7
Exoatmospheric Phase

7.1 Movement of the Center of Mass

7.1.1 Keplerian Trajectories

This topic is included in most mechanics textbooks, so it is difficult to bring anything new. However, it is a mandatory precondition before studying reentry phase. Also, we will try to approach it from a somewhat different point of view.

Let us consider now the classical problem of the movement of a point mass in a central attraction field according to r^{-2}.

7.1.1.1 Hypotheses and Nomenclature

- We assume the earth is a homogeneous spherical body.
- We consider an inertial frame K with origin at earth's center of mass G, whose accelerated movement is neglected.
- $\vec{r}$ denotes the radius vector from G to the center of mass M of a body of mass m.
- The gravitational field $\vec{\varphi} = -\frac{\mu \vec{r}}{r^2 r}$ is isotropic, and gravitational attraction force is $\vec{f} = m\vec{\varphi}$.
- The linear momentum of the body in K is $\vec{p} = m\dot{\vec{r}}$.
- The angular momentum is $\vec{l} = \vec{r} \wedge \vec{p}$.

7.1.1.2 Equations of Movement

For an observer fixed to K, fundamental principles of mechanics are written as,

$$\dot{\vec{p}} = \vec{f} \Leftrightarrow \ddot{\vec{r}} = \vec{\varphi}$$
$$\dot{\vec{l}} = \vec{r} \wedge \vec{f} \Leftrightarrow \vec{r} \wedge \ddot{\vec{r}} = \vec{r} \wedge \vec{\varphi}$$

Equations of movement are clearly independent of mass. Thus, the motion depends only on the kinematics initial conditions. We thus assume in the continuation of this chapter that m = 1 and that force, moment, and energy are related to the unit of mass.

The central force applied is along $\vec{r}$ and its moment relating to G is null. Consequently, the derivative of the angular momentum is null and the *angular momentum is constant in the movement*. Thus, *the movement is planar*, indeed $\vec{l} = \vec{r} \wedge \dot{\vec{r}} = \vec{l}_0$ implies $\vec{r}$ and $\dot{\vec{r}}$ are orthogonal to $\vec{l}_0$.

The movement takes place in the plane passing by G and orthogonal to $\vec{l}_0 = \vec{r}_0 \wedge \dot{\vec{r}}_0$, defined by $\vec{r}_0$ and $\dot{\vec{r}}_0$.

(i) Note: we eliminate the case where $\vec{l}_0 = 0$, which corresponds to an initial velocity directed toward the center of attraction and gives a rectilinear trajectory passing by G.

7.1.1.3 Movement in a Reference Frame Fixed with the Local Vertical

Let us choose frame K such that Gx is along $\vec{r}_0$ and Gz along $\vec{l}_0$. Gy axis is determined to complete the right-hand frame. So (Gx, Gy) is the trajectory plane. We consider $\vec{r} = \{x, y, o\}$ the instantaneous position of M in this plane, and we define a rotating frame E such that $Gz_E \equiv Gz$ and Gx_E is along $\vec{r}$ (Fig. 7.1).

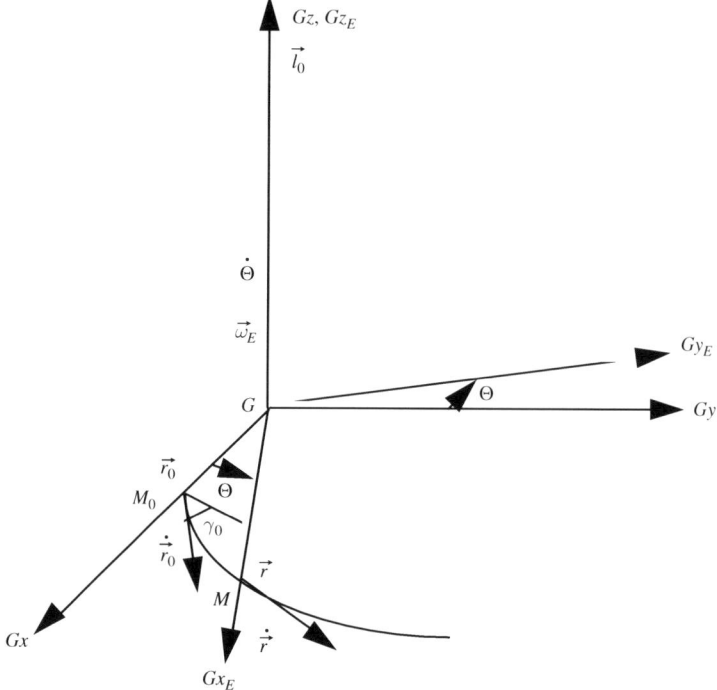

Fig. 7.1 Local frame E

7.1 Movement of the Center of Mass

We obtain E frame by rotating K with an angle θ around Gz. This is a noninertial frame, with a nonuniform angular rate $\vec{\omega}_E = \{0, 0, \dot{\theta}\}$ around Gz.

Instantaneous location and velocity of point M relating to E are $\vec{r}_E = \{r, 0, 0\}$ and $\dot{\vec{r}}_E = [\dot{r}, 0, 0]$, i.e., the motion of M relative to E is along the vertical. The components of attraction force are in this frame $\vec{f}_E = \left\{-\frac{\mu}{r^2}, 0, 0\right\}$, directed along $-\overrightarrow{Gx_E}$.

From (1.4), preceding equations established in K are written in the rotating frame E:

$$\vec{r}_E \wedge \left(\dot{\vec{r}}_E + \vec{\omega}_E \wedge \vec{r}_E\right) = \vec{l}_0 \tag{7.1}$$

$$\ddot{\vec{r}}_E = \vec{\varphi}_E - \left(2\vec{\omega}_E \wedge \dot{\vec{r}}_E + \dot{\vec{\omega}}_E \wedge \vec{r}_E + \vec{\omega} \wedge (\vec{\omega} \wedge \vec{r}_E)\right) \tag{7.2}$$

These equations are simplified by noting that $\vec{r}_E, \dot{\vec{r}}_E,$ and $\ddot{\vec{r}}_E$ are parallel and orthogonal to $\vec{\omega}_E$ and $\dot{\vec{\omega}}_E$. We may then obtain projections of these equations along Gx_E, Gy_E, and Gz_E. Equation (7.1) gives along Gz_E,

$$r^2\dot{\theta} = l_0 \tag{7.3}$$

Equation (7.2) gives along Gx_E,

$$\ddot{r} = -\frac{\mu}{r^2} + \dot{\theta}^2 r \tag{7.4}$$

The last term of the second member clearly represents the centrifugal force associated with the rotation motion of E.

Equation (7.2) gives along Gy_E,

$$2\dot{\theta}\dot{r} + \ddot{\theta}r = 0 \tag{7.5}$$

The first term corresponds to the Coriolis acceleration and the second to the tangential acceleration associated with the angular acceleration, both related with the rotational movement of E. After multiplication by r, (7.5) gives:

$$\frac{d}{dt}\left(r^2\dot{\theta}\right) = 0 \tag{7.6}$$

It is equivalent to (7.3), i.e., with the conservation of angular momentum $\vec{l}$.

This "mechanical" approach leads to the well-known system of differential equations:

$$\frac{1}{2}r^2\dot{\theta} = \dot{A} = \frac{l_0}{2} \tag{7.7}$$

$$\ddot{r} = -\frac{\mu}{r^2} + \dot{\theta}^2 r \tag{7.8}$$

Equation (7.7) clearly represents the second law of Kepler, "rate of swept area $\dot{A}$ is constant," $\dot{A}$ indicates the time derivative of the area swept by the radius vector.

Equation (7.8) can be transformed while using $\dot{\theta}$ from (7.7), and choosing θ as independent variable:

$$\frac{d}{dt} = \frac{d\theta}{dt}\frac{d}{d\theta} \Rightarrow \dot{\theta}\frac{d}{d\theta}\left(\dot{\theta}\frac{dr}{d\theta}\right) = -\frac{\mu}{r^2} + \dot{\theta}^2 \cdot r$$

which gives,

$$\frac{d}{d\theta}\left(\frac{1}{r^2}\frac{dr}{d\theta}\right) = -\frac{\mu}{\left(2\dot{A}\right)^2} + \frac{1}{r} \tag{7.9}$$

A last change of function $r = \frac{1}{u}$ allows obtaining a more convenient final form:

$$\frac{d^2u}{d\theta^2} + u = \frac{\mu}{\left(2\dot{A}\right)^2} \tag{7.10}$$

This can be integrated in,

$$\frac{1}{r} = u = \frac{\mu}{\left(2\dot{A}\right)^2} + c\cos(\theta - \theta_p) \tag{7.11}$$

Where the constants of integration are c, assumed positive, and the angle θ_p, which correspond to the maximum value of u, that is to say the minimum value of the distance to center of attraction (perigee of the trajectory).

While noting classically $\frac{1}{p} = \frac{\mu}{\left(2\dot{A}\right)^2}$ and $c = \frac{e}{p}$, we obtain:

$$r = \frac{p}{1 + e\cos(\theta - \theta_p)} \tag{7.12}$$

We leave to the reader to check that this corresponds to polar equation of a conic, with origin at focus, of parameters:

$$p = \frac{\left(2\dot{A}\right)^2}{\mu} \quad \text{"latus rectum"} \tag{7.13}$$

$$0 \le e \text{ "eccentricity"}$$

7.1 Movement of the Center of Mass

Table 7.1 Nature of conical curve depending on the eccentricity e

e = 0	D = 1	$r = r_c = p$		Circle
$0 < e < 1$	$0 < D \leq 1 + e$	$\dfrac{p}{1+e} \leq r \leq \dfrac{p}{1-e}$		Ellipse
$1 \leq e$	$0 \leq D \leq 1 + e$	$\dfrac{p}{1+e} \leq r \leq \infty$		Hyperbola or parabola ($e = 1$)

In the case of the ellipse, the polar angle θ relative to the initial point is classically named "true anomaly."

Examination of the denominator $D = 1 + e \cdot \cos(\theta - \theta_p)$ as a function of θ with the constraint $0 \leq r$ shows that we obtain various forms according to value of e (Table 7.1).

Whatever the value of e, the radius of the perigee is $r_p = \dfrac{p}{1+e}$, corresponding to $\theta = \theta_p$. In the case of closed trajectories ($e < 1$, with negative total energy), the radius of apogee is $r_a = \dfrac{p}{1-e}$, corresponding to $\theta = \theta_a = \theta_p \pm \pi$.

These results demonstrate the first law of Kepler: *"the trajectories of planets are ellipses whose one focus is located at sun mass center."*

7.1.1.4 Parameters of the Conical Curve

The initial conditions are:

$$\vec{r}_0 = \vec{r}(t=0) = \vec{r}(\theta = 0)$$

$$\vec{v}_0 = \dot{\vec{r}}_0 = \dot{\vec{r}}(t=0) = \dot{\vec{r}}(\theta = 0)$$

To determine the parameter e, we will use the conservation of energy in K,

$$E = T + U = \frac{1}{2}\vec{v}^2 - \frac{\mu}{r} \qquad (7.14)$$

It can be expressed as a function of r and θ using the expression of $\vec{v}$ in E:

$$\vec{v} = \vec{v}_E + \vec{\omega}_E \wedge \vec{r}_E$$

$$\frac{1}{2}\left(\dot{r}^2 + (r\dot{\theta})^2\right) - \frac{\mu}{r} = E \qquad (7.15)$$

The (7.7) and (7.12) make it possible to determine the time derivative of the radius vector $\dot{r} = \dot{\theta}\dfrac{dr}{d\theta}$, that is to say:

$$\dot{r} = \frac{2\dot{A}}{r^2}\frac{dr}{d\theta} = \frac{2\dot{A}e}{p}\sin(\theta - \theta_p) \qquad (7.16)$$

Then, while introducing (7.16) in (7.15), we obtain after some derivations:

$$e^2 - 1 = 2E\left(\frac{2\dot{A}}{\mu}\right)^2 = 2E \cdot \frac{p}{\mu}$$

$$E = \mu \frac{e^2 - 1}{2p}; \quad e = \sqrt{1 + 2E\frac{p}{\mu}} \qquad (7.17)$$

Having determined p and e, the parameter θ_p is obtained directly from r_0 and $\dot{r}_0 = v_{x0} = v_0 \sin \gamma_0$:

$$\cos \theta_p = \frac{p - r_0}{r_0 e} = \frac{1}{e}\left(\frac{p}{r_0} - 1\right)$$

$$\sin \theta_p = -\frac{p v_{x0}}{2\dot{A}e} = -\frac{p v_0 \sin \gamma_0}{2\dot{A}e} = -\frac{p \tan \gamma_0}{r_0 e}$$

In the case of closed trajectories, we can determine angular location of apogee r_a, which corresponds to $\theta_a = \theta_p \pm \pi$,

$$\cos \theta_a = \frac{1}{e}\left(1 - \frac{p}{r_0}\right)$$

$$\sin \theta_a = -\frac{p \tan \gamma_0}{r_0 e} \qquad (7.18)$$

From (7.17), we can express energy as a function of r_p, whatever the trajectory, and of r_p or r_a in the case of closed trajectories

$$E = \frac{\mu}{2}\frac{\left(e^2 - 1\right)}{p} = \frac{\mu}{2}\frac{(1 + e)(1 - e)}{p}$$

Taking into account the expression of the radius of perigee $r_p = \frac{p}{1+e}$, we obtain:

$$E = T_p + U_p = -(1 - e)\frac{\mu}{2r_p} = \frac{(1 - e)}{2}U_p \qquad (7.19)$$

In the same way,

$$E = T_a + U_a = -(e + 1)\frac{1\mu}{2r_a} = \frac{1 + e}{2}U_a \qquad (7.20)$$

In all cases,

$$T_p = -\frac{1 + e}{2}U_p \qquad (7.21)$$

7.1 Movement of the Center of Mass

And only in the case e < 1,

$$T_a = -\frac{1-e}{2} U_a \tag{7.22}$$

To analyze the relation between the energy and the nature of the trajectory, we consider all the trajectories having the same given radius of apogee or perigee $r_a = r_p = r_c$. By definition they all have, at this point, the same potential energy,

$$U_a = U_p = U_c = -\frac{\mu}{r_c} < 0 \tag{7.23}$$

From preceding expressions of kinetic energy at point c, we obtain the classification of Table 7.2, for a radius of apogee or perigee $r_c \approx r_t = 6371$ km close to the earth average radius.

Notice: for a given radius of perigee or apogee $r_c \neq r_t$ circular orbit velocity and escape velocity become $V_c' = V_c \sqrt{\frac{r_t}{r_c}}$ and $V_l' = V_l \sqrt{\frac{r_t}{r_c}}$

7.1.1.5 Elliptic Trajectories

From preceding results, we obtain geometrical parameters of the centered Cartesian representation:

– Semimajor axis

$$a = \frac{r_p + r_a}{2} = \frac{1}{2}\left(\frac{p}{1+e} + \frac{p}{1-e}\right) = \frac{p}{1-e^2} \tag{7.24}$$

Table 7.2 Classification of the Keplerian trajectories

Eccentricity	Nature of the conical curve	Total energy E and kinetic T at point c	Velocity at perigee or apogee
$0 < e < 1$	Elliptic: $r_p \leq r \leq r_a = r_c$	$U_c \leq E < \frac{U_c}{2}$ $0 \leq T_c < \frac{|U_c|}{2}$	$v_c < V_c$, circular orbit velocity
$e = 0$	Circular, radius r_c	$E = \frac{U_c}{2}$, $T_c = \frac{|U_c|}{2}$	Circular orbit velocity $v = V_c = \sqrt{\frac{\mu}{r_c}} \approx 7.91$ km/s
$0 < e < 1$	Elliptic: $r_p = r_c \leq r \leq r_a$	$\frac{|U_c|}{2} < E < 0$ $\frac{|U_c|}{2} < T_c < |U_c|$	$V_c < v_c < V_l$, escape velocity
$e > 1$	Parabolic, $r_p \leq r \to \infty$	$E = 0$, $T_c = |U_c|$	Escape velocity, $v_c = V_l = V_c \sqrt{2} \approx 11.$ km/s
$e > 1$	Hyperbolic, $r_p \leq r \to \infty$	$0 < E$ $|U_c| < T_c$	$V_l < v_c$

Taking into account (7.17), it results that energy of the trajectory is entirely determined by a:

$$E = -\frac{\mu}{2a} \qquad (7.25)$$

– Focus location

$$\pm c = \pm e \cdot a \qquad (7.26)$$

– Semiminor axis b results from the geometrical definition of the ellipse, i.e., the sum of the distances from a point M of the curve to focuses is constant:

$$(a - c) + (a + c) = 2d$$

where d is the hypotenuse of triangle OBC (Fig 7.2)

$$\Rightarrow b^2 = d^2 - c^2 = a^2 - c^2 = a^2 \cdot \left(1 - e^2\right)$$

$$\Rightarrow \text{semiminor axis } b = a\sqrt{1 - e^2} \qquad (7.27)$$

This leads us to the *third law of Kepler*
– Rate of swept area:

$$\frac{(2\dot{A})^2}{\mu} = p = \left(1 - e^2\right)a \Rightarrow \dot{A} = \frac{1}{2}\sqrt{\mu(1-e)^2 a} \Rightarrow \dot{A} = \frac{\sqrt{\mu}}{2}\frac{b}{\sqrt{a}}$$

– Area S of ellipse: $S = \pi a\, b$

From which the period of revolution T:

$$T = \frac{S}{\dot{A}} = \frac{2\pi}{\sqrt{\mu}} a^{\frac{3}{2}} \qquad (7.28)$$

"The square of time of revolution is proportional to the cube of the semimajor axis of the ellipse."

7.1.1.6 Chronology

We use the geometrical method having the constant rate of swept area and calculation of the area swept as function of θ, with origin at apogee.

The ellipse is the orthogonal projection of the corresponding circle (Fig. 7.2), of radius R = a located in the plane passing by the major axis, and forming an angle φ such that $\cos\varphi = b/a$.

We seek the value of area S swept by the radius vector since the passage to apogee CA until its instantaneous position CM. This area S is the projection of area

7.1 Movement of the Center of Mass

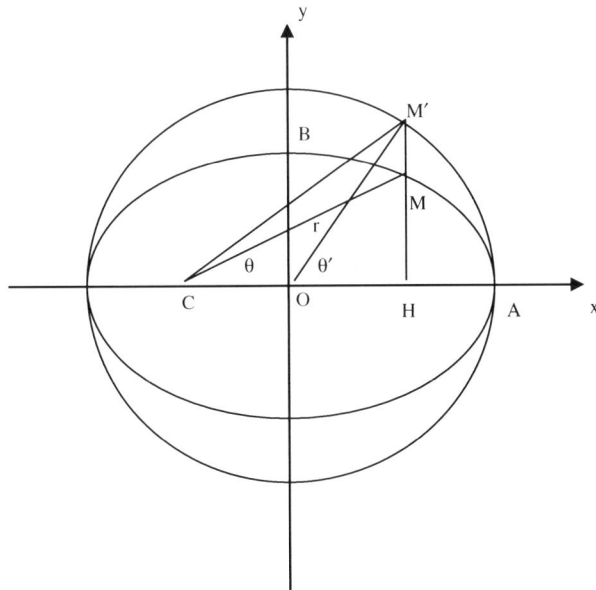

Fig. 7.2 Determination of swept area

S' of eccentric circular sector CAM', composed of the centered sector OAM of apex angle θ' and of the triangular element OCM'.

$$S' = \frac{a^2\theta'}{2} + \frac{1}{2}a\,c\sin\theta' \qquad (7.29)$$

$$S = S'\cos\varphi = \frac{a\,b}{2}(\theta' + e\sin\theta') \qquad (7.30)$$

We must then determine θ' (eccentric anomaly), as a function of θ:

$$HM' = \frac{HM}{\cos\varphi} \Rightarrow a\sin\theta' = \frac{a}{b}r\sin\theta$$

$$OH = a\cos\theta' = r\cos\theta - c$$

$$\sin\theta' = \sqrt{1-e^2}\,\frac{\sin\theta}{1-e\cos\theta} \qquad (7.31)$$

$$\cos\theta' = \frac{\cos\theta - e}{1 - e\cos\theta} \qquad (7.32)$$

We obtain finally the time since passage at apogee as a function of θ'

$$t = \frac{S}{\dot{A}} = \frac{T}{2\pi}(\theta' + e\sin\theta') \qquad (7.33)$$

Then as a function of θ:

$$\theta \in [0, \pi] \rightarrow t = \frac{T}{2\pi}\left[\text{Arc cos}\left(\frac{\cos\theta - e}{1 - e\cos\theta}\right) + \frac{e\sqrt{1-e^2}\sin\theta}{1 - e\cos\theta}\right] \quad (7.34)$$

$$\theta \in [\pi, 2\pi] \rightarrow t = \frac{T}{2\pi}\left[2\pi - \text{Arc cos}\left(\frac{\cos\theta - e}{1 - e\cos\theta}\right) + \frac{e\sqrt{1-e^2}\sin\theta}{1 - e\cos\theta}\right]$$

7.1.2 Ballistic Trajectories

Ballistic trajectories are closed Keplerian trajectories having an intersection with terrestrial sphere (Fig. 7.3). They are elliptical trajectories such that:

$$E < 0 \quad \Rightarrow \quad v_0 < \sqrt{2\frac{r_t}{r_0}}V_c \quad (7.35)$$

In order to intersect with the terrestrial surface, radii of perigee and apogee must verify:

$$r_p \leq r_t \leq r_a \quad \Leftrightarrow \quad (1-e)r_t \leq p \leq (1+e)r_t \quad (7.36)$$

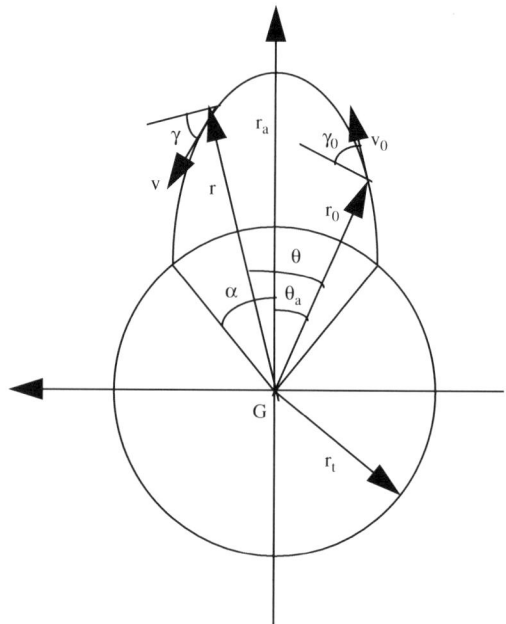

Fig. 7.3 Ballistic trajectories

7.1 Movement of the Center of Mass

For an initial point above the ground $r_0 \geq r_t$, the condition $r_t \leq r_a$ is automatically verified. The only remaining condition is that the radius of perigee is lower than earth radius:

$$p \leq r_t (1 + e)$$

This condition, which eliminates the trajectories of satellites being verified, the range measured at sea level is given by the relation:

$$r_t = \frac{p}{1 - e \cos \alpha}$$

where α indicates half angular range. This gives:

$$\cos \alpha = \frac{1}{e}\left(1 - \frac{p}{r_t}\right) \tag{7.37}$$

We must eliminate trajectories of angular range $\pi < 2\alpha < 2\pi$ such as $\cos \alpha < 0$, which are until now without practical interest (they correspond to passage by the antipodes):

$$0 \leq 2\alpha \leq \pi \Rightarrow \cos \alpha = \frac{1}{e}\left(1 - \frac{p}{r_t}\right) \geq 0 \Leftrightarrow p \leq r_t$$

$$\Rightarrow P = 2r_t \text{ arc cos}\left(\frac{1}{e}\left(1 - \frac{p}{r_t}\right)\right) \leq \pi r_t \approx 20000 \text{ km}$$

According to preceding results, parameters p and e are related to initial conditions by the relations:

$$p = \frac{(r_0 \, v_0 \, \cos \gamma_0)^2}{\mu} \tag{7.38}$$

$$e = \sqrt{1 + 2\left(\frac{v_0^2}{2} - \frac{\mu}{r_0}\right)\left(\frac{r_0 \, v_0 \cos \gamma_0}{\mu}\right)^2} \tag{7.39}$$

While $r_0 \geq r_t$, conditions necessary and sufficient to obtain a ballistic trajectory of range lower than 20000 km are finally reduced to the single inequality:

$$p \leq r_t \Leftrightarrow \left(\frac{r_0}{r_t}\right)^2 \left(\frac{v_0}{V_c}\right)^2 (\cos \gamma_0)^2 \leq 1$$

This leads to the following constraints on initial conditions:

- When initial velocity is lower than the circular orbit velocity $0 \leq v_0 \leq V_c\left(\frac{r_t}{r_0}\right)$, all the flight path angles $0 \leq \gamma_0 \leq \frac{\pi}{2}$ are possible.
- When initial velocity lies between the circular orbit and escape velocities $V_c\frac{r_t}{r_0} < v_0 < V_c\sqrt{2\frac{r_t}{r_0}}$, the flight path angle must satisfy $\gamma_{0\,min} \leq \gamma_0 \leq \frac{\pi}{2}$, with $\cos\gamma_{0\,min} = \left(\frac{V_c}{v_0}\right)\left(\frac{r_t}{r_0}\right)$.

For an initial point located at sea level ($r_0 = r_1$), these conditions become:
- for $v_0 \leq V_c$, complete interval $0 \leq \gamma_0 \leq \frac{\pi}{2}$ is possible.
- for $v_c \leq v_0 < v_l = V_c\sqrt{2}$, interval is restricted to $Arc\cos\left(\frac{V_c}{v_0}\right) \leq \gamma_0 \leq \frac{\pi}{2}$.

In other words, whatever the initial flight path angle, using an initial velocity lower than $v_c \approx 7910\,\text{m/s}$ guarantees an impact point of range lower than half earth's circumference (at least if we do not take into account constraints related to the existence of the relief and atmosphere!). A velocity in the interval $v_c \approx 7910\,\text{m/s}$ and $v_l \approx 11000\,\text{m/s}$ and a flight path angle higher than a limit equal to $Arc\cos\left(\frac{v_c}{v_0}\right)$ also makes this possible.

The angular location of the apogee relative to the initial point is given by:

$$\cos\theta_a = \frac{r_0 - p}{e r_0} \quad (7.40)$$

$$\sin\theta_a = \frac{p\tan\gamma_0}{e r_0} \quad (7.41)$$

7.1.2.1 Equations of the Trajectory

Principal useful relations, for an angular origin at the initial point are:

Instantaneous radius:

$$r = \frac{p}{1 - e\cos(\theta - \theta_a)} \quad (7.42)$$

Instantaneous velocity and flight path angle:

$$v = v_c\sqrt{\frac{r_t}{p}}\sqrt{1 + e^2 - 2e\cos(\theta - \theta_a)} \quad (7.43)$$

$$\sin\gamma = -\frac{e\sin(\theta - \theta_a)}{\sqrt{1 + e^2 - 2e\cos(\theta - \theta_a)}} \quad (7.44)$$

$$\cos\gamma = \frac{1 - e\cos(\theta - \theta_a)}{\sqrt{1 + e^2 - 2e\cos(\theta - \theta_a)}} \quad (7.45)$$

7.1 Movement of the Center of Mass

Velocity at apogee and perigee:

$$v_a = v_c \sqrt{\frac{r_t}{p}} (1 - e)$$

$$v_p = v_c \sqrt{\frac{r_t}{p}} (1 + e)$$

Total ballistic time (from sea level to sea level):

$$t = \frac{T}{\pi} \left[\arccos\left(\frac{\cos\alpha - e}{1 - e\cos\alpha}\right) + \frac{e\sqrt{1-e^2}\sin\alpha}{1 - e\cos\alpha} \right]$$

with $T = \frac{2\pi}{\sqrt{\mu}} a^{\frac{3}{2}}$ and $a = \frac{p}{1-e^2}$.

Ballistic time as a function of θ *is identical to the general case of the ellipse.*

Expression of the dynamic variables according to altitude: Relations as function to the true anomaly, θ can be used to determine the variables v, γ, t at a given altitude h. Indeed, we have according to (7.42):

$$\theta - \theta_a = \pm Arc\cos\left(\frac{1}{e}\left(1 - \frac{p}{r_t + h}\right)\right), \quad r_p \leq r_t + h \leq r_a \quad (7.46)$$

7.1.2.2 Maximum Range for a Given Energy

Total energy E of the trajectory being fixed at some negative value, we seek the optimal flight path angle γ_{opt}, which provides the maximum range P. To maximize α is equivalent to minimize $\cos\alpha$, obtained from (7.17) and (7.37):

$$\cos\alpha = \frac{r_t - p}{r_t\sqrt{1 + \frac{2}{\mu}Ep}} \quad \text{minimum} \Rightarrow \frac{\partial\cos\alpha}{\partial p}\frac{\partial p}{\partial \gamma_0} = 0 \quad (7.47)$$

The optimal flight path angle is obtained from $\frac{\partial\cos\alpha}{\partial p} = 0$. Taking into account (7.38) and (7.39),

$$p_{opt} = -(r_t + \frac{\mu}{E}) \quad \Rightarrow \quad \cos^2\gamma_{opt} = -\frac{\mu}{(r_0 v_0)^2}\left[r_t + \frac{\mu}{\frac{v_0^2}{2} - \frac{\mu}{r_0}}\right] \quad (7.48)$$

For an initial point at sea level ($r_0 = r_t$), the expression is simplified considerably:

$$\cos^2\gamma_{opt} = \frac{1}{2 - \frac{r_t v_0^2}{\mu}} = \frac{1}{2 - \left(\frac{v_0}{v_c}\right)^2} \quad (7.49)$$

The constraint $\cos^2 \gamma_{opt} \leq 1$ implies that the ground velocity must be lower than the circular orbit velocity V_c.

While carrying this value of γ_{opt} in the expression of $\cos \alpha$, we obtain the maximum range,

$$\cos \alpha_{opt} = \frac{1}{e_{opt}}\left(1 - \frac{p_{opt}}{r_t}\right)$$

While noting $\bar{v} = \frac{v_0}{V_c}$ (nondimensional velocity), according to (7.38), (7.39), and (7.49), we obtain:

$$p_{opt} = r_t \frac{\bar{v}^2}{2 - \bar{v}^2} \tag{7.50}$$

$$e_{opt} = \sqrt{1 - \bar{v}^2} \tag{7.51}$$

$$\cos \alpha_{opt} = \frac{2\sqrt{1 - \bar{v}^2}}{2 - \bar{v}^2} \quad \Rightarrow \quad \sin \alpha_{opt} = \frac{\bar{v}^2}{2 - \bar{v}^2} \tag{7.52}$$

Finally, from the expressions of γ_{opt} and $\sin 2\gamma_{opt}$, we obtain:

$$\cos \alpha_{opt} = \sin 2\gamma_{opt} \quad \Leftrightarrow \quad 2\alpha_{opt} = \pi - 4\gamma_{opt} \tag{7.53}$$

The theoretical maximum range corresponds to $\bar{v} = 1 \Rightarrow \gamma_{opt} = 0$, i.e., the circular trajectory, which is obviously unusable in practice:

$$P_{max} = \pi r_t \approx 20000 \, \text{km}$$

Remarks:

i) Relations (7.49) and (7.52) giving the optimal flight path angle and the optimal range are only a function of the parameter $\overline{T} = \frac{\bar{v}^2}{2}$ (nondimensional kinetic energy).
ii) Taking into account operational requirements and constraints related to missiles and atmospheric effects, the practical maximum ranges are about 12000 Km.

7.1.2.3 Minimal Energy for a Given Range

Path Angle Optimization for Range

The range being given, the preceding result (7.53) directly gives the flight path angle with minimum energy:

$$\gamma_{opt} = \frac{\pi}{4} - \frac{\alpha}{2} \tag{7.54}$$

7.1 Movement of the Center of Mass

We obtain the well-known optimum flight path angle equal to $\frac{\pi}{4}$ for the short ranges. Optimum flight path angle decreases linearly toward zero at the maximum angular range $2\alpha = \pi$.

Characteristics of Trajectories

For a fixed range $P = 2\alpha r_t \leq \pi r_t \approx 20000$ km, use of the optimal flight path angle corresponds to the minimum energy trajectory. According to (7.52) and (7.14), energy and sea-level velocity corresponding to $\gamma_{opt} = \frac{\pi}{4} - \frac{\alpha}{2}$ are:

$$E_{opt} = -\frac{V_c^2}{1+\sin\alpha} \leq E_{circ} = -\frac{V_c^2}{2} \tag{7.55}$$

$$\left(\frac{v_{opt}}{v_c}\right)^2 = \frac{2\sin\alpha}{1+\sin\alpha} \leq 1 \tag{7.56}$$

While carrying $\bar{v}_{opt} = \frac{v_{opt}}{V_c}$ in expressions (7.50) and (7.51), we obtain p_{opt} and e_{opt} as a function of α, then derived expressions:

$$r_{a,opt} = \frac{p_{opt}}{1-e_{opt}} \quad \text{(Radius of apogee)}$$

$$v_{a,opt} = v_c \sqrt{\frac{r_t}{p_{opt}}} \left(1-e_{opt}\right) \quad \text{(Velocity at apogee)}$$

$$t = \frac{T}{\pi}\left[\arccos\left(\frac{\cos\alpha - e_{opt}}{1 - e_{opt}\cos\alpha}\right) + \frac{e_{opt}\sqrt{1-e_{opt}^2}\sin\alpha}{1-e_{opt}\cos\alpha}\right] \quad \text{(Ballistic time), with:}$$

$$T = \frac{2\pi}{\sqrt{\mu}} \left(\frac{p_{opt}}{1-e_{opt}^2}\right)^{\frac{3}{2}}$$

Evolution with range of characteristic parameters of minimal energy trajectories is shown in Figs. (7.4–7.7).

7.1.2.4 Nonoptimal Trajectories

Nonoptimal Energy Conditions

Let us seek the conditions to obtain a given range $P = 2\alpha r_t \leq \pi r_t \approx 20000$ km while using more energy than the optimal trajectory:

$$E_{opt} < E < 0$$

$$v_{opt} < v_0 < \sqrt{2}v_c \quad (v_0 \text{ Sea-level velocity})$$

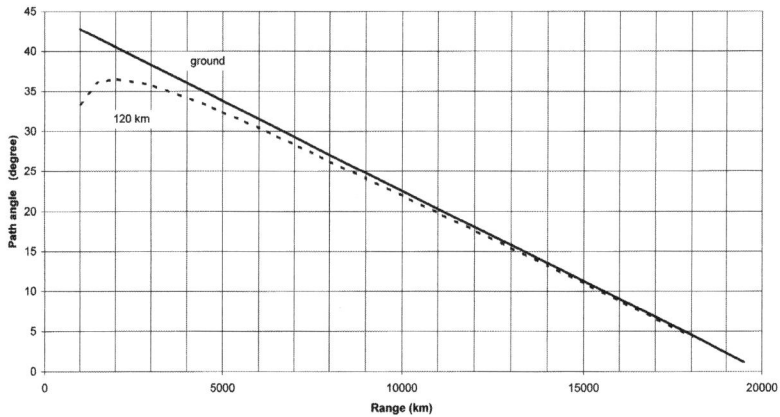

Fig. 7.4 Initial flight path angle as function of range

The range and energy being fixed, we have according to (7.37) and (7.17):

$$\frac{1}{e}\left(1 - \frac{p}{r_t}\right) = \cos\alpha$$

$$\frac{e^2 - 1}{2p} = \frac{E}{\mu}$$

While eliminating e between the two equations, we obtain an equation in p,

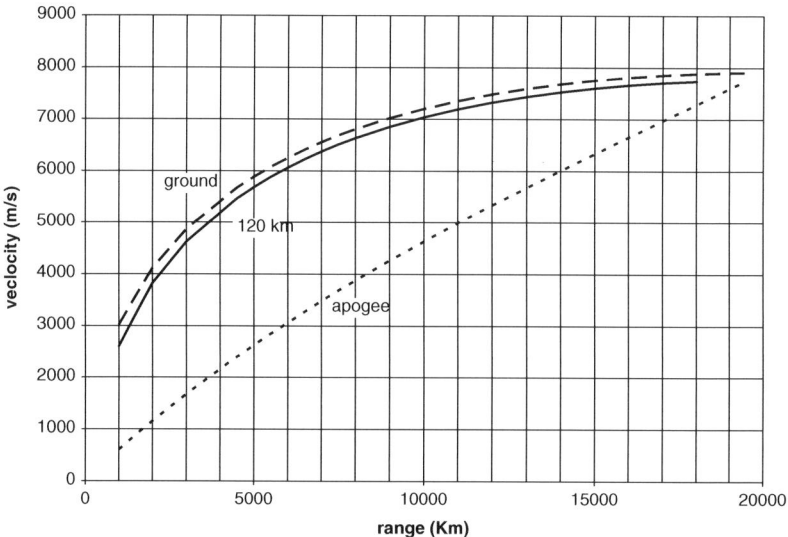

Fig. 7.5 Velocity as function of range

7.1 Movement of the Center of Mass

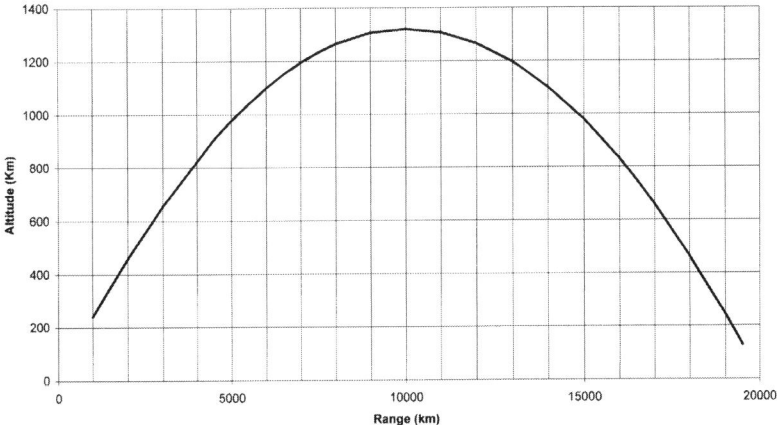

Fig. 7.6 Apogee altitude as function of range

$$2p\frac{E\cos^2\alpha}{\mu} = \left(1 - \frac{p}{r_t}\right)^2 - \cos^2\alpha \tag{7.57}$$

Then by using relations:

$$E = \frac{v_0^2}{2} - \frac{\mu}{r_t}$$

$$V_c^2 = \frac{\mu}{r_t}; \quad \bar{v} = \frac{v_0}{V_c}$$

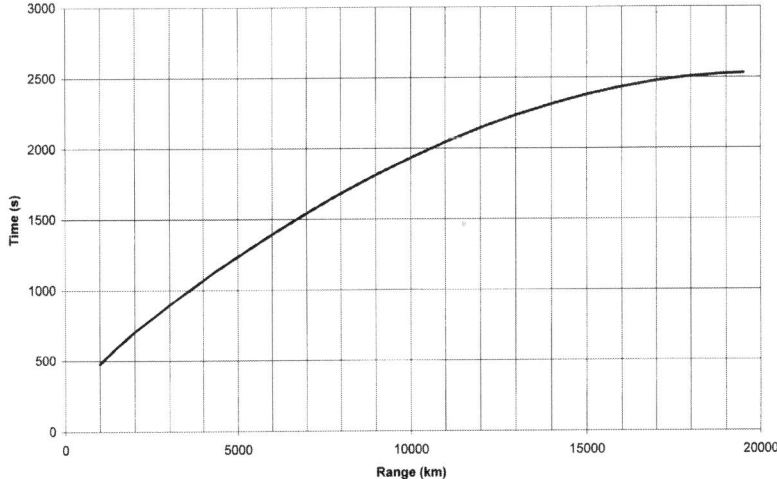

Fig. 7.7 Ballistic time as function of range

We obtain a quadratic equation in $\frac{p}{r_t}$,

$$\left(\frac{p}{r_t}\right)^2 - 2\left(\frac{p}{r_t}\right)\left(\sin^2\alpha + \frac{\cos^2\alpha}{2}\bar{v}^2\right) + \sin^2\alpha = 0 \quad (7.58)$$

which can be written in the canonical form:

$$\left(\frac{p}{r_t} - \left(\sin^2\alpha + \frac{\cos^2\alpha}{2}\bar{v}^2\right)\right)^2 = \left(\sin^2\alpha + \frac{\cos^2\alpha}{2}\bar{v}^2\right)^2 - \sin^2\alpha$$

It admits two solutions when the second member is positive, i.e., when $\bar{v}^2 \geq \frac{2\sin\alpha}{1+\sin\alpha} = \bar{v}_{opt}^2$, where $\bar{v}_{opt}$ is the velocity corresponding to minimal energy, that is to say:

$$\left(\frac{p}{r_t}\right) = \sin^2\alpha + \frac{\cos^2\alpha}{2}\bar{v}^2 \pm \sqrt{\left(\sin^2\alpha + \frac{\cos^2\alpha}{2}\bar{v}^2\right)^2 - \sin^2\alpha} \quad (7.59)$$

Finally, while noting that $\frac{p}{r_t} = \bar{v}^2\cos^2\gamma_0$, we obtain the solution of two flight path angles:

$$\cos^2\gamma_{0\pm} = \frac{1}{\bar{v}^2}\left\{\sin^2\alpha + \frac{\cos^2\alpha}{2}\bar{v}^2 \pm \sqrt{\left(\sin^2\alpha + \frac{\cos^2\alpha}{2}\bar{v}^2\right)^2 - \sin^2\alpha}\right\} \quad (7.60)$$

According to previous results:

a) For $\bar{v}_{opt} < \bar{v} \leq 1$ $(v_{opt} < v_0 \leq v_c)$, the two solutions exist whenever α (they are confused for $\bar{v} = \bar{v}_{opt}$).

b) For $1 < \bar{v} < \sqrt{2}$ $(v_c < v_0 < V_c\sqrt{2})$, we saw that this range was realizable with this initial velocity only if $\cos^2\gamma_0 \leq \frac{1}{\bar{v}^2}$, i.e., if

$$\left\{\sin^2\alpha + \frac{\cos^2\alpha}{2}\bar{v}^2 \pm \sqrt{\left(\sin^2\alpha + \frac{\cos^2\alpha}{2}\bar{v}^2\right)^2 - \sin^2\alpha}\right\} \leq 1 \,.$$

According to this condition, γ_{0+} never exists and γ_{0-} exists whenever α.

For $\alpha = \alpha_{opt}$ at fixed velocity $\bar{v} \leq 1$, we find again the flight path angle for maximum range:

$$\cos^2\gamma_{opt} = \frac{1}{2-\bar{v}^2} \quad \rightarrow \quad \gamma_{opt} = \frac{\pi}{4} - \frac{\alpha_{opt}}{2}$$

7.1 Movement of the Center of Mass

For $\bar{v}^2 = \bar{v}_{opt}^2$ at fixed range $2\alpha \leq \pi$, we find as previously for minimal energy:

$$\cos^2 \gamma'_{opt} = \frac{1 + \sin \alpha}{2} \quad \rightarrow \quad \gamma'_{opt} = \frac{\pi}{4} - \frac{\alpha}{2}$$

In summary, to reach a given range $P = 2\alpha r_t \leq \pi r_t \approx 20000$ km, there is a trajectory with minimal energy corresponding to the optimal flight path angle γ'_{opt}. For any value of the energy ranging between this minimal energy and a limit strictly lower than the energy of circular orbit (at sea level), there are two solutions for trajectories whose flight path are $\gamma_{0\pm}$:

- The higher flight path angle corresponds to γ_{0-}, called a lofted trajectory, because it has the highest altitude of apogee and longer ballistic time,
- The lower flight path angle corresponds to γ_{0+}, called a shallow trajectory, because it is most directly toward the target and of the shortest duration.

Then, we have:

$$\alpha < \alpha_{opt} \quad \rightarrow \quad \gamma_{0+} < \gamma_{opt} < \gamma'_{opt} < \gamma_{0-}$$

When energy is higher than the energy of circular orbit and strictly lower than the escape energy, only the lofted solution γ_{0-} exists.

For example, we obtain for the maximum useful range $2\alpha = \pi$:

$$\bar{v}_{opt} = 1; \; 1 \leq \bar{v} < \sqrt{2}$$
$$\cos^2 \gamma_{0+} = \frac{1}{\bar{v}^2}$$

In other words, we can theoretically reach this range with initial conditions from $\gamma_{opt} = 0 \rightarrow v_c = 7910$ m / s to $\gamma_{0+} = 45° \rightarrow v_l - \varepsilon \approx 11000$ m/s

Note that maximum range trajectories, being defined such that $\frac{\partial P}{\partial \gamma_0} = 0$, are most insensitive to error on the initial flight path angle γ_0. Consequently, using conditions different from minimal energy conditions result in degradation of the range accuracy.

Classification of Trajectories

Energy ratio is a method of classification of nonoptimal trajectories. This parameter has only a limited interest for missile or entry applications, but is worthy of discussion.

We seek to reach a range $P = 2\alpha r_t$ while using a nonoptimal trajectory. This implies that with same energy E, using optimal conditions, we are able to reach a range $P_{max} > P$. The energy ratio parameter is defined as the range loss with respect to this maximum range, expressed in % of the range P, when we use a nonoptimal flight path angle:

$$\frac{\Delta P}{P}(\%) = 100 \frac{P_{max} - P}{P} \tag{7.61}$$

Total energy, corresponding to $P_{max} = P\left(1 + \frac{1}{100}\frac{\Delta P}{P}\right)$, and the range P being determined, we are brought back to the preceding case, and we obtain two solutions for trajectories of flight path angles $\gamma_{0\pm}$.

By convention one defines the parameter Π, which takes the values:

$$\Pi = +\frac{\Delta P}{P} \quad \text{for the lofted trajectory } (\rightarrow \gamma_{0-})$$

$$\Pi = -\frac{\Delta P}{P} \quad \text{for the shallow trajectory } (\rightarrow \gamma_{0+})$$

which defines the energy ratio. However, notice that this classification applies only for energies where the maximum range trajectory exists, i.e., $P_{max} \leq 20000$ km and $v \leq V_c \approx 7910$ m/s. This is in fact largely superabundant for practical applications.

This involves, aside from range limitations due to the missile, that the maximum theoretical energy ratio corresponds to the maximum range of 20000 km. Thus we obtain for a range P:

$$\Pi_{max} = \pm 100 \frac{\pi r_t - P}{P} \approx \pm 100 \frac{20000 - P_{km}}{P_{km}}$$

Figures 7.8–7.11 give the characteristics of trajectories for 4000 km range and for energy ratio limited to $\pm 150\%$.

7.1.3 Influence of Earth Rotation

For a trajectory relative to the rotating earth, we define the initial conditions at a constant altitude corresponding to the end of launch phase. Altitude H_I depends on the missile and the mission (index I indicate the injection point on the ballistic trajectory). We assume here that the planet has a spherical symmetry and a constant rotation rate.

Depending on the frame used to observe the trajectory, the initial conditions can be defined relative to a nonrotating observer, located at planet mass center, or to an observer fixed to the latter.

Initial conditions at the point "I" of altitude H_I are entirely defined by (Fig. 7.12):

- The geocentric latitude λ_I
- The module of the relative velocity vector V_I

7.1 Movement of the Center of Mass

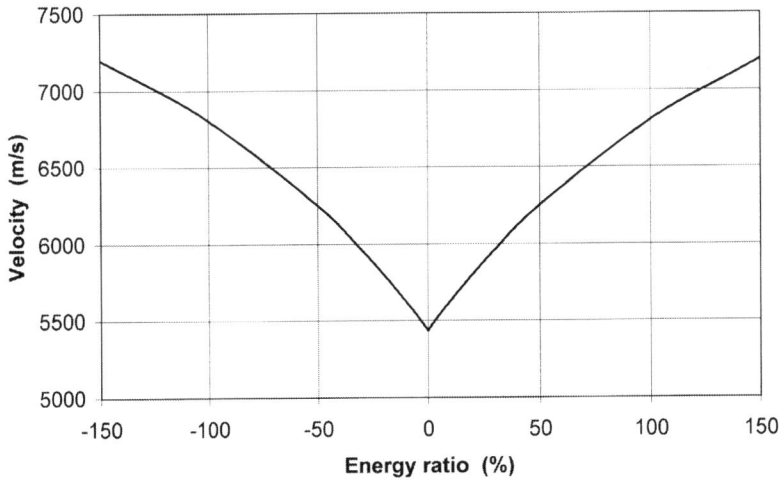

Fig. 7.8 Initial velocity as function of energy ratio

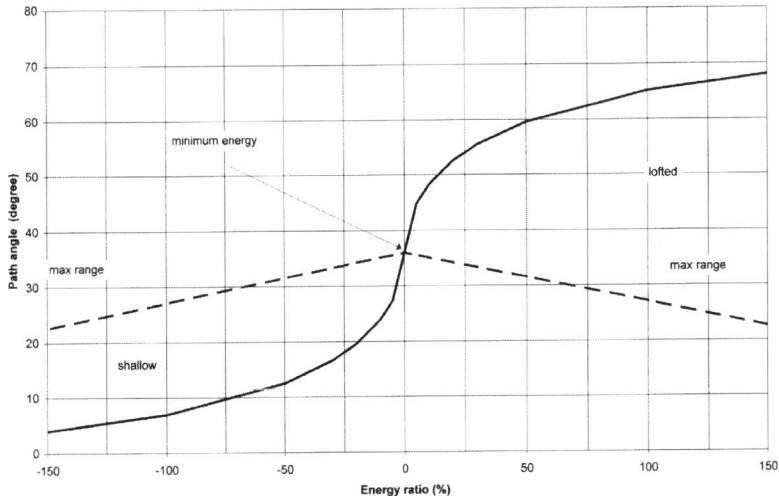

Fig. 7.9 Initial flight path angle as function of energy ratio

- Flight path angle, which corresponds to the angle between the relative velocity vector γ_I and the geocentric horizontal plane (normal with the initial radius vector $\vec{r}_I$ to the mass center of planet), $\gamma_I > 0$ for initial conditions of ballistic phase
- The azimuth angle of the relative velocity vector Az_I relative to north direction in the geocentric horizontal plane

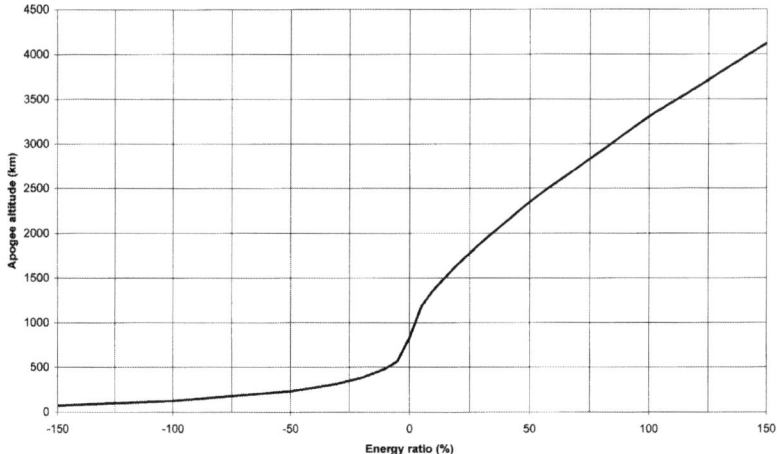

Fig. 7.10 Apogee altitude as function of energy ratio

The planet being assumed to have symmetry of revolution around pole axis, we choose the origin of longitudes at the meridian line of the initial point.

7.1.3.1 Effect of Initial Azimuth

For a revolving observer and a fixed relative velocity and flight path angle, trajectories corresponding to various azimuths are not symmetrical around the initial radius vector. This phenomenon does not correspond to our current experiment, but it is particularly important for the range of ballistic trajectories or for placing satellites

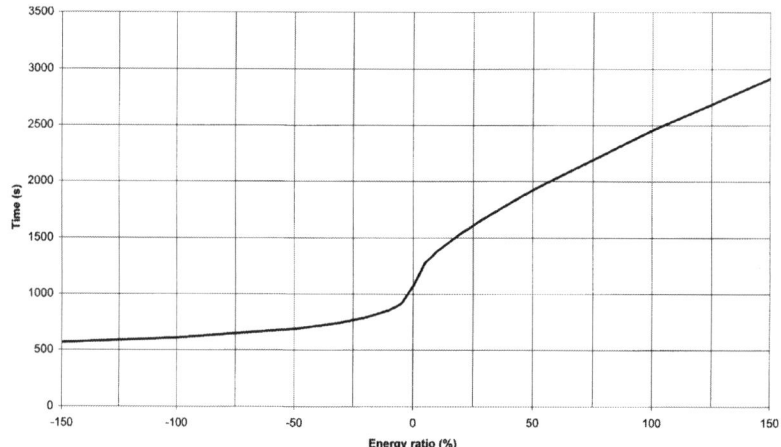

Fig. 7.11 Ballistic time as function of energy ratio

7.1 Movement of the Center of Mass

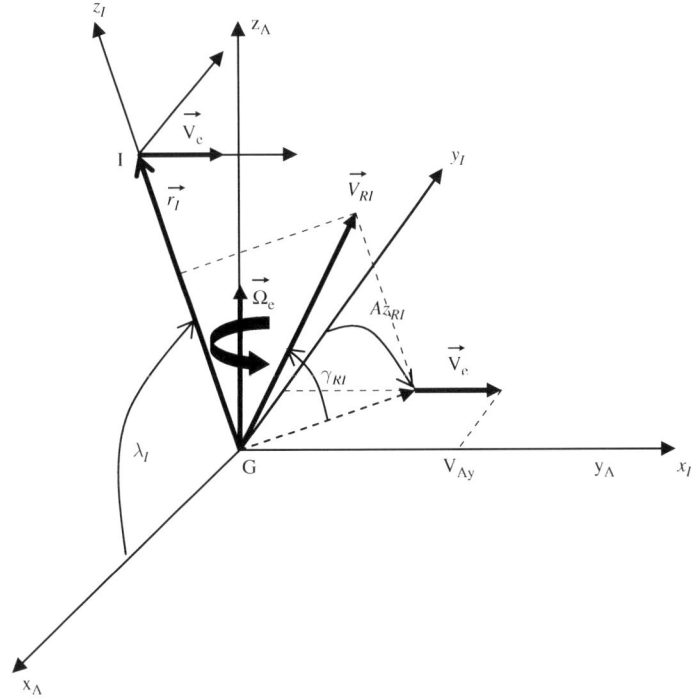

Fig. 7.12 Relative and absolute velocities

in orbit. Everyone knows that to place satellite in orbit, it is more advantageous to launch from the equator and toward east and that the range of ballistic missiles is greater for eastward trajectories. Indeed, for the same relative speed, thus roughly the same energy consumption, we profit from an increase in the absolute velocity and a reduction of the absolute flight path angle when one fires eastward. This effect of "sling" is more effective closer to equator. For a nonrotating observer, absolute range increases. Although, throughout flight, the points fixed on the earth's surface drift away eastward, the absolute range being greater, it results in a net increase of range measured point-to-point on the surface. For satellites, geometrical parameters of the orbit being determined by the absolute velocity, this effect is even more beneficial.

To quantify the range increase, there is no need to use a digital code. Indeed, starting from the relative conditions, we can determine the absolute velocity conditions in the frame $GX_IY_IZ_I$ of axes parallel with the local directions East, North, Zenith at point I (Fig. 7.12):

The drift velocity of a point fixed to earth at initial location I is:

$$\vec{V}_e = \vec{\Omega}_e \wedge \vec{r}_I \quad \rightarrow \quad V_{ex} = \Omega_e r_I \cos \lambda_I$$

Absolute velocity of a point of relative speed $\vec{V}_{RI}$ is thus:

$$\vec{V}_{AI} = \vec{V}_{RI} + \vec{V}_e$$
$$V_{AXI} = V_{AI} \cos \gamma_{AI} \sin Az_{AI} = V_{RI} \cos \gamma_{RI} \sin Az_{RI} + V_{ex}$$
$$V_{AYI} = V_{AI} \cos \gamma_{AI} \cos Az_{AI} = V_{RI} \cos \gamma_{RI} \cos Az_{RI}$$
$$V_{AZI} = V_{AI} \sin \gamma_{AI} = V_{RI} \sin \gamma_{RI}$$

The module, slope, and azimuth of absolute velocity are:

$$\frac{V_{AI}}{V_{RI}} = \sqrt{1 + x^2 + 2x \cos \gamma_{RI} \sin Az_{RI}} \tag{7.62}$$

$$\gamma_{AI} = Arc \sin \left(\frac{\sin \gamma_{RI}}{\sqrt{1 + x^2 + 2x \cos \gamma_{RI} \sin Az_{RI}}} \right) \tag{7.63}$$

$$\cos Az_{AI} = \frac{\cos Az_{RI} \cos \gamma_{RI}}{\sqrt{\cos^2 \gamma_{RI} + x^2 + 2x \cos \gamma_{RI} \sin Az_{RDi}}} \tag{7.64}$$

$$\sin Az_{AI} = \frac{\sin Az_{RI} \cos \gamma_{RI} + x}{\sqrt{\cos^2 \gamma_{RI} + x^2 + 2x \cos \gamma_{RI} \sin Az_{RI}}} \tag{7.65}$$

where $x = \frac{V_{ex}}{V_{RI}} = \frac{\Omega_e r_I \cos \lambda_I}{V_{RI}}$ indicates the ratio between the drift velocity at point I and initial relative speed modulus.

Thus maximum effect is at the equator, where the drift velocity at sea level is about 463 m/s.

To obtain a first-order estimate of range variation, we neglect the effect of atmosphere, and we assume an injection point located at sea level, $r_0 = r_t$.

Let us consider an initial point of latitude λ_I, a relative speed V_{RI}, a slope γ_{RI}, and an azimuth Az_{RI} (Fig. 7.13).

In the case of a nonrotating planet, the range of the elliptic trajectory is obtained from $V_0 = V_{RI}$, $\gamma_0 = \gamma_{RI}$ by (7.37), (7.38), and (7.39). The flight duration is obtained by (7.24), (7.28), (7.31), and (7.33):

$$p = \frac{(r_t v_0 \cos \gamma_0)^2}{\mu}$$

$$e = \sqrt{1 + 2 \left(\frac{v_0^2}{2} - \frac{\mu}{r_t} \right) \left(\frac{r_t v_0 \cos \gamma_0}{\mu} \right)^2}$$

$$\cos \alpha = \frac{1}{e} \left(1 - \frac{p}{r_T} \right); \quad a = \frac{p}{1 - e^2}$$

$$T = \frac{2\pi}{\sqrt{\mu}} a^{\frac{3}{2}}; \quad \sin \theta' = \sqrt{1 - e^2} \frac{\sin \alpha}{1 - e \cos \alpha}; \quad t = \frac{T}{\pi} (\theta' + e \sin \theta')$$

where α indicates half angular range $\alpha = \frac{P}{2 \cdot r_t}$.

7.1 Movement of the Center of Mass

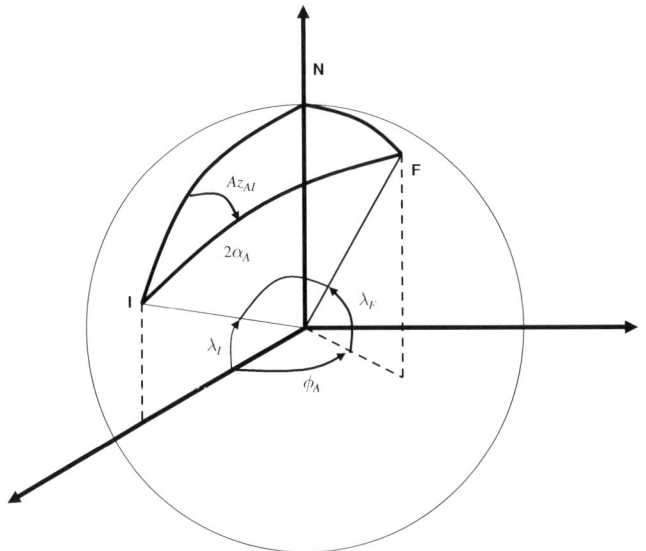

Fig. 7.13 Ground tracks in the absolute frame

For rotating planet, absolute angular range $2\alpha_A$ (seen by a nonrotating observer) and duration t_A are given by the same relations using absolute speed V_{AI} and slope γ_{AI} determined from relative conditions and relations (7.62)–(7.65).

One obtains the relative range P_R at sea level by observing that during flight duration absolute longitude (relative to a nonrotating angular origin corresponding to the initial meridian plan) of observers fixed to the planet surface increased uniformly with $\Delta\phi = \Omega_e \cdot t_A$. Latitude of the final point F is the same for motionless and revolving observers, $\lambda_{FA} = \lambda_{FR} = \lambda_F$. Longitude at arrival is deduced from $\Delta\phi$, i.e.,

$$\phi_R = \phi_A - \Omega_e \cdot t_A \tag{7.66}$$

We obtain the latitude λ_F and longitude ϕ_A of final point F in the inertial frame from λ_I, Az_{AI}, $2\alpha_A$ and spherical trigonometry relations in the spherical triangle NIF (Fig. 7.13):

$$\cos\left(\frac{\pi}{2} - \lambda_F\right) = \cos\left(\frac{\pi}{2} - \lambda_I\right)\cos(2\alpha_A) + \sin\left(\frac{\pi}{2} - \lambda_I\right)\sin(2\alpha_A)\cos Az_{AI}$$

$$\frac{\sin Az_{AI}}{\sin\left(\frac{\pi}{2} - \lambda_F\right)} = \frac{\sin\phi_A}{\sin(2\alpha_A)}$$

$$\cos(2\alpha_A) = \cos\left(\frac{\pi}{2} - \lambda_I\right)\cos\left(\frac{\pi}{2} - \lambda_F\right)$$
$$+ \sin\left(\frac{\pi}{2} - \lambda_I\right)\sin\left(\frac{\pi}{2} - \lambda_F\right)\cos\phi_A$$

This gives:

$$\begin{aligned}
\sin \lambda_F &= \sin \lambda_I \cos(2\alpha_A) + \cos \lambda_I \sin(2\alpha_A) \cos Az_{AI} \\
\cos \lambda_F &= \sqrt{1 - \sin^2 \lambda_F} \\
\sin \phi_A &= \frac{\sin(2\alpha_A) \sin Az_{AI}}{\cos \lambda_F} \\
\cos \phi_A &= \frac{\cos(2\alpha_A) - \sin \lambda_I \sin \lambda_F}{\cos \lambda_I \cos \lambda_F}
\end{aligned} \quad (7.67)$$

Finally, we obtain the apparent angular range $2\alpha_R$ for a rotating observer from $\phi_R = \phi_A - \Omega_e \cdot t_A$, λ_I, and λ_F by resolution of the spherical triangle NIF_R (Fig. 7.14):

$$\cos(2\alpha_R) = \sin \lambda_I \sin \lambda_F + \cos \lambda_I \cos \lambda_F \cos \phi_R \quad (7.68)$$

Figure 7.15 represents the increase in range compared to the nonrotating case according to the initial latitude and azimuth, for relative initial conditions

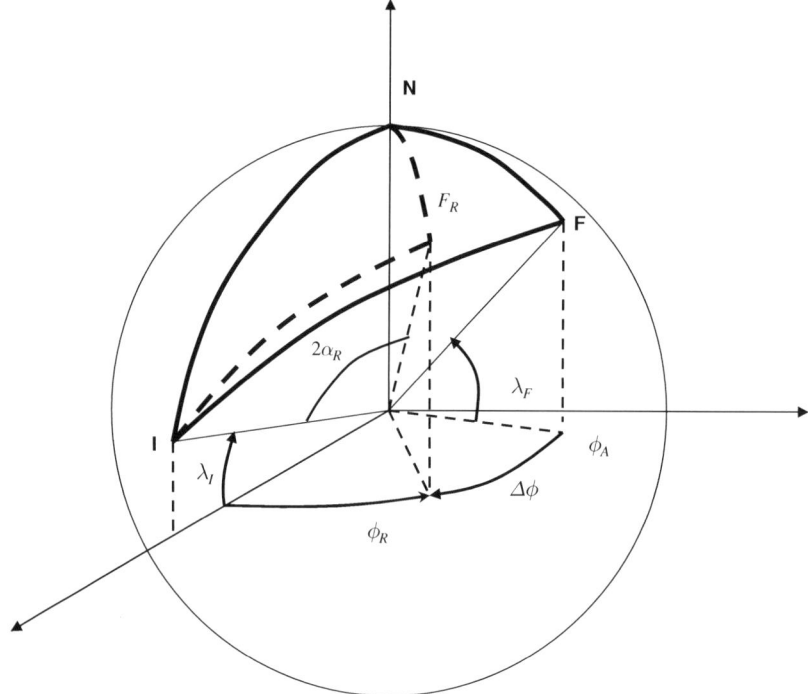

Fig. 7.14 Trajectory in rotating frame

7.1 Movement of the Center of Mass

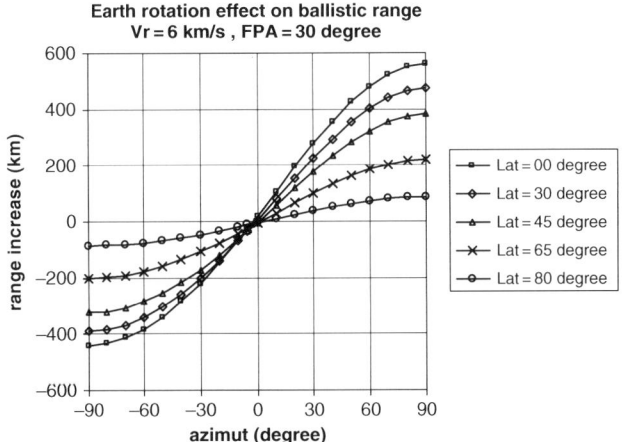

Fig. 7.15 Influence of azimuth on ballistic range

$V_{RI} = 6000$ m/s, $\gamma_{RI} = 30°$, which correspond to a range of 5264 Km for a nonrotating earth.

The variations of range reach -440 to $+550$ km at equator, -323 to $+382$ km at 45° of latitude.

Figure 7.16 displays results for a relative initial velocity 7000 m/s, and a 25° path angle, corresponding to a nonrotating ground range of 8910 km. In this case, the variations are more important and reach -1200 to $+1400$ km.

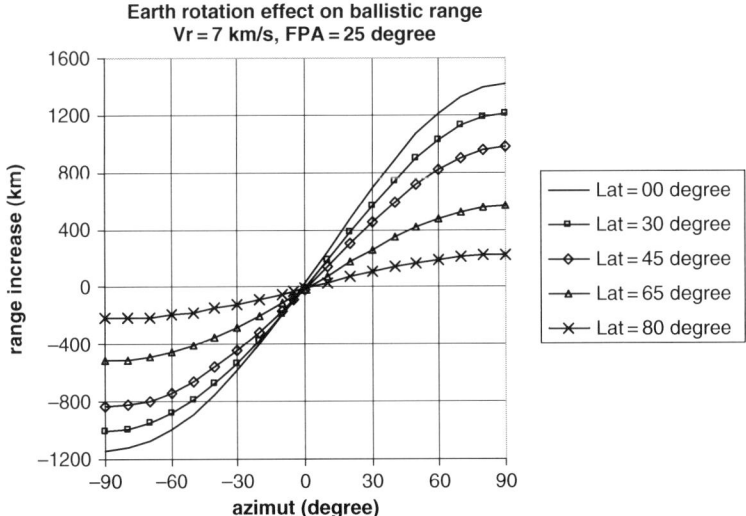

Fig. 7.16 Influence of azimuth on ballistic ranges

7.2 Movement Around Mass Center

The only external forces are the terrestrial forces of gravity, which are applied directly on the material points composing the solid. Resulting moment around the center of gravity is zero by definition. The center of gravity is distinct from the center of mass because the terrestrial field of gravitation is not homogeneous. This involves in general a small moment around the mass center:

$$\vec{M}_{ext/I} \approx M\left(\vec{R}_G - \vec{R}_I\right) \wedge \vec{g}\left(\vec{R}_I\right) \approx \sum_i m_i \vec{\mathfrak{R}}_i \wedge \frac{\partial \vec{g}\left(\vec{R}_I\right)}{\partial \vec{R}_I} \vec{\mathfrak{R}}_i$$

However, in the case of ballistic vehicles, trajectories durations are short compared with satellites, thus the effect of this moment can be neglected. This is a free rotational movement; thus, for an inertial observer or in accelerated translation, the angular momentum is constant in the movement:

$$\dot{\vec{H}} = 0 \quad \Leftrightarrow \quad \vec{H} = \vec{H}_0$$

7.2.1 Rotation Around a Principal Axis

We consider first the case where the initial rotation vector is along a principal axis of the solid. By definition, angular momentum is in the same direction. This involves for an observer fixed to the solid, according to the (1.36):

$$\frac{d\vec{H}}{dt} = -\vec{\omega} \wedge \vec{H} = 0 \Rightarrow \vec{H} = \vec{H}_0, \vec{\omega} = \vec{\omega}_0.$$

Thus, the movement is a uniform rotation around the initial axis, which keeps a constant inertial and relative direction.

In order to assess the stability of this motion, let us assume an initial rotation rate $t < 0 \rightarrow \vec{\omega}_0 = \{\omega_0 \ 0 \ 0\}$ and a small disturbance $\delta\vec{\omega}(0) = \{\varepsilon_1(0) \ \varepsilon_2(0) \ \varepsilon_3(0)\}$ such that $\|\delta\vec{\omega}(0)\| \ll \omega_0$ at t = 0. It results at t > 0 $\vec{\omega}(t) = \{\omega_0 + \varepsilon_1, \varepsilon_2, \varepsilon_3\}$. The resulting equation is written in the rotating frame as:

$$\begin{aligned} I_1\dot{\varepsilon}_1 + \varepsilon_2\varepsilon_3(I_3 - I_2) &= 0 \\ I_2\dot{\varepsilon}_2 + (\omega_0 + \varepsilon_1)\varepsilon_3(I_1 - I_3) &= 0 \\ I_3\dot{\varepsilon}_3 + (\omega_0 + \varepsilon_1)\varepsilon_2(I_2 - I_1) &= 0 \end{aligned}$$

While retaining only the first-order terms, we obtain:

$$\begin{aligned} I_1\dot{\varepsilon}_1 &\approx 0 \\ I_2\dot{\varepsilon}_2 + \omega_0\varepsilon_3(I_1 - I_3) &\approx 0 \\ I_3\dot{\varepsilon}_3 + \omega_0\varepsilon_2(I_2 - I_1) &\approx 0 \end{aligned}$$

7.2 Movement Around Mass Center

Thus, $\omega_1 \approx \omega_0 + \varepsilon_1(0)$ is a constant and

$$\begin{bmatrix} \dot{\varepsilon}_2 \\ \dot{\varepsilon}_3 \end{bmatrix} + \begin{bmatrix} 0 & \omega_0 \frac{I_1 - I_3}{I_2} \\ \omega_0 \frac{I_2 - I_1}{I_3} & 0 \end{bmatrix} \begin{bmatrix} \varepsilon_2 \\ \varepsilon_3 \end{bmatrix} \approx 0$$

This set of linear differential equations with constant coefficients classically admits a particular solutions $\vec{v}_i e^{\lambda t}$ such that,

$$\text{Det} \begin{bmatrix} \lambda & -\omega_0 \frac{I_1 - I_3}{I_2} \\ -\omega_0 \frac{I_2 - I_1}{I_3} & \lambda \end{bmatrix} = 0 \Leftrightarrow \lambda^2 - \omega_0^2 \frac{(I_1 - I_3)(I_2 - I_1)}{I_3^2} = 0$$

Two cases arise:

1) $\inf(I_2, I_3) < I_1 < \sup(I_2, I_3) \Rightarrow \lambda = \pm \omega_0 \sqrt{\frac{(I_1 - I_3)(I_2 - I_1)}{I_3^2}}$

2) $I_1 < \inf(I_2, I_3)$ ou $\sup(I_2, I_3) < I_1 \Rightarrow \lambda = \pm i\omega_0 \sqrt{\frac{(I_1 - I_3)(I_1 - I_2)}{I_3^2}}$

Free rotational movement around a principal axis of intermediate moment of inertia is unstable since the small disturbances are likely to diverge exponentially.

Free rotational movements around principal axes of maximum and minimum moment of inertia are stable since the small disturbances have a sinusoidal periodic evolution. This is the origin of the gyroscopic principle of stabilization widely used during the ballistic phase.

Let us note that when the vehicle has inertial symmetry of revolution $I_2 = I_3 = I_T \neq I_1$ the axis 1 always corresponds to a stable axis and the expression of the pulsation of the closed cycle is $\omega_0 \left| 1 - \frac{I_1}{I_T} \right|$.

In the case of a nonrigid body, the energy of rotation is likely to decrease under the effect of dissipative interior forces. It is easy to show that the only stable mode of free rotation corresponds to the principal axis of highest moment of inertia. Indeed, conservation of angular momentum and energy of rotation gives:

$$\vec{H}^2 = (I_1 \omega_1)^2 + (I_2 \omega_2)^2 + (I_3 \omega_3)^2 = H_0^2$$

$$T = \frac{1}{2} \left[I_1 \omega_1^2 + I_2 \omega_2^2 + I_3 \omega_3^2 \right]$$

Assuming axis 1 corresponds to the highest moment of inertia and using ω_1 from preceding equation, we can express the energy of rotation according to ω_2 and ω_3.

$$T = \frac{1}{2} \left[\frac{H_0^2}{I_1} + I_2 \left(1 - \frac{I_2}{I_1} \right) \omega_2^2 + I_3 \left(1 - \frac{I_3}{I_1} \right) \omega_3^2 \right]$$

With this choice of axis 1 coefficients of ω_2^2 and ω_3^2 are positive, and it results the kinetic energy is minimum when ω_2 and ω_3 are null, which is the proof of the proposal.

The asymptotic final state corresponds to minimum rotation rate compatible with the conservation of the angular moment, $\omega_1 = \frac{H_0}{I_1}$. This explains in particular the stability of earth rotational movement as it is close to an ellipsoid of revolution turning around its axis of maximum moment of inertia.

7.2.2 Coning Motion

Now we consider the case of a vehicle with inertial symmetry of revolution. The axis of symmetry is clearly a principal axis of inertia. In the case of a ballistic RV, it is the roll axis, which corresponds to the smallest moment of inertia. A ballistic reentry vehicle is very rigid by definition, designed to sustain very high aerodynamic loads during reentry. The free rotational motion around its roll axis is stable. At the time of separation from the missile, a spin is given to the vehicle, in order to stabilize its axis in a fixed inertial direction. This roll impulse is applied either using a mechanical device on the missile, or using roll thrusters on the RV. The roll rate is typically about one revolution per second. The choice of the direction generally corresponds to that of the velocity relative to earth at the beginning of reentry, in order to minimize initial angle of attack. Ideally, the spin device delivers a pure rolling moment and the angular momentum is along the roll axis. However, technological flaws results in small angular velocities in other axes, involving an initial misalignment of the angular momentum with the longitudinal axis. This misalignment is at the origin of the "coning motion," which corresponds, in the case of finite initial disturbances, to the sinusoidal evolution of small disturbances.

The symmetry of revolution of the body determines that its longitudinal axis and the angular velocity $\vec{\omega}$ are at all times in the same meridian plane around the angular momentum $\vec{H}_0$.

We use frame at the center of mass G, such that Gx_0 is along $\vec{H}_0$, and Gy_0, Gz_0 remain parallel to the inertial axes (Fig. 7.17). While using an Eulerian frame, having axes along principal inertia axes of the body and such that Gx_I is along the symmetry axis, we obtain the tensor of inertia:

$$[I] = \begin{bmatrix} I_x & 0 & 0 \\ 0 & I_T & 0 \\ 0 & 0 & I_T \end{bmatrix} ; I_y = I_z = I_T$$

In order to describe the motion, we use Euler angles (ψ, θ, φ) shown in Fig. 7.17 corresponding to the choice of axes (1, 2, 1) in Sect. 6.2.

These rotations transform the observation frame (Gx_0, Gy_0, Gz_0) into the Eulerian frame:

7.2 Movement Around Mass Center

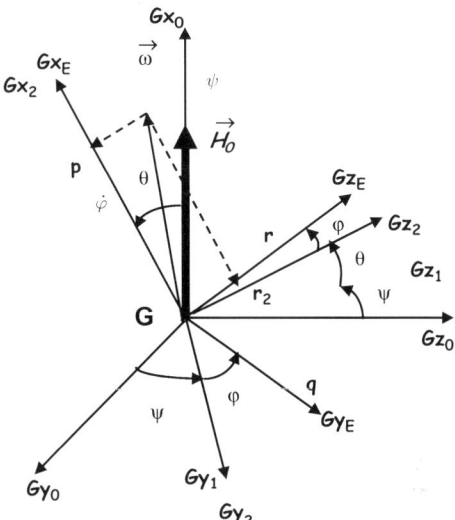

Fig. 7.17 Euler angles

- Rotation ψ around Gx_0 transforms (Gx_0, Gy_0, Gz_0) into (Gx_0, Gy_1, Gz_1).
- Rotation θ around Gy_1 transforms (Gx_0, Gy_1, Gz_1) into (Gx_E, Gy_1, Gz_2).
- Final rotation φ around Gx_E transforms (Gx_E, Gy_1, Gz_2) into the Eulerian frame (Gx_E, Gy_E, Gz_E).

It is easy to check that, except in the case where $\theta = 0$, the relation between (ψ, θ, φ) and the instantaneous position of (Gx_E, Gy_E, Gz_E) relative to (Gx_0, Gy_0, Gz_0) is reciprocal. When $\theta = 0$, angles ψ and φ are unspecified, and we can only define angle $\psi + \varphi$. Except in this last case, angular rate $\vec{\omega}$ can be written relative to the base (nonorthogonal) formed by the unit vectors of the three successive axes of rotations

$$\vec{\omega} = \dot{\psi}\vec{G}x_0 + \dot{\theta}\vec{G}y_1 + \dot{\varphi}\vec{G}x_E$$

From previous findings, $\vec{\omega}$, $\vec{G}x_E$ and $\vec{G}x_0$ are in the same plan. From the expression of $\vec{\omega}$, it is clear that the nutation rate $\dot{\theta}$ is null. This involve that angle θ is constant in the movement.

The motion is thus a precession motion $\dot{\psi}$ around the angular momentum $\vec{H}_0$ and a rolling motion $\dot{\varphi}$ around the symmetry axis $\vec{G}x_E$ (Fig. 7.18). It is now easy to determine these parameters in the transverse plane containing $\vec{\omega}$, $\vec{G}x_E$, $\vec{G}x_0$, and $\vec{G}z_2$.

It is clear that $\vec{H}$ and θ being constant, components Hx_E and Hz_2 of $\vec{H}$ along $\vec{G}x_E$ and $\vec{G}z_2$ are also constant. While using components p and r_2 of $\vec{\omega}$ on these last axes, this involves:

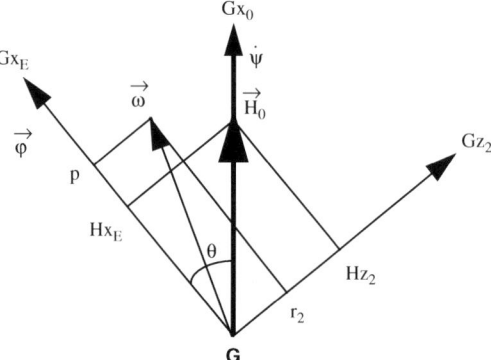

Fig. 7.18 Constants of coning motion

$$Hx_E = I_x p = H_0 \cos\theta_0 \Rightarrow p = p_0 = \frac{H_0 \cos\theta_0}{I_x}$$

$$Hz_2 = I_T r_2 = H_0 \sin\theta_0 \Rightarrow r_2 = r_{20} = \frac{H_0 \sin\theta_0}{I_T}$$

Thus, roll rate p and the component r_2 of angular velocity along $\vec{G}z_2$ are constants in the movement. Knowing the initial components of angular velocity in frame E, this allows us to determine θ_0,

$$\tg\theta_0 = \frac{Hz_2}{Hx_E} = \frac{I_T\, r_{20}}{I_x\, p_0} = \frac{I_T\sqrt{q_0^2 + r_0^2}}{I_x\, p_0}$$

We are now ready to determine all the unknown parameters:

$$\vec{\omega} = \dot{\psi}\,\vec{G}x_0 + \dot{\varphi}\,\vec{G}x_E = r_{20}\vec{G}z_2 + p_0\vec{G}x_E$$
$$\Rightarrow r_{20} = \dot{\psi}\sin\theta_0;\quad p_0 = \dot{\varphi} + \dot{\psi}\cos\theta_0$$
$$\Rightarrow \dot{\psi} = \dot{\psi}_0 = \frac{r_{20}}{\sin\theta_0} = \frac{\mu p_0}{\cos\theta_0};\quad \dot{\varphi} = p_0 - \dot{\psi}_0\cos\theta_0 = p_0(1 - \mu)$$

where $\mu = \frac{I_x}{I_T}$.

The precession rate $\dot{\psi}$ and the roll angle derivative $\dot{\varphi}$ are constants in the movement. They both have the same sign as the roll rate. Thus, evolution of Euler angles of a symmetrical spinning top in free rotation is:

$$\psi = \psi_0 + \dot{\psi}_0\, t$$
$$\theta = \theta_0$$
$$\varphi = \varphi_0 + \dot{\varphi}_0\, t$$

7.2 Movement Around Mass Center

The evolution of the angular velocity components in the Eulerian frame is then:

$$p = p_0$$
$$q = \sqrt{q_0^2 + r_0^2} \sin\left((1-\mu)p_0 t + \varphi_0\right)$$
$$r = \sqrt{q_0^2 + r_0^2} \cos\left((1-\mu)p_0 t + \varphi_0\right)$$

The pulsation $\omega_G = \dot{\varphi}_0 = (1-\mu)p_0$ is sometimes called gyrometric pulsation, because it is a characteristic of the measurements provided by lateral rate gyrometers on the vehicle. Taking into account present results the transverse angular velocity can be expressed using the complex form as:

$$q + i\,r = \sqrt{q_0^2 + r_0^2}\; e^{i\left(\frac{\pi}{2}-\varphi_0\right)}\; e^{-i\omega_G t}$$

which shows that the angular velocity has an apparent rotational movement around Gx at rate $-\omega_G$.

With typical values of ballistic RVs, $\mu \approx 0.1$, $p_0 \approx 1\;rps$, $\theta_0 \approx 5°$, we obtain:

$$\dot{\psi} \approx 0.1\;rps,\; \dot{\varphi} \approx 0.9\;rps$$

When the geometrical axis of symmetry of the body is aligned with the corresponding principal direction of inertia this describes a cone of half-angle θ_0 around the angular momentum with a 10-s period, the body turns around its axis of symmetry with a 1.1-s period. When the two axes are distinct, the principal axis of inertia describes the preceding movement and its geometrical axis of symmetry turns with a 1.1-second period.

Chapter 8
Six Degree-of-Freedom Reentry

8.1 General Equations of Motion

Let us consider an observation frame centered at a point O fixed to the earth surface of radius vector $\vec{r}_T$ from earth mass center. This frame has a uniform rotation rate $\vec{\Omega}_e$ and a centripetal acceleration $\vec{\Gamma}_e$ associated with the earth rotation movement around its mass center. According to (1.4), the linear acceleration of the center of mass relative to this frame is:

$$\dot{\vec{V}}_R = \vec{\gamma}_R = \frac{\vec{R}^A}{m} + \vec{\varphi} - \left\{ \vec{\Gamma}_e + 2\vec{\Omega}_e \wedge \vec{V}_R + \vec{\Omega}_e \wedge \left(\vec{\Omega}_e \wedge \vec{r} \right) \right\}$$

with:

$$\vec{\Gamma}_e = -\vec{\Omega}_e \wedge \left(\vec{\Omega}_e \wedge \vec{r}_T \right) = \Omega_e^2 \vec{r}_T + \left(\vec{\Omega}_e \cdot \vec{r}_T \right) \vec{\Omega}_e$$

where $\vec{\varphi}$ indicates the local value of the terrestrial gravitation field, a function of the location of the center of mass G of the vehicle (not to be confused with the gravity, which includes the centrifugal contribution associated with acceleration of the origin of reference frame, $\vec{g} = \vec{\varphi} - \vec{\Gamma}_e$).

In an Eulerian reference frame E bound to the vehicle, centered in G, the angular momentum theorem (1.32) and (1.37) provides the expression of angular acceleration:

$$\vec{I}\dot{\vec{\omega}} = \vec{M}_G^A - \vec{\omega} \wedge \vec{I}\vec{\omega} \quad \Rightarrow \quad \dot{\vec{\omega}} = \left[I^{-1} \right] \left[\vec{M}_G^A - \vec{\omega} \wedge [I]\vec{\omega} \right]$$

The force $\vec{R}^A$ and moment $\vec{M}_G^A$ represent the aerodynamic loads in G whose models were developed with Chap. 4. Aerodynamic loads are functions of variables of CG location, angular orientation, and associated velocities. Variables of linear location are naturally the coordinates of the vehicle center of mass in the reference frame of the observer. Angular variables of position may be according to the selected representation the cosine directors, Euler angles, or the quaternion associated with the relative orientation of the vehicle. Thus equations of evolution for the CG location are:

$$\dot{\vec{r}} = \vec{V}_R \quad \Leftrightarrow \quad \begin{bmatrix} \dot{x} \\ \dot{y} \\ \dot{z} \end{bmatrix} = \begin{bmatrix} v_{Rx} \\ v_{Ry} \\ v_{Rz} \end{bmatrix}$$

In the case of the choice {3 2 1} of Euler rotations, we obtain for the angular variables:

$$\dot{\psi} = \frac{q_R \sin \varphi + r_R \cos \varphi}{\cos \theta}$$

$$\dot{\theta} = q_R \cos \varphi - r_R \sin \varphi$$

$$\dot{\varphi} = p_R + \sin \theta \frac{q_R \sin \varphi + r_R \cos \varphi}{\cos \theta},$$

When using quaternion:

$$\dot{r}_0 = -\frac{1}{2}(r_1 p_R + r_2 q_R + r_3 r_R)$$

$$\dot{r}_1 = -\frac{1}{2}\{r_0 p_R + r_2 r_R - r_3 q_R\}$$

$$\dot{r}_2 = -\frac{1}{2}\{r_0 q_R + r_3 p_R - r_1 r_R\}$$

$$\dot{r}_3 = -\frac{1}{2}\{r_0 r_R + r_1 q_R - r_2 p_R\}$$

with $\vec{\omega} = \begin{bmatrix} p & q & r \end{bmatrix}$, absolute angular velocity and $\vec{\omega}_R = \begin{bmatrix} p_R & q_R & r_R \end{bmatrix} = \vec{\omega} - \vec{\Omega}_e$, angular velocity of the vehicle relative to the earth rotating frame.

To express the resultant of the aerodynamic loads $\vec{R}_A$ in the rotating reference frame of the observer, we will use the rotation transform matrix derived from Euler angles or from quaternion.

$$[A(\psi, \theta, \varphi)] = \begin{bmatrix} \cos \theta \cos \psi & \cos \theta \sin \psi & -\sin \theta \\ -\cos \varphi \sin \psi + \sin \varphi \sin \theta \cos \psi & \cos \varphi \cos \psi + \sin \varphi \sin \theta \sin \psi & \sin \varphi \cos \theta \\ \sin \varphi \sin \psi + \cos \varphi \sin \theta \cos \psi & -\sin \varphi \cos \psi + \cos \varphi \sin \theta \sin \psi & \cos \varphi \cos \theta \end{bmatrix}$$

or,

$$[A(r_0, r_1, r_2, r_3)] = \begin{bmatrix} r_0^2 + r_1^2 - r_2^2 - r_3^2 & 2(r_0 r_3 + r_1 r_2) & 2(r_1 r_3 - r_0 r_2) \\ 2(r_1 r_2 - r_0 r_3) & r_0^2 + r_2^2 - r_1^2 - r_3^2 & 2(r_0 r_1 + r_2 r_3) \\ 2(r_0 r_2 + r_1 r_3) & 2(r_2 r_3 - r_0 r_1) & r_0^2 + r_3^2 - r_1^2 - r_2^2 \end{bmatrix}$$

The general expression of $\vec{R}_A$ along vehicle axes is:

8.1 General Equations of Motion

$$X^A = -\bar{q} \cdot S_{ref} \cdot C_A$$
$$Y^A = \bar{q} \cdot S_{ref} \cdot C_Y^A$$
$$Z^A = -\bar{q} \cdot S_{ref} \cdot C_N$$

Resulting in the expression of $\vec{R}^A$ along observation axes,

$$\begin{bmatrix} R_x^A \\ R_y^A \\ R_z^A \end{bmatrix} = [A]^T \begin{bmatrix} X^A \\ Y^A \\ Z^A \end{bmatrix}$$

The equation of evolution of the angular momentum in a vehicle reference frame (Eulerian), having axes parallel to the aerodynamic frame, uses the expression of the aerodynamic moments in this reference frame,

$$L = \bar{q} \cdot S_{ref} \cdot L_{ref} \cdot \left(C_{l/G} + C_{lp/G} \left(\frac{p_R \, L_{ref}}{V_R} \right) \right)$$
$$M = \bar{q} \cdot S_{ref} \cdot L_{ref} \cdot \left(C_{m/G} + C_{mq/G} \left(\frac{q_R \, L_{ref}}{V_R} \right) \right)$$
$$N = \bar{q} \cdot S_{ref} \cdot L_{ref} \cdot \left(C_{n/G} + C_{nr/G} \left(\frac{r_R \, L_{ref}}{V_R} \right) \right)$$

Let us recall that the aerodynamic loads are functions of the motion of the vehicle relative to atmosphere gaseous mixture. While taking into account the wind velocity $\vec{W}$, the relative motion is obtained from $\vec{V}_R = \vec{V}_{R/gas} + \vec{W}$. We must replace $\vec{V}_R$ (velocity of the vehicle relating to earth) by $\vec{V}_{R/gas} = \vec{V}_R - \vec{W}$ in the above expressions of the aerodynamic loads. In addition, we saw that the aerodynamic coefficients are also functions of angular velocity relative to gas, direction, and instantaneous state of the upstream flow. Thus, dynamic evolution of the vehicle obeys a differential system of state equations such that the derivatives of state vector depend only on instantaneous state vector. This system is highly nonlinear, primarily because of the aerodynamic model and the equations of evolution of angular state. The dimension of the state vector is 12 when we use Euler angles, 13 when we use quaternion:

$$\vec{X}^T = \begin{bmatrix} x & y & z & \psi & \theta & \varphi & V_{Rx} & V_{Ry} & V_{Rz} & p & q & r \end{bmatrix}$$

or,

$$\vec{X}^T = \begin{bmatrix} x & y & z & r_0 & r_1 & r_2 & r_3 & V_{Rx} & V_{Ry} & V_{Rz} & p & q & r \end{bmatrix}$$

8.2 Solutions of General Equations

It is clear that in the general case, the evolution problem is not accessible by analytical methods. We must use a numerical approach using an algorithm of integration for first-order nonlinear differential equations. A Runge Kutta algorithm of order higher or equal to 4 (see Fig. 8.1 a very simple FORTRAN 77 subroutine) with constant time steps is quite suitable provided we use a time step adapted to the dynamics of the vehicle. However, it is preferable to use an automatic method with variable time step to obtain a good compromise between accuracy, computing time, and ease of use of the code.

The numerical approach is essential when we assess accurately the dynamic behavior of a particular vehicle in the design phase of definition or justification. However, the comprehension of fundamental dynamic phenomena, as in other fields

```
        SUBROUTINE DRK4(T,X,XS,XI,DERIV,XPRIM,DT,NDIM)
C****************************************************************
C       RUNGE KUTA 4TH ORDER INTEGRATION SUBROUTINE
C       FOR FIRST ORDER DIFFERENTIAL SYSTEM
C       T : TIME
C       DT : TIME STEP
C       NDIM : STATE VECTOR DIMENSION
C       X(NDIM) : STATE VECTOR INPUT/OUTPUT
C       XS(NDIM), XI(NDIM) : WORKING VECTORS
C       XPRIM(NDIM) : DERIVATIVE OF STATE VECTOR
C       INPUT :  T, X(T), DERIV
C       OUTPUT : T+DT, X(T+DT)
C       DERIV , SUBROUTINE COMPUTING DERIVATIVE OF
C       STATE VECTOR :
C       SUBROUTINE DERIV(T,X,XPRIM)
C****************************************************************
        IMPLICIT REAL*8 (A-H,O-Z)
        DIMENSION X(1),XS(1),XI(1),XPRIM(1),RKI(4),RKA(4)
        DATA RKI/0.D0,0.5D0,0.5D0,1.D0/,RKA/0.166666666667D0,
     S  2*0.333333333333D0,0.166666666667D0/
        DO 1 I=1,NDIM
        XS(I)=X(I)
1       XI(I)=X(I)
        TK=T
        DO 2 NRK=1,4
        IF(NRK.EQ.1)GO TO 3
        DO 4 I=1,NDIM
4       XI(I)=XS(I)+RKI(NRK)*XPRIM(I)*DT
        TK=T+RKI(NRK)*DT
3       CALL DERIV(TK,XI,XPRIM)
        DO 2 I=1,NDIM
2       X(I)=X(I)+RKA(NRK)*XPRIM(I)*DT
        T=T+DT
        RETURN
        END
```

Fig. 8.1 Fortran subroutine for Runge Kutta algorithm

8.2 Solutions of General Equations

of physics, needs idealized studies simplified to a point to be exploitable where analytical methods are tractable. In addition to the teaching virtue, this approach makes it possible to easily assess orders of magnitude of various dynamics reentry phenomena or to explain some unexpected behavior. It is also necessary to emphasize that, even in the case of sophisticated numerical solutions, the theoretical accuracy is somewhat illusory in comparison with the uncertainties in the aerodynamic model and the state of the atmosphere.

In the continuation of this work, we will simplify the most important phenomena using an analytical approach, while comparing these simplified results with those of sophisticated digital codes. We will first study the reentry motion of the center of mass with zero angle of attack hypotheses (the exact approach corresponds to a three degree-of-freedom digital code). Then we will assess the movement around the center of gravity in the case of a reentry with incidences:

- When the vehicle has symmetry of revolution together with initial incidence (with or without spin)
- When the vehicle has low asymmetries and spin

The exact approach requires use of a six degree-of-freedom digital code. To obtain analytical solutions, we will assume that the motion of the mass center is decoupled from the motion around the center of mass. The results will be quantitatively more or less exact but qualitatively correct most of the time, except in the case of strong increase in the incidence at the end of the flight. In this last case, only the numerical study using an adapted code can give realistic dynamic loads (provided that the aerodynamic models were correctly evaluated).

Chapter 9
Zero Angle of Attack Reentry

This assumption represents the case of a perfectly symmetrical vehicle, statically and dynamically stable, with a negligible incidence at the beginning of reentry, penetrating a calm atmosphere. In this case, the axis of symmetry of the vehicle remains practically aligned with the Flight Path Vector relative to the earth and the aerodynamic load is simply the axial force, which is also the drag. To obtain a realistic three degree-of-freedom solution with rigor, we must consider the effect of gravity and terrestrial geoid's model, a standard atmosphere model, effect of earth rotation on the axes of observation and on movement relative to air. However, H. J. Allen has developed a very useful set of approximate solutions for ballistic vehicles with high initial velocity and high ballistic coefficient [ALL]:

(h1) zero incidence,
(h2) constant ballistic coefficient

$$\beta = \frac{m}{S_{ref} C_A},$$

(h3) zero gravity,
(h4) homogeneous atmosphere with an exponential vertical density profile,

$$\rho(z) = \rho_s \, e^{-\frac{z}{H_R}},$$

(h5) flat, nonrotating earth (Fig. 9.1).

This approximation applies to the decelerated phase (under 50 km altitude) when the drag force becomes higher than the weight of the vehicle. Indeed, between 120 and 60 km, the trajectory of the center of mass is still very close to the outer atmosphere ballistic trajectory (it is the accelerated reentry phase).

9.1 Allen's Reentry Results

(h3) + (h5) ⇒ rectilinear trajectory, constant flight path angle γ

P. Gallais, *Atmospheric Re-Entry Vehicle Mechanics.*
© Springer 2007

Fig. 9.1 H. Julian Allen with blunt body theory (1957, NASA picture)

$$(h1) + (h2) + (h3) + (h5) \Rightarrow \dot{V} = -\frac{1}{2}\frac{\rho V^2}{\beta}$$

$$\dot{z} = -V \sin \gamma$$

with convention $\gamma > 0$ for the reentry.

In order to solve this set of differential equations, first let us choose z as the independent variable instead of t:

$$\Rightarrow \dot{V} = \frac{dV}{dz}\frac{dz}{dt} = -V \sin \gamma \frac{dV}{dz}$$

$$\Rightarrow \frac{dV}{dz} = \frac{\rho V}{2\beta \sin \gamma}$$

Then considering (h4), we choose ρ as a new independent variable:

$$\frac{dV}{dz} = \frac{dV}{d\rho}\frac{d\rho}{dz} = -\frac{\rho}{H_R}\frac{dV}{d\rho}$$

$$\Rightarrow \frac{1}{V}\frac{dV}{d\rho} = -K$$

with $K = \frac{H_R}{2\beta \sin \gamma}$.

9.1 Allen's Reentry Results

This differential equation is integrated to give the velocity according to the density of air,

$$V = V_0 e^{-K(\rho - \rho_0)} \approx V_0 e^{-K\rho} \tag{9.1}$$

This expression provides the order of magnitude of the environment undergone by the vehicle during the reentry:

9.1.1 Axial Load Factor and Dynamic Pressure

Longitudinal load factor: $n_x = \frac{A}{mg} = \frac{1}{2}\rho \frac{V^2}{\beta g} = \frac{V_0^2}{2\beta g} \rho e^{-2K\rho}$

Maximum longitudinal load factor: n_x maximum $\Leftrightarrow \frac{d}{d\rho}\left(\rho e^{-2K\rho}\right) = 0$

$$\Leftrightarrow \rho = \frac{1}{2K}; \quad z = H_R \ln(2K\rho_s)$$

$$n_{x,\max} = \frac{V_0^2}{2g H_R} \frac{\sin \gamma_0}{e}$$

The value of the maximum is independent of β.

When $H_R \rho_s \leq \sin \gamma_0$, impact on the ground occurs before meeting the theoretical maximum. As β is assumed constant, this altitude apply as well to the maximum dynamic pressure. Unlike maximum axial load, maximum dynamic pressure depends on β.

In the case of an atmosphere with parameters $H_{\text{ref}} = 7000$ m, $\rho_0 = 1.39$ kg/m³, for $\beta = 10^4$ kg/m² and $V_0 = 6000$ m/s typical of a ballistic missile, it results:

$$\gamma_0 = 30° \quad \rightarrow \quad z \approx 6100 \text{ m}; \quad n_{x,\max} \approx 40 \text{ g}$$
$$\gamma_0 = 70° \quad \rightarrow \quad z \approx 800 \text{ m}; \quad n_{x,\max} \approx 75 \text{ g}$$

9.1.2 Heat Flux

The correlation suggested by Sutton and Graves [SUT] for convective heat flux at stagnation point in laminar flow (cold wall) corresponds to:

$$\Phi = C\sqrt{\frac{\rho}{R_N}} V^3 \text{ W/m}^2$$

where R_N is the radius of curvature of the wall, C a constant, which depends on the composition of the planetary atmosphere considered. Respective values $C = 1.83 \cdot 10^{-4}$ kg$^{\frac{1}{2}}$.m^{-1} and $C = 1.89 \cdot 10^{-4}$ kg$^{\frac{1}{2}}$.m^{-1} of the constant are well adapted to the earth's atmosphere and the Martian atmosphere (95% CO_2).

Maximum heat flux is reached when:

$$\frac{d}{d\rho}\left(\sqrt{\rho}e^{-3K\rho}\right) = 0 \iff \rho = \frac{1}{6K} \iff z = H_R \ln\left(6K\rho_s\right),$$

The corresponding maximum value is:

$$\Phi_{max} = C \frac{V_0^3}{\sqrt{R_N}} \sqrt{\frac{\beta \sin \gamma_0}{3eH_R}}$$

Under preceding conditions in earth's atmosphere and for a nose radius $R_N = 25$ mm representative of ballistic RV, we obtain:

$$\gamma_0 = 30° \rightarrow z \approx 15400 \text{ m}; \quad \Phi_{max} \approx 68 \, Mw/m^2$$
$$\gamma_0 = 70° \rightarrow z \approx 10100 \text{ m}; \quad \Phi_{max} \approx 93 \, Mw/m^2$$

It should be noted that the maximum load factor is independent of the vehicle (at least within the framework of this approximation, when the maximum is reached before impact). The maximum heat flux is not only a function of initial velocity and flight path angle but also of the ballistic coefficient and nose radius. Thus, vehicles with high ballistic coefficient and small nose radius undergo considerable heat fluxes.

9.1.3 Thermal Energy at Stagnation Point

In order to assess thermal energy per unit of area at the stagnation point, we will first establish a generalized expression of the Allen formula, in the case of a more general atmosphere (i.e., not isothermal), which is homogeneous and in vertical equilibrium. For the static pressure we have:

$$\frac{dp}{dz} = -\rho g$$

Using otherwise the same assumptions, if we use pressure instead of the density as the independent variable, we obtain:

$$\frac{dV}{dz} = \frac{\rho V}{2\beta \sin \gamma_D}$$
$$\frac{dV}{dz} = \frac{dV}{dp}\frac{dp}{dz} = -\rho g \frac{dV}{dp}$$

9.1 Allen's Reentry Results

That is to say,

$$\Rightarrow \frac{dV}{dp} = \frac{V}{2\beta g \sin \gamma_D}$$

$$V \approx V_D e^{-\int_{p_D}^{p} \frac{dp}{p_c}}$$

where the variable $p_c = 2\beta g \sin \gamma_D$, homogeneous with a pressure, is characteristic of the trajectory and vehicle. When p_c is constant, we obtain:

$$V \approx V_D e^{-\frac{(p-p_D)}{p_c}} \approx V_D e^{-\frac{p}{p_c}}$$

When the atmosphere is isothermal, we have $p = \rho \frac{R}{M} T_S$. Taking into account the expression of the reference height $H_R = \frac{RT_S}{Mg}$ established in Chap. 3, we find the Allen solution.

From the expression of the heat flux, we can determine surface thermal energy in the vicinity of the stagnation point:

$$\Phi = \frac{dE}{dt} = C\sqrt{\frac{\rho}{R_N}} V^3 \Rightarrow E(t) = \frac{C}{\sqrt{R_N}} \int_D \sqrt{\rho} V^3 \, dt$$

By carrying out a change of variable using the static pressure:

$$dt = \frac{dp}{\left(\frac{dp}{dz} \cdot \frac{dz}{dt}\right)} = \frac{dp}{\rho g V \sin \gamma_D}$$

$$E(p) = \frac{C}{g \sin \gamma_D \sqrt{R_N}} \int_{p_D}^{p} \frac{V^2}{\sqrt{\rho}} dp = \frac{C}{g \sin \gamma_D \sqrt{R_N}} \sqrt{\frac{RT_S}{M}} V_D^2 \int_{p_D}^{p} \frac{e^{-2\frac{p}{p_c}}}{\sqrt{p}} dp$$

Finally, while posing $u = \sqrt{2\frac{p}{p_c}}$, we obtain:

$$E(u) = \frac{C}{g \sin \gamma_D \sqrt{R_N}} \sqrt{\frac{p_c RT_S}{2M}} V_D^2 2 \int_{u_D}^{u} e^{-u^2} du$$

Taking into account $u_D \approx 0$, the surface thermal energy received until an altitude corresponding to an atmospheric pressure p is close to:

$$E \approx \frac{C}{\sqrt{R_N}} \sqrt{\frac{\pi \beta H_R}{\sin \gamma_D}} V_D^2 \left[Erf\left(\sqrt{2\frac{p}{p_c}}\right) \right] \quad \text{watts/m}^2$$

In the case of objects with low ballistic coefficient such as space probes or small meteorites, the parameter p_c is small in front of the sea level pressure p_s. The maximum value 1 of Erf function as well as maximum thermal energy is reached before impact. If the object is not vaporized, maximum surface energy received is then:

$$E_{max} \approx \frac{C}{\sqrt{R_N}} \sqrt{\frac{\pi \beta H_R}{\sin \gamma_D}} V_D^2 \cdot \text{watts/m}^2.$$

9.1.4 Duration of Reentry

Knowing instantaneous vertical velocity, we can determine the variable time as function of altitude:

$$dt = \frac{dz}{V_Z} = -\frac{dz}{V \sin \gamma_D}$$

$$t - t_D = -\int_{z_0}^{z} \frac{dz}{V \sin \gamma_D} = -\frac{1}{V_D \sin \gamma_D} \int_{z_0}^{z} e^{K\rho} dz$$

Using an exponential atmosphere and the variable ρ instead of Z, we have

$$dz = -H_R \frac{d\rho}{\rho}$$

Time can be written as:

$$t - t_D \approx \frac{H_R}{V_D \sin \gamma_D} \int_{K\rho_D}^{K\rho} \frac{e^{K\rho} d(K\rho)}{K\rho}$$

While noting $x = K\rho$, the integral can be approximated by:

$$I(x, x_D) = \int_{x_D}^{x} \frac{e^x dx}{x} \approx \left[Ln |x| + x + \frac{x^2}{4} + \frac{x^3}{18} + \frac{x^4}{96} + \ldots \right]_{x_0}^{x}$$

The lower limit of integration $x_D = K\rho_D$ is close to zero, the integral can be written as:

$$I(x, x_D) \approx \left[Ln \left| \frac{x}{x_D} \right| + x + \frac{x^2}{4} + \frac{x^3}{18} + \frac{x^4}{96} + \ldots \right]$$

That is to say,

$$t - t_D \approx \frac{H_R}{V_D \sin \gamma_D} \left[\frac{z_D - z}{H_R} + K\rho \left(1 + \frac{(K\rho)}{4} + \frac{(K\rho)^2}{18} + \frac{(K\rho)^3}{96} \right) \right]$$

Figures 9.2–9.4 show comparison between the results of Allen approximation generalized to a nonisothermal atmosphere and the answers of a three degree-of-freedom code. The assumptions are: $\gamma_D = 30.4°$, $V_D = 6061$ m/s $\beta = 10^2$ and 10^4 kg/m^2, and US66 standard atmosphere. Heat fluxes are given for nose radius $R_N = 1$ m.

9.1 Allen's Reentry Results

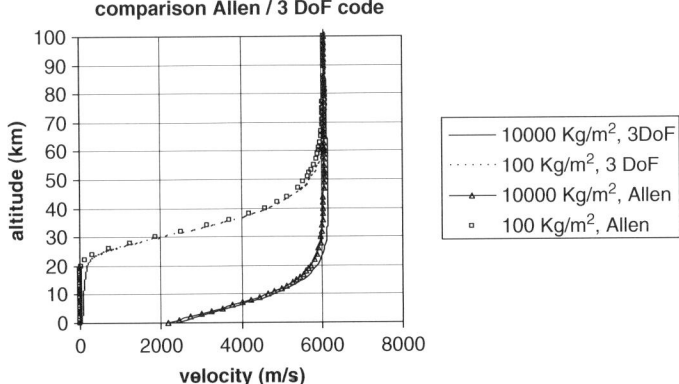

Fig. 9.2 Evolution of velocity

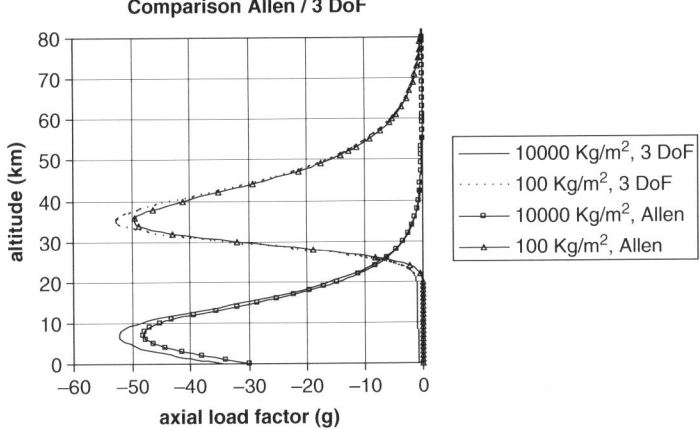

Fig. 9.3 Evolution of axial load factor

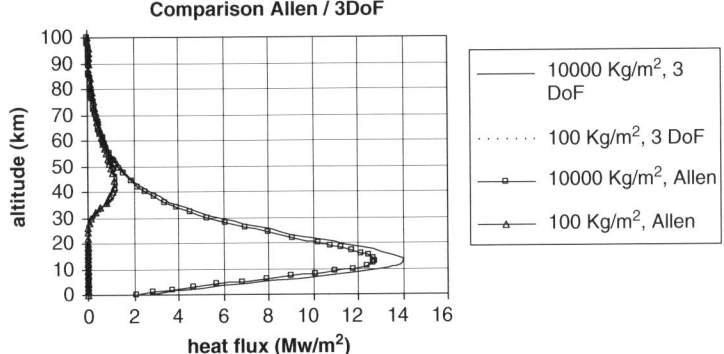

Fig. 9.4 Evolution of heat flux

These results show a good agreement between analytical calculations and three degree-of-freedom digital code. The undervaluation of maximum of deceleration and heat flux can be explained by not taking into account gravity, which leads to overestimating the loss of speed. The Allen approximation is thus quite able to give excellent orders of magnitude of the mechanical and thermal constraints during the reentry.

9.2 Influence of Ballistic Coefficient and Flight Path Angle

Figures 9.5–9.7 show the results of three degree-of-freedom code with sensitivity of reentry conditions to ballistic coefficient and initial flight path angle, in the same velocity conditions as the preceding case.

We can observe that the objects with low ballistic coefficient decelerate at relatively high altitude and quickly reach a subsonic limit velocity with a vertical flight

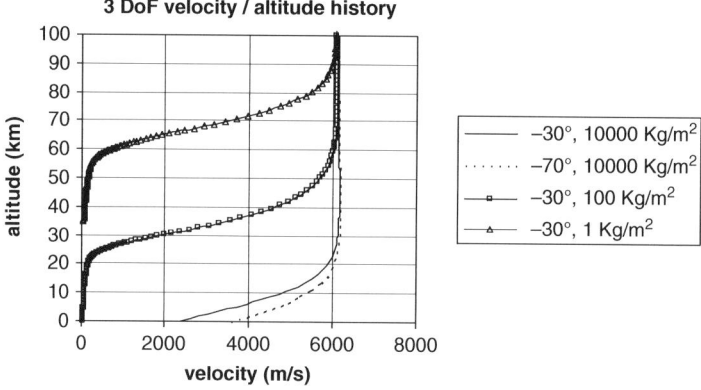

Fig. 9.5 Sensitivity of velocity on flight path and ballistic coefficient

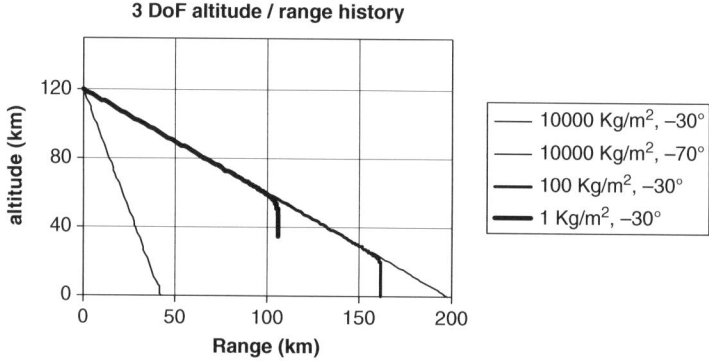

Fig. 9.6 Sensitivity of altitude/range on flight path and ballistic coefficient

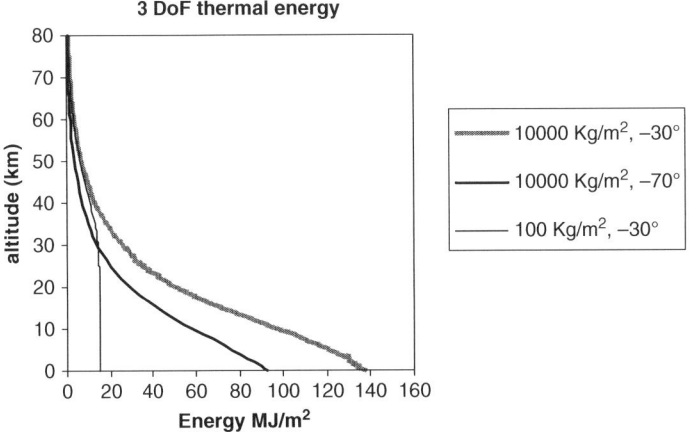

Fig. 9.7 Sensitivity of thermal energy on flight path and ballistic coefficient

path (55 Km and 20 Km for $\beta = 1$ and 100 Kg/m², respectively). Of course, the Allen assumptions of rectilinear of trajectory and negligible gravity are no longer valid in this case. In fact, meteorites of small size are vaporized before reaching their limit velocity, because their balance of thermal energy per unit of volume is very unfavorable. Even when we use three degree-of-freedom code, accurate calculation would require modeling the drag coefficient in free molecular and rarefied modes. On the other hand, massive meteorites reach the surface with high speed retention and a low thermal energy per unit volume, thus they reach the surface with devastating kinetic energies. In the case $\beta = 10^4$ kg/m², final surface energy (Fig. 9.7) and that estimated using the analytical formula are 136.9 and 151.5 MJ/m², respectively. This shows again that the Allen approximation provides useful orders of magnitude of the mechanical and thermal reentry phenomena.

9.3 Influence of Range

Figures 9.8–9.15 give the influence of range on reentry conditions for a ballistic coefficient $\beta = 10^4$ kg/m², obtained using three degree-of-freedom code.

Initial conditions at 120 km are summarized in Table 9.1, for elliptic trajectories and a nonrotating earth.

Table 9.2 summarizes the values of the parameters and corresponding altitudes of various ranges of minimal energy trajectories, $\beta = 10^4$ kg/m² and 25 mm nose radius.

Increased range corresponds to increased initial velocity V_0, thus also increased axial and lateral load factors, pressure, and stagnation enthalpy proportional to V_0^2, as well as heat flux proportional to V_0^3. Increased range corresponds to lower flight

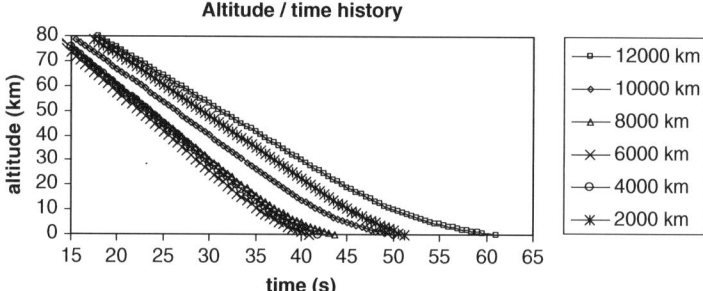

Fig. 9.8 Effect of range on time: altitude history

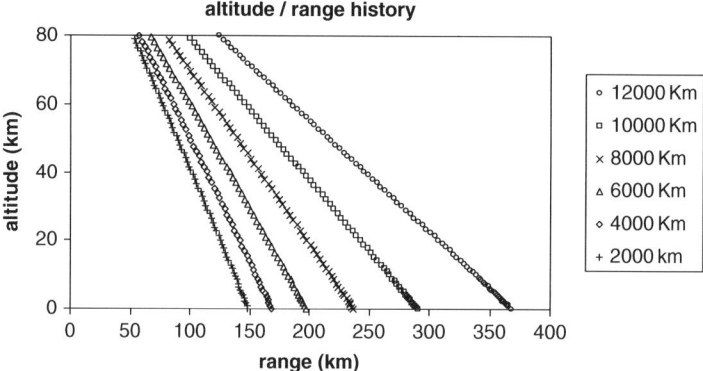

Fig. 9.9 Effect of range on Flight Path

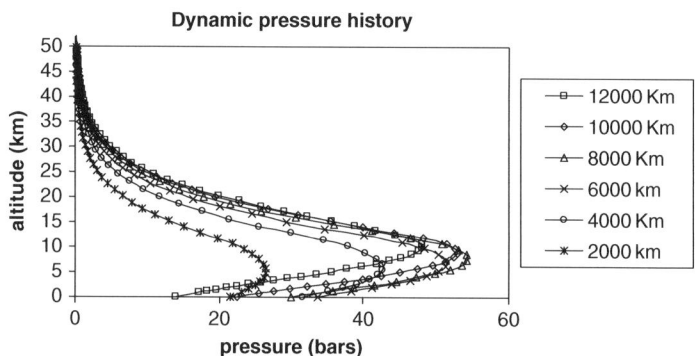

Fig. 9.10 Effect of range on dynamic pressure

9.3 Influence of Range

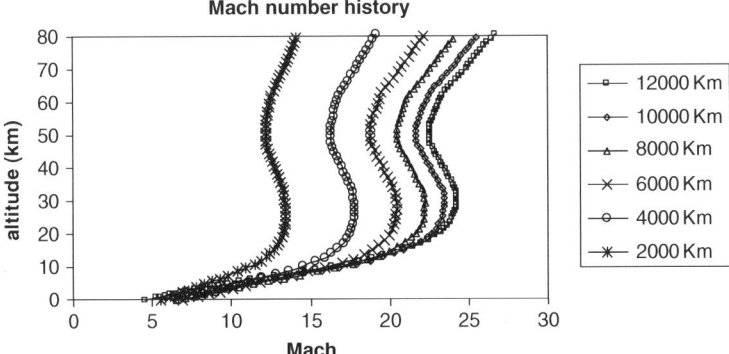

Fig. 9.11 Effect of range on Mach number history

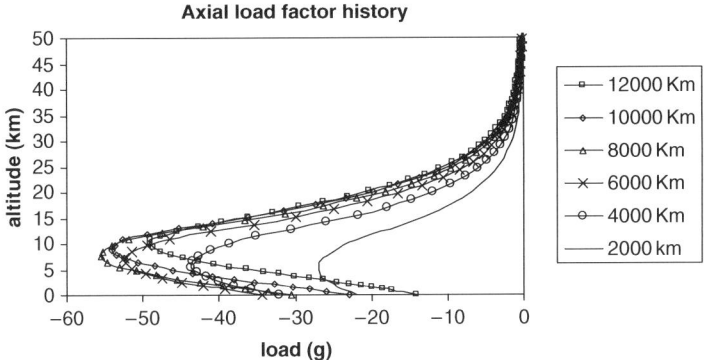

Fig. 9.12 Effect of range on axial load factor

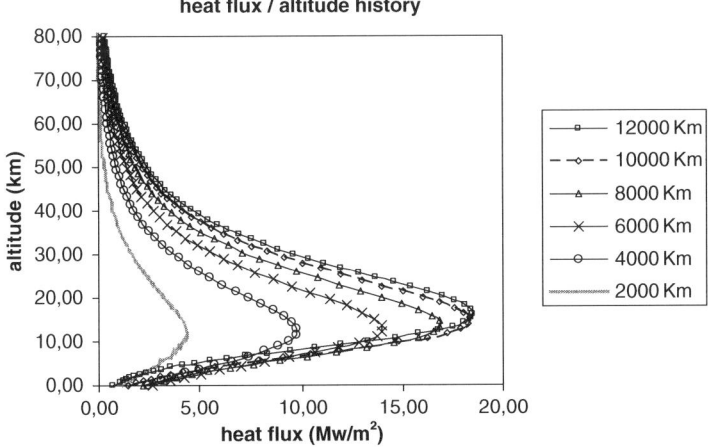

Fig. 9.13 Effect of range on heat flux/altitude history

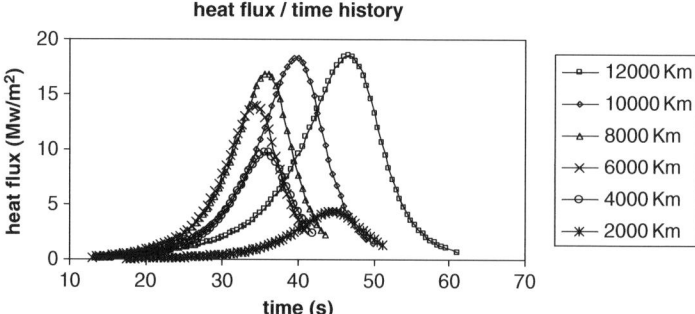

Fig. 9.14 Effect of range on heat flux/time history

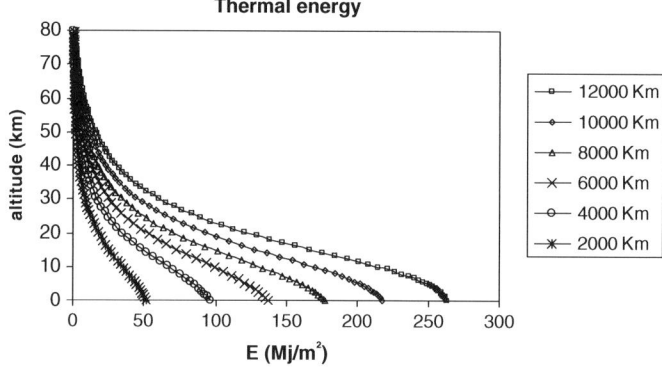

Fig. 9.15 Effect of range on thermal energy

path angle γ_0, which increases the duration of the reentry and thus the total heat received by the vehicle, although maximum heat flux is proportional to $(\sin \gamma_0)^{\frac{1}{2}}$.

Table 9.3 summarizes influence of ballistic coefficient on the same parameters for a 6000-km range minimum energy trajectory.

These results show that increased ballistic coefficient β increases the aerothermal loads (stagnation pressure, heat flux, quantity of heat received, and ablation). We can observe in addition that the approximation of Allen for an exponential atmosphere, which predicts a maximum value of the axial load factor independent of the ballistic coefficient, is faulty. The main origin of this variation is due to the differences

Table 9.1 Initial conditions of elliptic trajectories and a nonrotating earth at 120 km

Range (Km)	2000	4000	6000	8000	10000	12000
V_D (m/s)	3821	5216	6061	6633	7036	7323
γ_D (degrees)	−36.6	−34.2	−30.4	−26.2	−21.9	−17.6

9.3 Influence of Range

Table 9.2 Values of the parameters and corresponding altitudes of various ranges of minimal energy trajectories ($\beta = 10^4$ kg/m^2 and 25 mm nose radius)

Range (km)	Final Mach	Final velocity (m/s)	Maximum axial load factor (g/km)		Maximum stagnation pressure (bars/km)		Maximum heat flux (MW.m^{-2}/km)		Final thermal energy (MJ.m^{-2})
12000	4.4	1508	49.1	10.2	88.7	10.2	117	16.1	1670
10000	5.6	1916	54.0	8.9	97.5	8.9	116	15.1	1320
8000	6.5	2211	55.4	7.5	100	7.5	106	13.2	1120
6000	6.9	2347	52.4	7.3	94.6	7.3	89	12.3	870
4000	6.7	2273	43.5	5.5	78.7	5.5	61	11.4	610
2000	5.5	1876	22.0	5.6	49	5.6	28	11.2	320

Table 9.3 Influence of ballistic coefficient on the parameters of a 6000-km range minimum energy trajectory

β(kg/m^2)	Final Mach	Final velocity (m/s)	Maximum axial load factor (g/km)		Maximum stagnation pressure (bars/km)		Maximum heat flux (MW.m^{-2}/km)		Final thermal energy (MJ.m^{-2})
20000	11.2	3818	49.1	1.4	167	1.4	122	10.3	1087
10000	6.9	2347	52.4	7.3	94.6	7.3	89	12.3	870
5000	2.7	926	59.4	11.02	53.7	11.02	63	17.1	636

Table 9.4 Comparison between the durations of reentry from 120 km for minimum energy trajectories [calculated using three degree-of-freedom code (3DoF) and the Allen approximation]

Range (Km)	2000	4000	6000	8000	10000	12000
Duration (3DoF code)	51.22 s	41.77 s	40.81 s	43.59 s	49.82 s	61.06 s
Duration (Allen)	55.8 s	43.5 s	42 s	44.6 s	50.7 s	62.3 s

between the exponential atmosphere and a more realistic model such as the US66 standard atmosphere.

Finally, Table 9.4 shows comparison between the durations of reentry from 120 km for minimum energy trajectories, calculated using three degree-of-freedom code and the Allen approximation.

The agreement is quite good, considering that the approximation of Allen assumes zero gravity.

Chapter 10
Decay of Initial Incidence

This phase proceeds between 120 and 60 km, initially in rarefied flow then in a continuous mode. The aerodynamic resultant at the center of gravity is still negligible in front of the weight of the vehicle, and the aerodynamic moment becomes comparable with the gyroscopic moment. The velocity of the center of mass increases under the effect of gravity, thus this is the *accelerated phase* of reentry. The incidence starts to evolve under the effect of the aerodynamic moment; for a stable vehicle it is the beginning of convergence.

To obtain an approximate solution for the motion around mass center, we will assume the trajectory is rectilinear in an inertial frame related to a motionless earth and atmosphere. We choose this reference frame (Fig. 10.1) such that the $o\vec{x}$ axis is along the relative velocity. The choice of normal and lateral axes is arbitrary. We will use the {3, 2, 1} set of Euler angles to represent the orientation of vehicle relative to the inertial frame. We assume the vehicle is axisymmetric. In order to benefit from this aerodynamic and inertial symmetry, we write the equation for the angular momentum in a reference frame $Ox_2y_2z_2$ such that the $o\vec{x}_2$ axis is along the symmetry axis, with $\varphi = \dot{\varphi} = 0$. This observation frame follows the motion in ψ and θ of the vehicle, but its axis Oy_2 remains in the Oxy plane of the inertial frame.

Angular velocity $\vec{\omega} = \begin{bmatrix} p & q & r \end{bmatrix}$ of the vehicle and angular velocity $\vec{\omega}_e = \begin{bmatrix} p_e & q_e & r_e \end{bmatrix}$ of the frame $Ox_2y_2z_2$ are written in these axes as:

$$p = -\dot{\psi} \sin\theta + \dot{\varphi}$$
$$p_e = -\dot{\psi} \sin\theta$$
$$q = q_e = \dot{\theta}$$
$$r = r_e = \dot{\psi} \cos\theta$$

We obtain the angular momentum derivative,

$$\frac{d\vec{H}}{dt} = \begin{bmatrix} I_x \dot{p} \\ I_T \dot{q} \\ I_T \dot{r} \end{bmatrix} + \begin{bmatrix} p_e \\ q_e \\ r_e \end{bmatrix} \wedge \begin{bmatrix} I_x p \\ I_T q \\ I_T r \end{bmatrix} = \begin{bmatrix} L \\ M \\ N \end{bmatrix}$$

Fig. 10.1 Quasi inertial frame

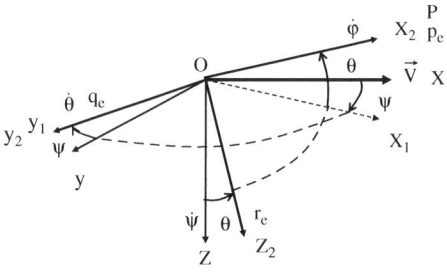

where L, M, and N are the components of the aerodynamic moment.

This gives:

$$I_x \dot{p} = L$$
$$I_T q_e + (I_x p - I_T p_e) r_e = M$$
$$I_T r_e + (I_T p_e - I_x p) q_e = N$$

Then,

$$I_x \dot{p} = L$$
$$\ddot{\theta} + \left(\frac{I_x}{I_T} p + \dot{\psi} \sin\theta\right) \dot{\psi} \cos\theta = \frac{M}{I_T}$$
$$\frac{d}{dt}(\dot{\psi} \cos\theta) - (\dot{\psi} \sin\theta + \frac{I_x}{I_T} p) \dot{\theta} = \frac{N}{I_T}$$

Let us assume that the angles θ and ψ are small, and the pitching moment is a linear function of angle of attack. The object being axi symmetrical, we can choose arbitrary transverse axes such as incidence and sideslip angles are $\alpha = \theta$ and $\beta = -\psi$. Then aerodynamic moment is written as:

$$M = \bar{q} S_{ref} L_{ref} (C_{m\alpha/G} \theta + C_{mq/G} \frac{q L_{ref}}{V})$$
$$N = -\bar{q} S_{ref} L_{ref} (C_{m\alpha/G} \beta + C_{mq/G} \frac{r L_{ref}}{V})$$
$$= \bar{q} S_{ref} L_{ref} (C_{m\alpha/G} \psi + C_{mq/G} \frac{r L_{ref}}{V})$$

Thus we obtain, while neglecting the second-order terms:

$$I_x \dot{p} = L$$
$$\ddot{\theta} + \mu p \dot{\psi} \approx -\omega^2 \theta - m_q \dot{\theta}$$
$$\ddot{\psi} - \mu p \dot{\theta} \approx -\omega^2 \psi - m_q \dot{\psi}$$

with:

$$\omega^2 = -\frac{\bar{q} S_{ref} L_{ref} C_{m\alpha/G}}{I_T}$$

$$m_q = -\frac{\bar{q} S_{ref} L_{ref}^2 C_{mq/G}}{I_T V}$$

These equations are valid within our approximation for vehicles of revolution, with or without roll velocity. Obviously, for a statically and dynamically stable vehicle, we have:

$$C_{m\alpha/G} < 0 \quad \Rightarrow \quad \omega^2 > 0$$
$$C_{mq/G} < 0 \quad \Rightarrow \quad m_q > 0$$

The parameter ω represents the natural pulsation of pitching and yawing oscillations, m_q is the aerodynamic damping term.

10.1 Zero Spin Rate

When the roll moment L and the "spin" rate p are null, the preceding system is simplified:

$$\ddot{\theta} + m_q \dot{\theta} + \omega^2 \theta \approx 0$$
$$\ddot{\psi} + m_q \dot{\psi} + \omega^2 \psi \approx 0$$

First, we notice that the motions in ψ and θ are uncoupled and their equations are identical. We need to study only one of the modes ψ or θ. The instantaneous movement is an elliptic vibration resulting from the addition of two linear orthogonal modes. In addition, when coefficients m_q and ω^2 are independent of time, we obtain the equations of the damped linear oscillator. It is obviously not the case of a ballistic RV, which enters in the atmosphere. During the accelerated phase, velocity varies little, but the density of air increases exponentially, as well as the dynamic pressure, m_q, and ω^2 coefficients. We will see that the influence of the increase in density is an essential factor in the dynamic behavior of the vehicle during this phase, namely, "*density damping*." In order to show this result analytically, we simplify the preceding model by assuming the aerodynamic damping moment is null (we will see in Chap. 13 that during this phase, it is negligible compared to density damping effect). Now, let us consider the simplified equation of the θ mode:

$$\ddot{\theta} + \omega^2 \theta \approx 0$$

10.1.1 First Approximate Solution

We look for an approximate solution, while assuming that the relative variation of natural pulsation $\frac{\Delta \omega}{\omega}$ is very small over one period T of oscillation:

$$\varepsilon = \frac{\Delta \omega}{\omega} = \frac{\dot{\omega} T}{\omega} = 2\pi \frac{\dot{\omega}}{\omega^2} \ll 1$$

Let us consider the instantaneous power of an angular movement with constant pulsation:

$$\dot{E} = I_T \frac{d}{dt}\left(\frac{1}{2}\dot{\theta}^2 + \frac{1}{2}\omega^2\theta^2\right) = I_T \left(\ddot{\theta} + \omega^2\theta\right)\dot{\theta} = 0.$$

$$\Rightarrow E = \frac{1}{2}I_T\dot{\theta}^2 + \frac{1}{2}I_T\omega^2\theta^2 = \frac{1}{2}I_T\dot{\theta}^2 + \frac{1}{2}K_\theta\theta^2 = E_0$$

where K_θ is the coefficient of the restoring moment. In this case, the power received by the oscillator is null and energy is constant.

When the pulsation is variable, the derivative of energy is:

$$\dot{E} = \frac{d}{dt}\left(\frac{1}{2}I_T\dot{\theta}^2 + \frac{1}{2}I_T\omega^2\theta^2\right) = I_T\left(\ddot{\theta} + \omega^2\theta\right)\dot{\theta} + I_T\theta^2\omega\dot{\omega}.$$

The term on the right between brackets being null, we obtain:

$$\dot{E} = I_T\theta^2\omega\dot{\omega}$$

The instantaneous power is not zero for the motion with variable pulsation. Rotational energy E is not a constant in the movement. In fact, this linear oscillator is not an isolated system and, when $\dot{\omega} > 0$, is receiving energy. Clearly, energy received come from the initial kinetic energy of the center of mass.

Using the assumption, $\varepsilon = \frac{2\pi}{\omega}\frac{\dot{\omega}}{\omega}$ is small and constant, we write now $\dot{\omega}$ in the form[1],

$$\dot{\omega} = \frac{\omega^2}{2\pi}\varepsilon.$$

Let us calculate the average power received (or removed) to the oscillating system over one period:

---

[1] Notice: According to our assumption, evolution of the pulsation is not unspecified, but follows,

$$\frac{d}{dt}\left(\frac{1}{\omega}\right) = \frac{\varepsilon}{2\pi} \Rightarrow \omega = \frac{\omega_0}{1 - \frac{t}{\tau_0}},$$

with $\tau_0 = \frac{1}{2\pi\varepsilon}T_0$.

10.1 Zero Spin Rate

$$P_m = \langle \dot{E} \rangle = \frac{1}{T} \int_t^{t+T} \dot{E} d\tau$$

Because of our assumption, on a one period scale, the motion is very close to a sinusoidal motion with constant pulsation. Thus we obtain:

$$\langle \dot{E} \rangle = \frac{1}{T} \int_t^{t+T} I_T \theta^2 \omega \dot{\omega} d\tau \approx I_T \frac{\omega^3}{2\pi} \varepsilon \frac{1}{T} \int_t^{t+T} \theta^2 d\tau$$

The mean value of the θ^2 term for a sinusoidal evolution is equal to $\frac{1}{2} \theta_{max}^2$, that is to say:

$$\langle \dot{E} \rangle \approx \frac{1}{2} I_T \omega^2 \theta_{max}^2 \left(\frac{\varepsilon \omega}{2\pi} \right) = E \cdot \left(\frac{\dot{\omega}}{\omega} \right)$$

We conclude with a quite simple approximate result, applicable to any linear oscillating system with a slowly varying frequency:

$$\frac{\langle \dot{E} \rangle}{E} = \frac{\dot{\omega}}{\omega} \quad \Rightarrow \quad \frac{E}{E_0} = \frac{\omega}{\omega_0}$$

Thus total energy of the oscillator increases linearly with pulsation. Moreover, as instantaneous energy can be derived from the instantaneous maximum amplitude of oscillations, we have:

$$\frac{E}{E_0} = \frac{\omega^2 \cdot \theta_{max}^2}{\omega_0^2 \cdot \theta_{0,max}^2}$$

We obtain finally:

$$\frac{\theta_{max}}{\theta_{0,max}} = \left(\frac{\omega_0}{\omega} \right)^{\frac{1}{2}}$$

During the accelerated phase of reentry, velocity is approximately constant and we have:

$$\left(\frac{\omega_0}{\omega} \right)^{\frac{1}{2}} = \left(\frac{\bar{q}_0}{\bar{q}} \right)^{\frac{1}{4}} \approx \left(\frac{\rho_0}{\rho} \right)^{\frac{1}{4}}$$

$$\Rightarrow \frac{\theta}{\theta_0} \approx \left(\frac{\rho_0}{\rho} \right)^{\frac{1}{4}}$$

Although the initial assumption is not verified (especially for the first periods), we can note from Fig. 10.2 that this expression of the amplitude is in good agreement with 6 DoF assessment of the evolution of incidence at the beginning of reentry.

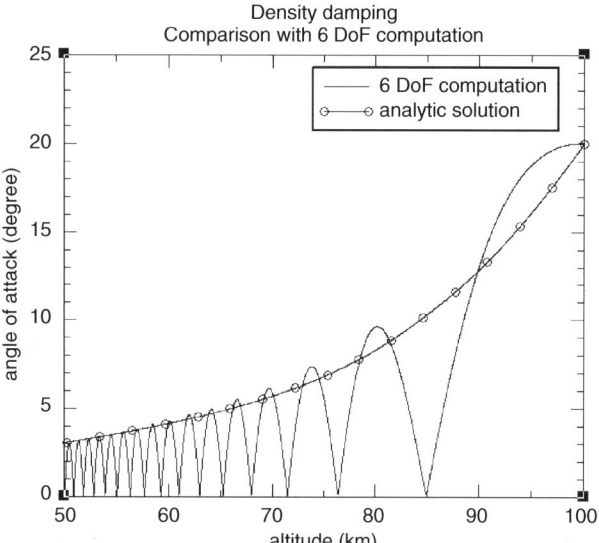

Fig. 10.2 Density damping

This result clearly highlights the phenomenon of *"density damping"*. Because of exponential increase of density, we obtain an exponential growth of pulsation accompanied by an exponential decay of the amplitude of oscillations (this applies, of course, only to statically stable vehicles). Rigorously, the term damping used for this phenomenon is not suitable because, unlike a traditional damping, it is not directly related to a dissipation of energy. Paradoxically, when the frequency increases, which is the case at the beginning of reentry and corresponds to an increase in energy of the oscillator, we obtain a decrease of the amplitude. The reverse phenomenon possibly happens at the end of the reentry on trajectories where the maximum dynamic pressure occurs before impact. In this case, a decrease of dynamic pressure is accompanied by aerodynamic pulsation decreases, which corresponds to an increase in the amplitude of the incidence. The phenomenon is in general much less marked, because the variation of frequency is much slower and initial incidence very low. In fact, the dissipative aerodynamic damping moment, which is not negligible at this time, decreases or cancels the incidence buildup.

10.1.2 Second Approximate Solution

Now, we seek a more exact solution, using the assumption of a reentry at constant velocity and gradient of pitching moment coefficient in an exponential atmosphere. Under these conditions, we can write:

10.1 Zero Spin Rate

$$z = z_D - V_D \cdot \sin\gamma_D \cdot t\;;\quad \frac{d}{dt} = -V_D \cdot \sin\gamma_D \cdot \frac{d}{dz}$$

$$\omega^2 = \omega_D^2 e^{-\frac{z-z_D}{H_R}}$$

The equation in θ becomes,

$$\frac{d^2\theta}{dz^2} + k^2\theta = 0$$

where $k = \frac{\omega_D e^{-\frac{z-z_D}{2H_R}}}{V_D \sin\gamma_D} = k_D e^{-\frac{z-z_D}{2H_R}}$ is the wave number associated with the altitude wavelength of the incidence oscillation such that $\lambda = \frac{2\pi}{k}$.

We now consider the variable φ built from the wave number $\frac{H_R}{\lambda}$ relative to the reference height of the atmosphere, $\varphi = 2 \cdot 2\pi \frac{H_R}{\lambda} = 2 \cdot k H_R$. We have:

$$\varphi = 2k_D H_R e^{-\frac{z-z_D}{2H_R}}$$

With this new variable, space derivatives of θ are transformed:

$$\frac{d}{dz} = \frac{d\varphi}{dz}\frac{d}{d\varphi} = -k_D e^{-\frac{z-z_D}{2H_R}}\frac{d}{d\varphi} = -\frac{\varphi}{2H_R}\frac{d}{d\varphi}$$

$$\left(\frac{d}{dz}\right)^2 = -\frac{\varphi}{2H_R}\frac{d}{d\varphi}\left(-\frac{\varphi}{2H_R}\frac{d}{d\varphi}\right) = \left(\frac{\varphi}{2H_R}\right)^2\left(\frac{d}{d\varphi}\right)^2 + \frac{\varphi}{(2H_R)^2}\frac{d}{d\varphi}$$

It is easy to check that:

$$\frac{d\theta}{dz} = -k\frac{d\theta}{d\varphi} \Rightarrow d\varphi = -k\,dz$$

$$\frac{d\theta}{dt} = \omega\frac{d\theta}{d\varphi} \Rightarrow d\varphi = \omega\,dt$$

This shows that φ represents the phase of the oscillation. We obtain finally a differential equation relative to variable φ,

$$\varphi^2\frac{d^2\theta}{d\varphi^2} + \varphi\frac{d\theta}{d\varphi} + \varphi^2\theta = 0$$

The observant reader will certainly identify a Bessel equation of zero order in its simplest form:

$$\varphi^2\frac{d^2\theta}{d\varphi^2} + \varphi\frac{d\theta}{d\varphi} + (\varphi^2 - n^2)\theta = 0$$

For the detailed study of the solutions of Bessel equations, we invite the reader to read to specialized course [ARF]. The general solution of zero order is a linear combination of Bessel function of first species, $J_0(\varphi)$, and of Bessel function of second species, $N_0(\varphi)$ (or Neumann function, noted $Y_0(\varphi)$):

$$\theta(\varphi) = A\, J_0(\varphi) + B\, N_0(\varphi)$$

Using Bessel functions properties, we have:

$$\frac{d\theta(\varphi)}{d\varphi} = -A\, J_1(\varphi) - B\, N_1(\varphi)$$

Coefficients A and B can be determined from initial conditions θ_D and $\dot{\theta}_D$:

$$t = 0 \quad \rightarrow \quad z_D \quad \rightarrow \quad \varphi_D = \frac{2\omega_D H_{\text{ref}}}{V_D \sin \gamma_D}$$

$$A J_0(\varphi_D) + B\, N_0(\varphi_D) = \theta_D$$

$$-A J_1(\varphi_D) - B\, N_1(\varphi_D) = \left.\frac{d\theta}{d\varphi}\right|_D = \frac{\dot{\theta}_D}{\omega_D}$$

which gives:

$$A = \frac{\theta_D N_1(\varphi_D) + \frac{\dot{\theta}(D)}{\omega_D} N_0(\varphi_D)}{J_0(\varphi_D)\, N_1(\varphi_D) - J_1(\varphi_D)\, N_0(\varphi_D)}$$

$$B = -\frac{\theta_D J_1(\varphi_D) + \frac{\dot{\theta}(D)}{\omega_D} J_0(\varphi_D)}{J_0(\varphi_D)\, N_1(\varphi_D) - J_1(\varphi_D)\, N_0(\varphi_D)}$$

The behavior of the Bessel functions of first and second species of orders 0 and 1 is shown in Fig. 10.3. Evolutions of the incidence, angular rate and pulsation calculated for a typical reentry vehicle are also represented. The examples are as follows:

Atmosphere	$\rho_s = 1.39\,\text{kg/m}^3;\, H_R = 7000\,\text{m}$
Vehicle coefficients	$S_{\text{ref}} = .12\,\text{m}^2\,;\, L_{\text{ref}} = 1.3\,\text{m}\,;\, I_T = 6\,\text{m}^2\cdot\text{kg}$
	$C_{m\alpha/G} = -0.2\,\text{radian}^{-1}$
Initial conditions	$z_D = 120\,\text{km}\,;\, V_D = 6100\,\text{m/s},\, \gamma_D = 30°$
	$\theta(Z_D) = 10°;\quad \dot{\theta}(Z_D) = 0;\quad \varphi_D \approx 0$

10.1 Zero Spin Rate

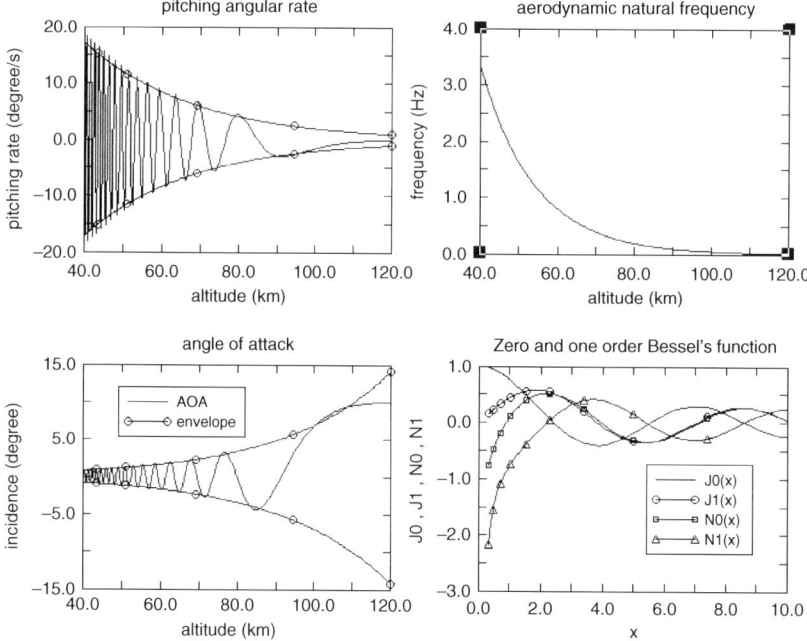

Fig. 10.3 Angular rate, angle of attack and aerodynamic frequency histories

Asymptotic Form

- For $\varphi \to 0$, we have:

$$J_0(\varphi) \to 1 \; ; \; J_1(\varphi) \approx \frac{\varphi}{2}$$
$$N_0(\varphi) \approx \frac{2}{\pi}\left(\text{Ln}\left(\frac{\varphi}{2}\right) + \gamma\right) \; ; \; N_1(\varphi) \approx -\frac{2}{\pi\varphi}$$

with $\gamma \approx \frac{228}{395}$

- For $\varphi \to \infty$, we have:

$$J_n(\varphi) \approx \sqrt{\frac{2}{\pi\varphi}} \cos\left(\varphi - \left(n + \frac{1}{2}\right)\frac{\pi}{2}\right)$$
$$N_n(\varphi) \approx \sqrt{\frac{2}{\pi\varphi}} \sin\left(\varphi - \left(n + \frac{1}{2}\right)\frac{\pi}{2}\right)$$

By using the preceding expressions with the initial condition $A = \theta_D$, and $B = 0$, we obtain the asymptotic form of θ and its derivative for large φ:

$$\theta \approx \theta_D \sqrt{\frac{2}{\pi\varphi}} \cos\left(\varphi - \frac{\pi}{4}\right)$$

$$\dot{\theta} \approx -\theta_D \omega_D \sqrt{\frac{2}{\pi\varphi}} \sin\left(\varphi - \frac{\pi}{4}\right)$$

$$\ddot{\theta} \approx \theta_0 \omega_D^2 \sqrt{\frac{2}{\pi\varphi}} \cos\left(\varphi - \frac{\pi}{4}\right)$$

Taking into account the definition of φ, we find a new behavior of the amplitude of θ in $\omega^{-\frac{1}{2}} \approx \rho^{-\frac{1}{4}}$. The phase is proportional to $\omega \approx \rho^{\frac{1}{2}}$.

The approximate envelopes of incidence, angular pitching rate, and acceleration are given by:

$$\frac{\theta}{\theta_D} \approx \frac{1}{\sqrt{\pi H_R k_D}} e^{\frac{z-z_D}{4H_R}} \; ; \quad \frac{\dot{\theta}}{\theta_D} \approx \frac{\omega_D}{\sqrt{\pi H_R k_D}} e^{\frac{z-z_D}{4H_R}} \; ; \quad \frac{\ddot{\theta}}{\theta_D} \approx \frac{\omega_D^2}{\sqrt{\pi H_R k_D}} e^{\frac{z-z_D}{4H_R}}$$

These parameters are represented in Fig. 10.3 together with the exact theoretical solutions for θ and $\dot{\theta}$ calculated previously.

10.2 Nonzero Spin

Although the equations of motion established at the beginning of this chapter apply, we will use more convenient formulation to analyze the behavior of a vehicle with spin. We use the {1, 2, 1} set of Euler angles described in Sect. 6.2.1. We use again an inertial observation frame, such that axis Ox is along the center-of-mass velocity vector and transverse axes Oy, and Oz. Thus, θ represents the angle between the axis of symmetry of the vehicle and the flight path vector (total incidence). As previously stated, we will write the equation of the angular momentum in a reference frame $Ox_2y_2z_2$ (Fig. 10.4) such that $\varphi = \dot{\varphi} = 0$ (this reference frame, which is not Eulerian, follows the ψ and θ motion of the vehicle, but its axis Oy_2 remains in the plane normal to flight path vector).

Angular rates $\vec{\omega} = \begin{bmatrix} p & q & r \end{bmatrix}$ of vehicle and $\vec{\omega}_e = \begin{bmatrix} p_e & q_e & r_e \end{bmatrix}$ of observation frame are,

$$p = \dot{\psi} \cos\theta + \dot{\varphi}$$
$$p_e = \dot{\psi} \cos\theta$$
$$q = q_e = \dot{\theta}$$
$$r = r_e = \dot{\psi} \sin\theta$$

We import these values into the equation of the angular momentum,

Fig. 10.4 Euler angles definition

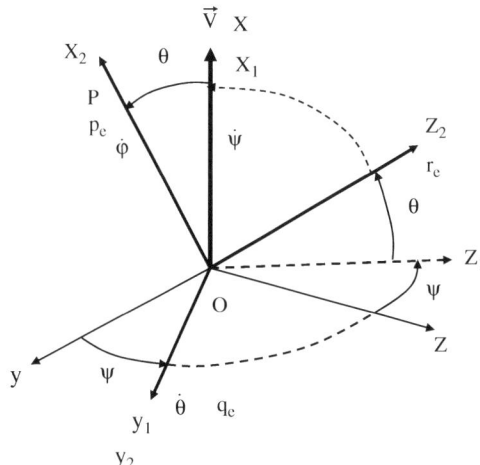

$$\frac{d\vec{H}}{dt} = \begin{bmatrix} I_x \dot{p} \\ I_T \dot{q} \\ I_T \dot{r} \end{bmatrix} + \begin{bmatrix} p_e \\ q_e \\ r_e \end{bmatrix} \wedge \begin{bmatrix} I_x p \\ I_T q \\ I_T r \end{bmatrix} = \begin{bmatrix} L \\ M \\ N \end{bmatrix},$$

Then we obtain,

$$I_x \dot{p} = L$$
$$\ddot{\theta} + \mu p \dot{\psi} \sin\theta - \dot{\psi}^2 \cos\theta \sin\theta = \frac{M}{I_T}$$
$$\frac{d}{dt}(\dot{\psi}\sin\theta) + \dot{\theta}\dot{\psi}\cos\theta - \mu p\dot{\theta} = \frac{N}{I_T}$$

If we neglect the damping moment, the aerodynamic moment is limited to the static pitching moment, which we assume to be,

$$M \approx \bar{q} S_{ref} L_{ref} C_{m\alpha/G} \sin\theta$$
$$N = 0$$

When roll rate is constant, the system is written as:

$$\ddot{\theta} + \mu p_0 \dot{\psi} \sin\theta - \dot{\psi}^2 \cos\theta \sin\theta = -\omega^2 \sin\theta$$
$$\ddot{\psi} \sin\theta + 2\dot{\psi}\dot{\theta}\cos\theta - \mu p_0 \dot{\theta} = 0$$

with $\omega^2 = -\frac{\bar{q} S_{ref} L_{ref} C_{m\alpha/G}}{I_T}$.

Multiplying the second equation by $\sin\theta$ and integrating, we obtain a constant in the movement. Indeed, as the aerodynamic moment component along the flight path vector is null, corresponding component of the angular momentum is constant:

$$\frac{d}{dt}\left(\dot{\psi}\sin^2\theta + \mu p_0 \cos\theta\right) = 0$$

$$\dot{\psi}\sin^2\theta + \mu p_0 \cos\theta = \frac{H_{\vec{V}}}{I_T} = C.$$

Now, we assume that the initial conditions correspond to a free coning motion. We can evaluate the constant C in two modes of motion.

- Precession mode: $\vec{H}_0$ is along $\vec{V}$ and $\sqrt{q_0^2 + r_0^2} \neq 0 \Leftrightarrow \tau = 0; \quad \varepsilon \neq 0$

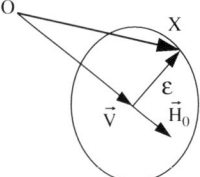

From results on free coning motion (7.2),

$$\dot{\psi}_0 = \frac{\mu p_0}{\cos\theta_0} \Rightarrow C = \frac{\mu p_0}{\cos\theta_0}$$

$$\dot{\psi} = \frac{\mu p_0 \left[\frac{1}{\cos\theta_0} - \cos\theta\right]}{\sin^2\theta}$$

- Nutation mode: $\vec{H}_0$ is along Ox, and not parallel with $\vec{V} \Leftrightarrow \tau \neq 0; \quad \varepsilon = 0$

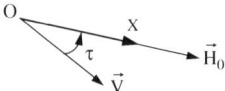

$$\dot{\psi}_0 = 0 \Rightarrow C = \mu p_0 \cos\theta_0$$

$$\dot{\psi} = \frac{\mu p_0(\cos\theta_0 - \cos\theta)}{\sin^2\theta}$$

This shows in the first case precession that rate is always the same sign as roll rate p_0. In the second case, sign is always opposed to roll rate.

We can synthesize these results:

$$\dot{\psi}_\pm = \frac{\mu p_0(\cos\theta_0^{\mp 1} - \cos\theta)}{\sin^2\theta}$$

10.2 Nonzero Spin

Now we assume the nutation motion θ is quasi-static, i.e., $\ddot{\theta} \approx 0$ (it is not the case for the precession motion ψ, because $\ddot{\psi} \neq 0$). Multiplying the θ equation by θ, we obtain a quadratic equation in $\dot{\psi}$,

$$(\dot{\psi}\cos\theta)^2 - \mu p_0 \dot{\psi}\cos\theta = \omega^2 \cos\theta.$$

Hence, while posing $p_r = \frac{\mu p_0}{2}$, we derive two solutions for $\dot{\psi}$:

$$\dot{\psi}_{\pm} = \frac{p_r \pm \sqrt{p_r^2 + \omega^2 \cos\theta}}{\cos\theta}$$

As in the preceding case, one of the solutions is always the same sign as roll rate and the other is of opposite sign.

In order to determine the evolution of θ, let us consider again the expression of the constant in the movement C resulting from the conservation of H_y. Using the assumption of small angles of attack, $\sin\theta \approx \theta$; $\cos\theta \approx 1 - \frac{\theta^2}{2}$, this expression takes the form:

$$\theta^2 \left(\dot{\psi} - p_r\right) \approx \pm \theta_0^2 \, p_r$$

When we square this equation and replace the bracketed term by its expression derived from the relation $\left(\dot{\psi}\cos\theta - p_r\right)^2 = p_r^2 + \omega^2 \cos\theta$ established previously, we obtain:

$$\theta^4 \left(p_r^2 + \omega^2\right) \approx \theta_0^4 \, p_r^2$$

while naming $\xi = \left(\frac{\omega}{p_r}\right)^2$, we have finally:

$$\frac{\theta}{\theta_0} \approx (1+\xi)^{-\frac{1}{4}}$$

$$\dot{\psi}_{\pm} \approx p_r \left(1 \pm \sqrt{1+\xi}\right)$$

In the precession mode of initial coning motion ($\vec{H}_0$ along $\vec{V}$), the X axis of the vehicle converges toward the flight path vector while turning in the same direction as roll rate. The precession rate $\dot{\psi}$ is initially equal to the free coning value and further increases in absolute value.

In the nutation mode of initial coning motion ($\vec{H}_0$ along symmetry axis), the Ox axis of the vehicle is initially motionless and converges toward the flight path vector while turning in the direction opposite to rolling motion.

We obtain the same type of decay of the incidence as for a vehicle without spin, i.e., a decrease of the amplitude related to the increase of air density. However, the initial decrease is slower because of the gyroscopic moment, and the roll rate is high.

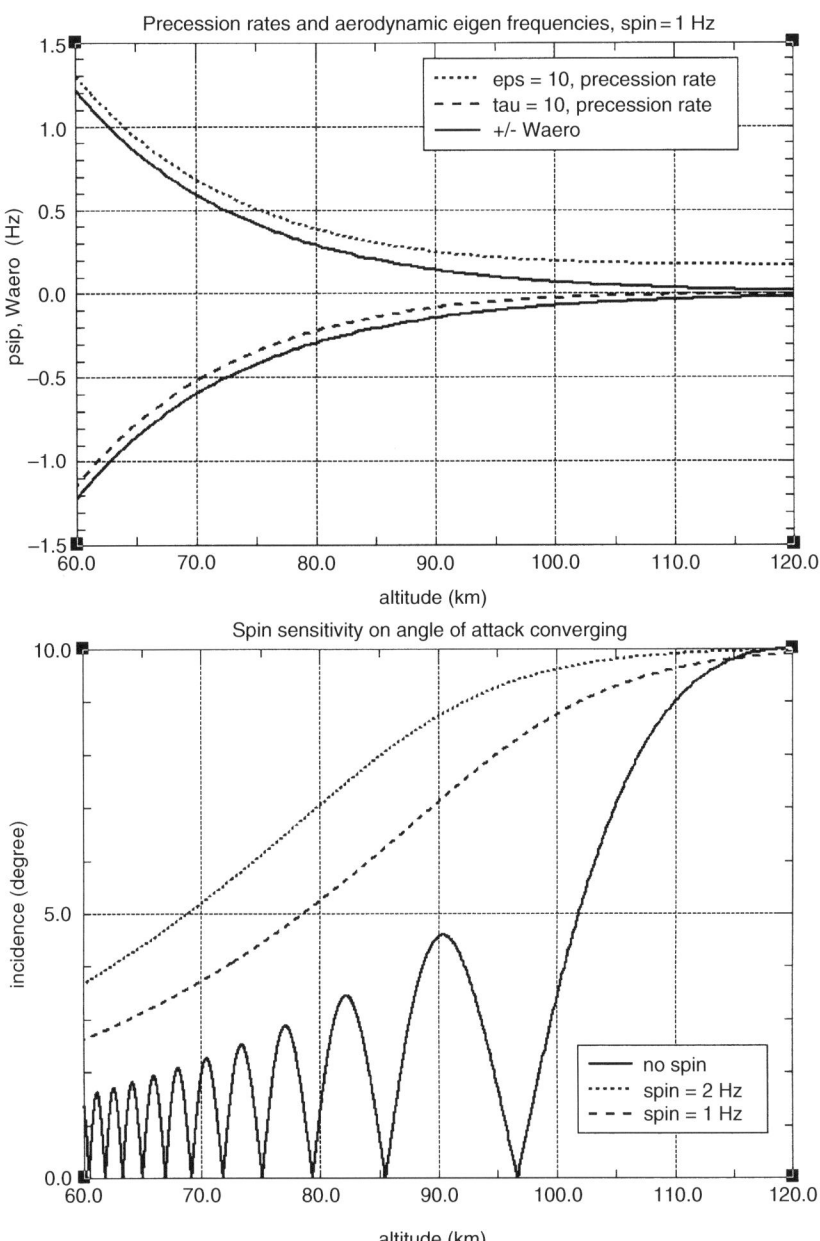

Fig. 10.5 Angle of attack and aerodynamic frequencies histories

10.2 Nonzero Spin

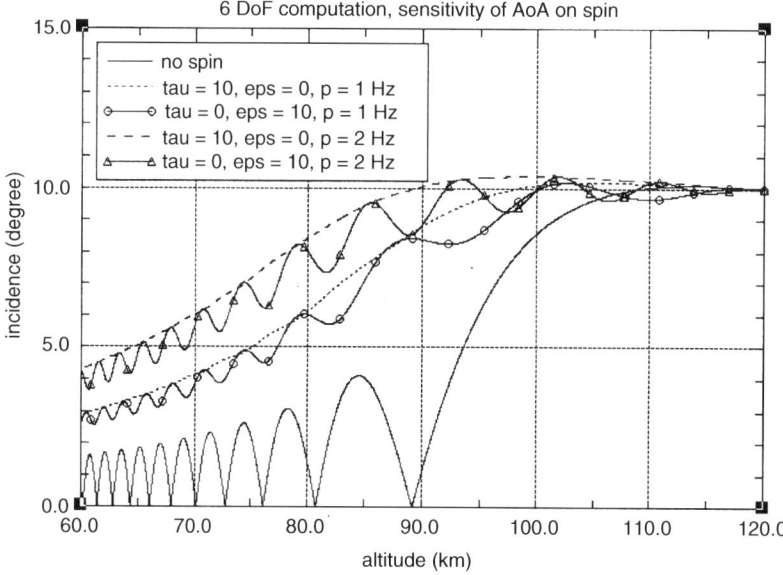

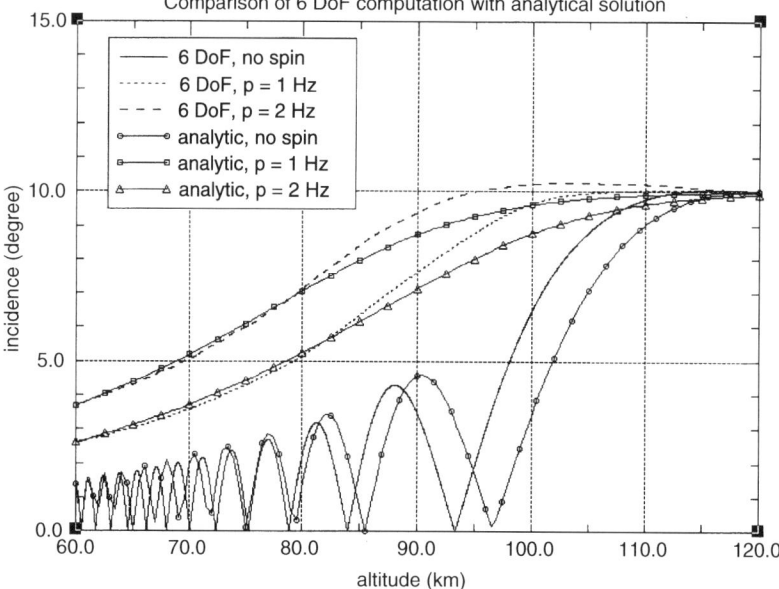

Fig. 10.6 Angle of attack histories

In fact, one finds the same rate of decrease in $\left(\frac{\rho_0}{\rho}\right)^{\frac{1}{4}}$ only when $\xi = \left(\frac{\omega}{p_r}\right)^2 \gg 1$, i.e., when the aerodynamic torque becomes prevalent.

Thus the gyroscopic moment related to roll rate has in general an unfavorable effect with respect to the initial convergence of the incidence. However, in certain situations this couple can be useful, for example, to control the divergence of the incidence at the encounter of a transitory static instability in rarefied flows, as happens to some space probes with high cone angles.

Figure 10.5 shows the behavior of a vehicle according to the present approximate method (Lref = 1.624 m, Sref = 0.196 m^2, $C_{N\alpha}$ = 2.2345 radian^{-1}, other assumptions identical to the vehicle used in the preceding paragraph). Figure 10.6 give a comparison between these approximate results with six degree-of-freedom calculations. Exact calculation confirms the behavior predicted by the quasi-static approximation. The differences observed come primarily from the different atmosphere models, and not accounting for the rotation of the flight path vector under the effect of gravity. We note indeed that the real six degree-of-freedom motion has oscillations in the case of the initial conditions of precession ($\tau = 0, \varepsilon \neq 0$). This is the effect of the rotation of the flight path vector due to gravity. Thus, there appears a term $\delta\tau$ of second mode with precession rate $\dot{\psi}_-$ of sign opposite to p_r, which is added to the initial mode ε with rate $\dot{\psi}_+$ of same sign as p_r. With initial conditions corresponding to the nutation mode ($\tau \neq 0, \varepsilon = 0$), the term $\delta\tau$ also appears but its precession rate is the same. This explains why the difference between six degree-of-freedom and approximate calculations is not oscillatory.

Chapter 11
End of the Convergence of the Incidence

During this phase, between 60 Km and sea level, we are in continuum flow and the aerodynamic drag becomes comparable, then higher than the weight of the vehicle. For the motion of the center of mass, it is the decelerated phase, which was previously analyzed with the Allen approximation. Under the effect of the aerodynamic moment, incidence oscillations continue to diminish until acquisition of the static trim in low altitude (generally in the vicinity of 10–15 Km). Depending on the level of inertial and aerodynamic asymmetries of the vehicle, this static trim can vary from zero to a highly amplified incidence associated with potentially catastrophic transverse loads.

We consider here the case of a vehicle of revolution, with spin, having low aerodynamic and inertial asymmetries.

We assume that it enters an atmosphere at rest above a nonrotating earth, flat and without gravity. The analysis will be made with the assumption of low incidence and sideslip angles. The aerodynamic model is assumed to be linear with respect to incidence and sideslip angles.

Aerodynamic asymmetries are represented by disturbance of this symmetrical basic model.

These asymmetries originate in defects of the external surface, either preexistent by construction or appearing during reentry by ablation under the effects of high heat flux. They are modeled by adding constant coefficients of forces and moments at zero incidence and sideslip to the basic aerodynamic coefficients of the symmetrical model.

Inertial asymmetries are either a lateral displacement of the center of gravity (CG offset) away from the aerodynamic axis of symmetry (corresponding to the nondisturbed model) or an angular deviation of the principal axis of roll inertia from the axis (principal axis misalignment). The amplitude of the misalignment θ_I is assumed sufficiently low to allow linear approximation of the sine and cosine lines and the simplification of the tensor of inertia. The effect of a lateral displacement of the aerodynamic center of pressure is similar to that of the CG offset and will not be studied specifically. The transverse moments of inertia are assumed equal (in practice the effect of small differences is negligible).

The equations of motion will be defined in an Eulerian reference frame fixed to the vehicle, with origin at the mass center and axis Gx parallel with aerodynamic axis of symmetry. The transverse axes direction (fixed relative to the vehicle) will be chosen in order to simplify modeling. The aerodynamic reference frame has its lateral axes parallel to the precedents and is centered at O on the aerodynamic axis of symmetry (O is the point of origin of defining the moments).

These assumptions allow the derivation of an approximate analytical solution. It is quite obvious that use of six degree-of-freedom code is required to study the general case.

11.1 Linear Equations

The general equations of evolution of the angular and linear momentum are written in the Eulerian reference frame of vehicle:

$$\dot{\vec{H}} + \vec{\Omega} \wedge \vec{H} = \vec{M}^A_{/G}$$

$$\dot{\vec{V}} + \vec{\Omega} \wedge \vec{V} = \frac{\vec{F}^A}{m}$$

We assume that the off-diagonal terms of the tensor of inertia are small compared with the diagonal terms:

$$(I_T - I_x)\sin\theta_I \cos\phi_I \approx I_T(1-\mu)\theta_y = \varepsilon_y \ ; \ (I_T - I_x)\sin\theta_I \sin\phi_I \approx I_T(1-\mu)\theta_z = \varepsilon_z$$

Thus the tensor of inertia is,

$$[I] = \begin{bmatrix} I_x & -\varepsilon_y & -\varepsilon_z \\ -\varepsilon_y & I_T & 0 \\ -\varepsilon_z & 0 & I_T \end{bmatrix}$$

The equation of the angular momentum is written as:

$$[I]\begin{bmatrix}\dot{p}\\ \dot{q}\\ \dot{r}\end{bmatrix} = \begin{bmatrix}L_{/G}\\ M_{/G}\\ N_{/G}\end{bmatrix} - \begin{bmatrix}p\\ q\\ r\end{bmatrix} \wedge [I]\begin{bmatrix}p\\ q\\ r\end{bmatrix}$$

By using expression of the matrix of inertia, we obtain:

$$\dot{p} = \frac{L_{/G}}{I_x} + \frac{(1-\mu)}{\mu}\left[\theta_y \dot{q} + \theta_z \dot{r} + p(\theta_z q - \theta_y r)\right] \qquad (11.1)$$

$$\dot{q} = \frac{M_{/G}}{I_T} + (1-\mu)\left\{pr + \theta_y \dot{p} - \theta_z(p^2 - r^2) + \theta_y qr\right\}$$

$$\dot{r} = \frac{N_{/G}}{I_T} - (1-\mu)\left\{pq - \theta_z \dot{p} - \theta_y(p^2 - q^2) + \theta_z qr\right\}$$

11.1 Linear Equations

While θ_y and θ_z are small, we neglect the roll acceleration terms in the pitch and yaw equations:

$$\dot{q} \approx \frac{M_{/G}}{I_T} + (1-\mu)\left\{pr - \theta_z\left(p^2 - r^2\right) + \theta_y qr\right\}$$
$$\dot{r} \approx \frac{N_{/G}}{I_T} - (1-\mu)\left\{pq - \theta_y\left(p^2 - q^2\right) + \theta_z qr\right\} \quad (11.2)$$

The expressions of the force and the aerodynamic moment around the origin O of the aerodynamic reference frame are:

$$F_x^A = -\bar{q} S_{ref} C_A$$
$$F_y^A \approx \bar{q} S_{ref}\left(-C_{N\alpha}\beta + C_{Y0}^A\right) = -\bar{q} S_{ref} C_{N\alpha}(\beta - \beta_0)$$
$$F_z^A \approx -\bar{q} S_{ref}\left(C_{N\alpha}\alpha + C_{N0}\right) = -\bar{q} S_{ref} C_{N\alpha}(\alpha - \alpha_0)$$
$$L_{/o} = \bar{q} S_{ref} L_{ref}\left(C_{l0} + C_{lp/o}\frac{p L_{ref}}{V}\right)$$
$$M_{/o} \approx \bar{q} S_{ref} L_{ref}\left(C_{m\alpha/o}\alpha + C_{m0/o} + C_{mq/o}\frac{q L_{ref}}{V}\right)$$
$$= \bar{q} S_{ref} L_{ref}\left(C_{m\alpha/o}(\alpha - \alpha_{e,o}) + C_{mq/o}\frac{q L_{ref}}{V}\right)$$
$$N_{/o} \approx \bar{q} S_{ref} L_{ref}\left(-C_{m\alpha/o}\beta + C_{n0/o} + C_{mq/o}\frac{r L_{ref}}{V}\right)$$
$$= \bar{q} S_{ref} L_{ref}\left(-C_{m\alpha/o}(\beta - \beta_{e,o}) + C_{mq/o}\frac{r L_{ref}}{V}\right)$$

Angles $\alpha_{e,o}$ and $\beta_{e,o}$ are the equilibrium trim values of incidence and sideslip, respectively, dependent on aerodynamic asymmetries, for a center of mass in O. α_0 and β_0 are the incidence at zero normal force and sideslip at zero lateral force, dependent on these same asymmetries. These angles are small compared to static trim angles and they will be neglected.

Note that the force and transverse moment related to roll rate (Magnus effect) are negligible in the case of the ballistic reentry vehicles, because roll rate is low.

The static moment around the center of gravity G located at $\vec{r}_G = \begin{bmatrix} x_G & y_G & z_G \end{bmatrix}$ in the aerodynamic frame is:

$$\vec{M}_{s/G} = \vec{M}_{s/O} - \vec{r}_G \wedge \vec{F}^A$$

Dynamic moment (damping) is transformed using the special rules defined in Sect. 4.3.2.2.

Thus we obtain the expression of the total aerodynamic moment around G:

$$L_{/G} \approx \bar{q} S_{ref} L_{ref} \left(C_{l0} + C_{lp/G} \frac{pL_{ref}}{V} \right) + \bar{q} S_{ref} C_{N\alpha} \left[y_G \alpha - z_G \beta \right]$$

$$M_{/G} \approx \bar{q} S_{ref} L_{ref} \left(C_{m\alpha/G}(\alpha - \alpha_e - \alpha_G) + C_{mqG} \frac{qL_{ref}}{V} \right)$$

$$N_{/G} \approx \bar{q} S_{ref} L_{ref} \left(-C_{m\alpha/G}(\beta - \beta_e - \beta_G) + C_{mqG} \frac{rL_{ref}}{V} \right)$$

where

$$\alpha_G = -z_G C_A \Big/ L_{ref} C_{m\alpha G} \quad ; \quad \beta_G = -y_G C_A \Big/ L_{ref} C_{m\alpha G} \quad (11.3)$$

are the static trim incidence and sideslip induced by CG offset, and

$$\alpha_e = \alpha_{e,G} = -\frac{C_{m0/G}}{C_{m\alpha/G}} \quad ; \quad \beta_e = \beta_{e,G} = \frac{C_{n0/G}}{C_{m\alpha/G}}$$

are the static trim incidence and sideslip corresponding to aerodynamic asymmetries.

With:

$C_{m\alpha/G}$, coefficient of gradient of pitching moment around the center of mass $\quad C_{m\alpha/G} = -\dfrac{x_G - x_{CP}}{L_{ref}} C_{N\alpha} < 0$

We now introduce these expressions into the equation of the angular momentum, while noting:

Coefficient of the pitch damping term	$m_q = -\bar{q} S_{ref} \dfrac{L_{ref}^2}{I_T V} C_{mq/G}$
Gradient of normal force	$N_\alpha = \bar{q} S_{ref} C_{N\alpha}$
Roll acceleration due to pure aerodynamic couples	$\dot{p}_0 = \bar{q} S_{ref} L_{ref} \dfrac{C_{l0}}{I_x}$
Coefficient of the roll damping term	$l_p = -\bar{q} S_{ref} \dfrac{L_{ref}^2}{I_T V} C_{lp/G}$

We obtain from (11.1) and (11.2):

11.1 Linear Equations

$$\dot{p} \approx \dot{p}_0 - l_p p + \frac{N_\alpha}{I_x} \left[y_G \alpha - z_G \beta \right] + \frac{(1-\mu)}{\mu} \left[\theta_y \dot{q} + \theta_z \dot{r} + p \left(\theta_z q - \theta_y r \right) \right] \tag{11.4}$$

$$\dot{q} \approx -\omega_n^2 (\alpha - \alpha_e - \alpha_G) - m_q q + (1-\mu) pr + (1-\mu) \left[-\theta_z \left(p^2 - r^2 \right) + \theta_y qr \right]$$

$$\dot{r} \approx \omega_n^2 (\beta - \beta_e - \beta_G) - m_q r - (1-\mu) pq + (1-\mu) \left[\theta_y \left(p^2 - q^2 \right) - \theta_z qr \right] \tag{11.5}$$

We will note the trim incidence and sideslip associated with principal axis misalignment, and the total trim incidence and sideslip:

$$\alpha_\theta = -\frac{(1-\mu)\theta_z p^2}{\omega_n^2}; \quad \beta_\theta = -\frac{(1-\mu)\theta_y p^2}{\omega_n^2} \tag{11.6}$$

We will restrain the influence of principal axis misalignment on the roll equilibrium state such that

$$\dot{\alpha} = \dot{\beta} = \dot{q} = \dot{r} = 0 \rightarrow \alpha_E, \beta_E, q_E, r_E$$

Thus we will neglect the influence of the quadratic terms in q and r in the last two equations, because equilibrium values of q and r are usually small with respect to roll rate p (this assumption will be checked *a posteriori*).

Using preceding assumptions, we obtain a linear system in α, β, q, r, $\dot{\alpha}$, $\dot{\beta}$, $\dot{q}$, and $\dot{r}$ for the angular momentum equation (according to what has just been said, these equations of evolution are correct from the point of view of principal axis misalignment only with respect to the equilibrium state):

$$\dot{q} = -\omega_n^2 (\alpha - \alpha_T) - m_q q + (1-\mu) pr$$
$$\dot{r} = \omega_n^2 (\beta - \beta_T) - m_q r - (1-\mu) pq \tag{11.7}$$

In the same way, the linear momentum equation is written as:

$$\begin{bmatrix} \dot{V}_x \\ \dot{V}_y \\ \dot{V}_z \end{bmatrix} + \begin{bmatrix} p \\ q \\ r \end{bmatrix} \wedge \begin{bmatrix} V_x \\ V_y \\ V_z \end{bmatrix} = -\frac{\bar{q} S_{ref}}{m} \begin{bmatrix} C_A \\ C_{N\alpha} (\beta - \beta_0) \\ C_{N\alpha} (\alpha - \alpha_0) \end{bmatrix}$$

Further derivations are,

$$\dot{V}_x + q V_z - r V_y = -\frac{\bar{q} S_{ref} C_A}{m}$$

$$\dot{V}_y + r V_x - p V_z = -\frac{\bar{q} S_{ref} C_{N\alpha}}{m} (\beta - \beta_0)$$

$$\dot{V}_z + p V_y - q V_x = -\frac{\bar{q} S_{ref} C_{N\alpha}}{m} (\alpha - \alpha_0)$$

Using small angles approximations:

$$\tan\alpha = \frac{V_z}{V_x} \approx \alpha; \quad \tan\beta = \frac{V_y}{V_x} \approx \beta$$

$$\dot\alpha \approx \frac{\dot V_z}{V_x} - \alpha \frac{\dot V_x}{V_x}; \quad \dot\beta \approx \frac{\dot V_y}{V_x} - \beta \frac{\dot V_x}{V_x}$$

$$V_x \approx V$$

When we divide the three initial equations by V_x, taking into account the approximations above, we obtain by neglecting the small terms of order higher than one,

$$\frac{\dot V}{V} + q\alpha - r\beta \approx -\frac{\bar q S_{ref} C_A}{mV} \tag{11.8}$$

$$\dot\beta - \frac{\bar q S_{ref} C_A}{mV}\beta + r - p\alpha \approx -\frac{\bar q S_{ref} C_{N\alpha}}{mV}(\beta - \beta_0)$$

$$\dot\alpha - \frac{\bar q S_{ref} C_A}{mV}\alpha + p\beta - q \approx -\frac{\bar q S_{ref} C_{N\alpha}}{mV}(\alpha - \alpha_0) \tag{11.9}$$

In order to obtain a more convenient form, we introduce complex numbers for incidence $\xi = \beta + i\alpha$ and transverse angular rate $\Omega = q + i\,r$. We obtain by linear combination of the pitch and yaw derivative equations,

$$\dot\Omega = i\omega_n^2 (\xi - \xi_T) - m_q \Omega - i(1-\mu)p\Omega \tag{11.10}$$

with

$$\xi_T = \beta_T + i\alpha_T; \quad \xi_0 = \beta_0 + i\alpha_0$$

In the same way we obtain by linear combination of the linear momentum equations along normal and lateral axes:

$$\dot\xi - \frac{A}{mV}\xi - i\Omega + ip\xi = -\frac{N_\alpha}{mV}(\xi - \xi_0) \tag{11.11}$$

While denoting $\ell_\alpha = \frac{N_\alpha - A}{mV}$ and $n_\alpha = \frac{N_\alpha}{mV}$, we arrange these two equations in a linear inhomogeneous differential system with variable complex coefficients:

$$\begin{bmatrix}\dot\xi \\ \dot\Omega\end{bmatrix} = \begin{bmatrix} -(\ell_\alpha + ip) & i \\ i\omega_n^2 & -(m_q + i(1-\mu)p) \end{bmatrix}\begin{bmatrix}\xi \\ \Omega\end{bmatrix} + \begin{bmatrix} n_\alpha \xi_0 \\ -i\omega_n^2 \xi_T \end{bmatrix} \tag{11.12}$$

Such is the form of the approximate linear differential system for the angular motion of a ballistic reentry vehicle with spin and small asymmetries.

To obtain a solution, we must combine the equations of roll evolution (11.4) together with those of velocity and altitude of the center of mass (11.8)

To determine the evolution of the center of mass location, we must project the components of velocity in the Eulerian frame $\vec{V} = \begin{bmatrix} v_x \approx V & v_y \approx \beta V & v_z \approx \alpha V \end{bmatrix}$ on the earth's observation frame and integrate them. To do this, we need to know the orientation of the Eulerian frame relative to the terrestrial axes and thus determine Euler angles by integration of angular rates.

At this stage, the system is obviously not simple enough to achieve an analytical solution and we need a last simplification step.

11.2 Instantaneous Angular Movement

Now we assume linear velocity, dynamic pressure, and roll rate are constant. This academic case represents the approximation of the motion of a model with spin fired horizontally with a gun in hyperballistic tunnel filled with a homogeneous gas, during which velocity variation can be considered negligible. Moreover, it allows to analysis of a short period of time for a reentry vehicle with constant roll rate. Let us note that constant roll rate requires, among other conditions, the absence of center of gravity offset.

In this case, coefficients of the first-order inhomogeneous differential system (11.12) giving the evolution of ξ and Ω are constant.

Eigen frequencies λ of the homogeneous system with constant coefficients are classically solution of the zero determinant equation:

$$\Delta = \begin{vmatrix} -(\ell_\alpha + ip) - \lambda & i \\ i\omega_n^2 & -(m_q + i(1-\mu)p) - \lambda \end{vmatrix} = 0$$

$$\Leftrightarrow \lambda^2 + F\lambda + G = 0 \qquad (11.13)$$

with,

$$F = \ell_\alpha + m_q + i(2-\mu)p$$
$$G = \omega_n^2 - (1-\mu)p^2 + m_q \ell_\alpha + ip\left((1-\mu)\ell_\alpha + m_q\right)$$

Note: Frequency or angular pulsation is used interchangeably. It obviously represents of the same physical entity, expressed either in Hertz or in radians/second.

While neglecting $\frac{\mu}{4}\left|p(\ell_\alpha - m_q)\right| << \omega_n^2$, we obtain the Eigen frequencies:

$$\begin{aligned} \lambda_+ &= \Lambda_+ + i\omega_+ \approx -\frac{\ell_\alpha + m_q}{2} + \frac{\mu p(m_q - \ell_\alpha)}{4\omega_a} + i\left[-\left(1 - \frac{\mu}{2}\right)p + \omega_a\right] \\ \lambda_- &= \Lambda_- + i\omega_- \approx -\frac{\ell_\alpha + m_q}{2} - \frac{\mu p(m_q - \ell_\alpha)}{4\omega_a} + i\left[-\left(1 - \frac{\mu}{2}\right)p - \omega_a\right] \end{aligned} \qquad (11.14)$$

with

$$\omega_a = \sqrt{\omega_n^2 + \left(\frac{\mu p}{2}\right)^2 - \left(\frac{\ell_\alpha - m_q}{2}\right)^2} \approx \sqrt{\omega_n^2 + \left(\frac{\mu p}{2}\right)^2}$$

11.2.1 Epicyclic Movement

The differential system being linear, for an observer fixed to the rotating axes, the general expression of the apparent motion is the weighted sum of the two frequency modes (in the absence of asymmetries):

$$\xi = \xi_+ e^{\Lambda_+ t} e^{i\omega_+ t} + \xi_- e^{\Lambda_- t} e^{i\omega_- t}$$

Now, let us consider an observer fixed to the vehicle axis of symmetry but motionless in roll (the reference frame of the observer has the same Gx axis as the vehicle, its transverse axes are such that $p_{obs} = 0$). These axes are called "aeroballistic"[VAU]. They turn at angular rate $\vec{\omega}_{obs} = \begin{bmatrix} -p & 0 & 0 \end{bmatrix}$ with respect to the vehicle axes. Thus one passes from the vehicle axes to the aeroballistic axes by a rotation ϕ around Ox such that $\dot{\phi} = -p$. Relations for a vector v in the transverse plane from the aeroballistic frame (index obs) to the vehicle frame are thus:

$$v = e^{i\phi} v_{obs}$$
$$\dot{v} = e^{i\phi} \dot{v}_{obs} + i\dot{\phi} e^{i\phi} v_{obs} = e^{i\phi} \left[\dot{v}_{obs} - i p v_{obs} \right]$$

Remarks:

1. We must not confuse ϕ with $-\varphi$ the usual Euler angle of the vehicle. Indeed we have $\dot{\phi} = -p$, whereas $\dot{\varphi} = p - \dot{\psi} \cos\theta$.
2. In terms of Euler angles of the aeroballistic frame, we have:

$$\theta_{obs} = \theta \quad ; \quad \psi_{obs} = \psi$$
$$\dot{\theta}_{obs} = \theta \quad ; \quad \dot{\psi}_{obs} = \dot{\psi}$$
$$p_{obs} = \dot{\varphi}_{obs} + \dot{\psi}_{obs} \cos\theta_{obs} = \dot{\varphi}_{obs} + \dot{\psi} \cos\theta = 0$$
$$\dot{\varphi} + \dot{\psi} \cos\theta = p$$
$$\Rightarrow \dot{\phi} = \dot{\varphi}_{obs} - \dot{\varphi} = -p$$
$$\Rightarrow \dot{\varphi}_{obs} = -\dot{\psi} \cos\theta$$

3. At low incidences $\cos\theta \approx 1$, $\dot{\varphi}_{obs} \approx -\dot{\psi} = -\dot{\psi}_{obs}$, the transverse planes of the vehicle and aeroballistic frame are very close to the plane normal to the center of mass velocity vector. Thus the transverse axes of the aeroballistic frame, which rotate at rate $\dot{\psi}_{obs} + \dot{\varphi}_{obs} \approx 0$ are quasi inertial, they represent approximately the point of view of a motionless observer.

Taking into account $\phi = -pt$, we obtain evolution of relative velocity and transverse angular rate in the aeroballistic frame:

$$\begin{bmatrix} \dot{\xi}_{obs} \\ \dot{\Omega}_{obs} \end{bmatrix} = \begin{bmatrix} -\ell_\alpha & i \\ i\omega_n^2 & -m_q + i\mu p \end{bmatrix} \begin{bmatrix} \xi_{obs} \\ \Omega_{obs} \end{bmatrix} + \begin{bmatrix} n_\alpha \xi_0 \\ -i\omega_n^2 \xi_T \end{bmatrix} e^{ipt}$$

11.2 Instantaneous Angular Movement

While neglecting the term $n_\alpha m_q \xi_0$, the system can be arranged in the more usual second-order differential form, similar to a dampened oscillator:

$$\ddot{\xi}_{obs} + F'\dot{\xi}_{obs} + G'\xi_{obs} \approx \lfloor \omega_n^2 \xi_T + i(1-\mu) p n_\alpha \xi_0 \rfloor e^{ipt} \quad (11.15)$$

with,

$$F' = \ell_\alpha + m_q - i\mu p$$
$$G' = \omega_n^2 + m_q \ell_\alpha - i\mu p \ell_\alpha$$

The new characteristic root equation is written as:

$$\lambda^2 + F'\lambda + G' = 0 \quad (11.16)$$

which gives the Eigen frequencies:

$$\lambda'_+ = \Lambda_+ + i\omega'_+ \approx -\frac{\ell_\alpha + m_q}{2} + \frac{\mu p (m_q - \ell_\alpha)}{4\omega_a} + i\left[\frac{\mu}{2}p + \omega_a\right]$$

$$\lambda'_+ = \Lambda_- + i\omega'_- \approx -\frac{\ell_\alpha + m_q}{2} - \frac{\mu p (m_q - \ell_\alpha)}{4\omega_a} + i\left[\frac{\mu}{2}p - \omega_a\right] \quad (11.17)$$

Thus, real parts of the Eigen frequencies (damping) are identical to the preceding ones. The imaginary parts are:

$$\omega' = \omega + p$$

11.2.1.1 Precession and Nutation Modes

While noting $p_r = \frac{\mu p}{2}$ as before, the preceding results give:

$$\omega'_+ \approx \frac{\mu}{2}p + \omega_a \approx p_r + \sqrt{\omega_n^2 + p_r^2} = p_r\left(1 + \text{sgn}(p_r)\sqrt{1 + \left(\frac{\omega_n}{p_r}\right)^2}\right)$$

$$\omega'_- \approx \frac{\mu}{2}p - \omega_a \approx p_r\left(1 - \text{sgn}(p_r)\sqrt{1 + \left(\frac{\omega_n}{p_r}\right)^2}\right) \quad (11.18)$$

At high altitudes, when p is positive and $\omega_n \to 0$, ω'_+ et ω'_- are, respectively, equal to the precession rates $\dot{\psi}_+$ and $\dot{\psi}_-$ determined before in the quasi-static approximation of initial decay of the incidence of a symmetrical reentry vehicle. Pulsation ω'_+ is positive and corresponds to precession initial conditions $\tau = 0$; $\varepsilon \neq 0$ of coning motion.

Pulsation ω'_- is negative and corresponds to nutation initial conditions $\tau \neq 0$; $\varepsilon = 0$ of coning motion.

As continuation of the precession coning mode, the corresponding aerodynamic mode at Eigen frequency ω'_+ should be called (when p is positive) precession mode. In the same way the mode with ω'_- Eigen frequency should be called nutation mode. According to this terminology, when p is negative, ω'_- becomes the precession frequency and ω'_+ the nutation frequency.

In fact, the most widely accepted aerodynamic definition assigns the nutation mode at highest Eigen frequency (in absolute value) in the aeroballistic axes and the precession to the lower frequency mode. This last definition is adopted here, which is in contradiction with gyroscopic terminology.

11.2.1.2 Evolution of the Eigen Frequencies

During the reentry, the Eigen frequencies (or pulsations) evolve with the dynamic pressure. Let us note $\Delta\omega = \omega_a - |p_r| \approx \sqrt{\omega_n^2 + p_r^2} - |p_r| \geq 0$, which is zero at the beginning of reentry. The pulsations of precession and nutation in the aeroballistic frame are written according to $\Delta\omega$:

$$\omega'_{nut} = 2p_r + \text{sgn}(p)\Delta\omega$$
$$\omega'_{prec} = -\text{sgn}(p)\Delta\omega \qquad (11.19)$$

Thus, in this reference frame, the *precession* frequency is always of *sign opposed to the roll rate*; it is *null initially* and growing with dynamic pressure. The nutation frequency always has same sign as roll rate; it is initially equal to the precession frequency of the coning motion and its module increases with dynamic pressure.

In the vehicle axes, the frequencies are:

$$\omega_{nut} = \omega'_{nut} - p = -p(1-\mu) + \text{sgn}(p)\Delta\omega$$
$$\omega_{prec} = \omega'_{prec} - p = -p - \text{sgn}(p)\Delta\omega$$

The pulsation of precession is always opposed sign to the roll rate and increases in absolute value; its initial value is $-p$. The pulsation of nutation is also initially of sign opposed to rolling, of initial value $-p(1-\mu)$, but its sign changes when:

$$\Delta\omega = \sqrt{\omega_n^2 + p_r^2} - |p_r| = |p|(1-\mu)$$
$$\Leftrightarrow p = p_c = \pm \frac{\omega_n}{\sqrt{1-\mu}} \qquad (11.20)$$

We will see later that p_c value, called critical roll frequency, plays an important role in the roll resonance and lock-in theory.

Figure 11.1 shows a typical evolution of the aeroballistic Eigen frequencies, with $p = 1$ Hz roll rate.

11.2 Instantaneous Angular Movement

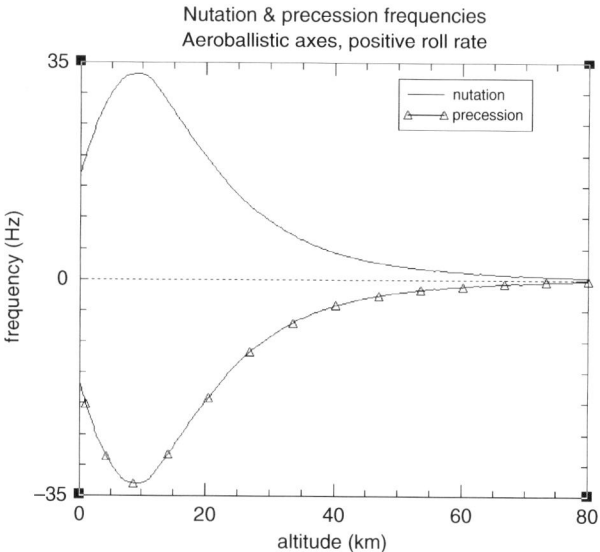

Fig. 11.1 Evolution of the Eigen frequencies with altitude

11.2.1.3 Representation of Angular Motion

For an observer fixed to the aeroballistic frame, the expression of the complex angle attack is epicycle:

$$\xi' = \xi'_+ e^{\Lambda_+ t} e^{i\,\omega'_+ t} + \xi'_- e^{\Lambda_- t} e^{i\,\omega'_- t}$$

Taking into account the approximate expressions of ω'_+ and ω'_-, this can be written:

$$\xi' = e^{i\,p_r\,t} \left[\xi'_+ e^{\Lambda_+ t} e^{i\omega_a t} + \xi'_- e^{\Lambda_- t} e^{-i\omega_a t} \right]$$

Incidence oscillation for short periods of time appears as elliptic oscillation slowly damped, whose axes turn slowly at angular rate p_r. We can also notice that the damping rates of the two modes are slightly different. This difference is negligible below 60 km altitude and also in current applications where $\frac{\mu\,p}{2\omega_a} \ll 1$.

For an observer fixed to vehicle axes, the evolution of the complex angle of attack becomes,

$$\xi = e^{-i\,p\left(1-\frac{\mu}{2}\right)t} \left[\xi_+ e^{\Lambda_+ t} e^{i\omega_a t} + \xi_- e^{\Lambda_- t} e^{-i\omega_a t} \right]$$

According to initial conditions of free coning motion at beginning of reentry (τ, ε), we obtain pure precession motion ($\tau \neq 0, \varepsilon = 0$), pure nutation ($\tau = 0, \varepsilon \neq 0$), or mixed ($\tau \neq 0, \varepsilon \neq 0$), represented in Figs. 11.2 and 11.3.

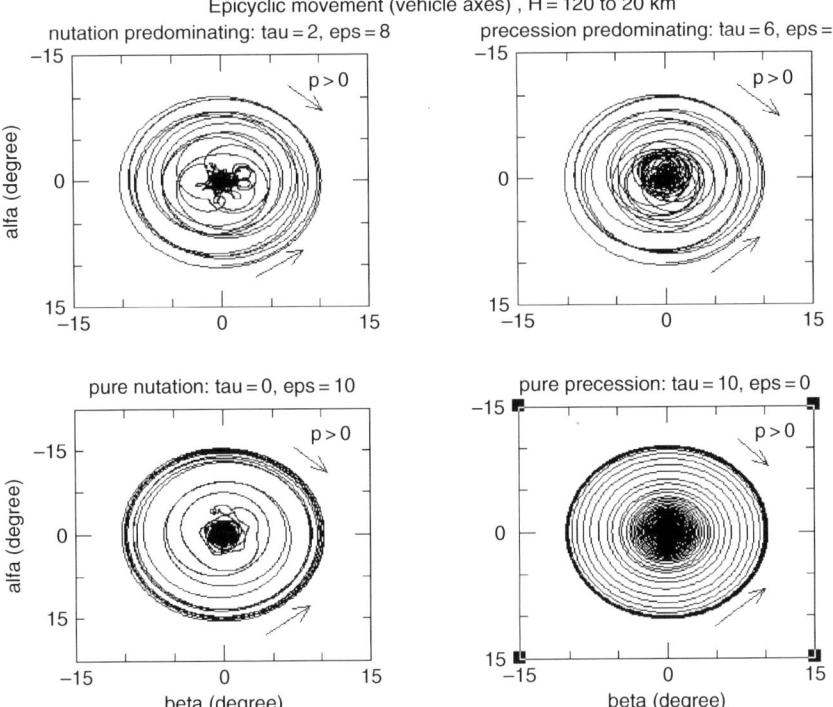

Fig. 11.2 Epicyclic motion (120–20 Km)

As we saw in the preceding paragraph, the apparent rotation movement of relative velocity vector in the vehicle axes is always initially opposed to roll rate. However, when the initial component of nutation is higher than the component of precession ($\varepsilon > \tau$), the later motion is accompanied by a change of direction of the apparent rotation of the complex angle of attack, as in the case of a pure nutation motion. When initial conditions are of precession type ($\varepsilon < \tau$), the motion proceeds without change of the direction of apparent rotation.

11.2.2 Tricyclic Movement

Now we study the general case of a vehicle with small aerodynamic and/or inertial asymmetries.

In the case of a symmetrical and dynamically stable vehicle ($\Lambda_+, \Lambda_- < 0$), instantaneous motion is centered on zero angle of attack and the amplitude tends asymptotically toward zero. When an asymmetry exists, an additional term of excitation is necessary in the second member of the equation of evolution.

11.2 Instantaneous Angular Movement

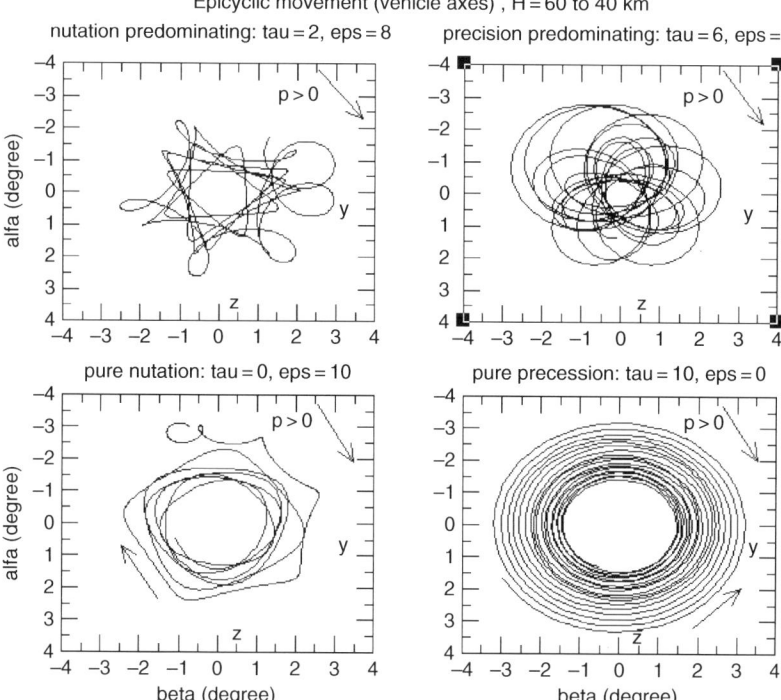

Fig. 11.3 Epicyclic motion (60–40 Km)

11.2.2.1 Trim Angle of Attack

Let us examine the asymptotic response to this excitation, referenced to an observer fixed to the vehicle. For him, asymmetries are fixed and result in constant force and moment terms (excitation at zero frequency). While posing $\dot{\xi} = 0, \dot{\Omega} = 0$ in the equation of evolution (11.12), the asymptotic response is a solution of the linear system:

$$\begin{bmatrix} -(\ell_\alpha + i\,p) & i \\ i\omega^2 & -(m_q + i\,(1-\mu)\,p) \end{bmatrix} \begin{bmatrix} \xi \\ \Omega \end{bmatrix} = \begin{bmatrix} -n_\alpha \xi_0 \\ i\omega^2 \xi_T \end{bmatrix}$$

The determinant of the system is equal to the coefficient G of the characteristic equation. Thus we obtain the equilibrium state of the rolling vehicle,

$$\xi_E = \frac{\omega_n^2 \xi_T + (m_q + i(1-\mu)p) n_\alpha \xi_0}{G} \approx \frac{\omega_n^2 \xi_T}{G}$$

$$\Omega_E = \frac{i\omega_n^2 [n_\alpha \xi_0 - (\ell_\alpha + ip)\xi_T]}{G} \approx \frac{p\omega_n^2 \xi_T}{G} = p\xi_E \quad (11.21)$$

The complex rolling trim angle can be arranged:

$$\xi_E \approx \frac{\xi_T}{1 - \text{sgn}(1-\mu)\left(\frac{p}{p_{cr}}\right)^2 + iD\left(\frac{p}{p_{cr}}\right)} \quad (11.22)$$

with

$$\xi_T = \xi_e + \xi_G + \xi_\theta \quad (11.23)$$

$$p_{cr} = \frac{\omega_n}{\sqrt{|1-\mu|}}, \text{ critical roll rate} \quad (11.24)$$

$$D = \frac{(1-\mu)\ell_\alpha + m_q}{\omega_n \sqrt{|1-\mu|}} \text{ damping parameter} \quad (11.25)$$

This can be expressed in the form:

$$\frac{\xi_E}{\xi_T} = A e^{i\Delta\Phi} \quad (11.26)$$

with

$$A = \frac{1}{\sqrt{(1-\varepsilon x^2)^2 + (Dx)^2}} \quad (11.27)$$

$$x = \frac{p}{p_{cr}}; \quad \varepsilon = \text{sgn}(1-\mu); \quad \sin\Delta\Phi = -ADx; \quad \cos\Delta\Phi = A\left(1 - \varepsilon x^2\right) \quad (11.28)$$

We must observe that $\xi_\theta = -x^2(\theta_y + i\theta_z)$ in ξ_T depends on roll rate, but we will continue to formally treat it like other static trim incidences.

- When $\mu \geq 1$, which is usually the case of entry capsules, $\varepsilon = -1$, $A \leq 1$, the effect of rolling motion is always to decrease the static trim. There is no problem of roll resonance for such vehicles. For this reason the further developments are devoted to vehicles having $\mu < 1$.
- When $\mu < 1$, in general the case of ballistic reentry vehicles, the behavior of rolling trim angle of attack with respect to roll rate is typically that of a resonant system.

11.2 Instantaneous Angular Movement

Resonance happens when $x = \frac{p}{p_{cr}} = \pm\sqrt{1 - \frac{D^2}{2}} \approx \pm 1$. Amplification factor A of the static trim angle ξ_T is then maximum, and its phase lag is $\mp\frac{\pi}{2}$, depending on the sign of roll rate:

$$\xi_{E,\max} = \frac{\xi_T}{D\sqrt{1 - \left(\frac{D}{2}\right)^2}} e^{-i\operatorname{sgn}(p)\frac{\pi}{2}}$$

When $p \ll p_{cr}$, amplification is close to unity and phase lag is negligible, $\xi_E \approx \xi_T$.

When $p \gg p_{cr}$, amplification is close to zero and phase lag is close to $180°$.

Figures 11.4 and 11.5 give the shape of the amplification and phase lag curves for the rolling trim according to x, for various damping D.

The physical origin of resonance appears more clearly with the point of view of an inertial observer, fixed to the aeroballistic frame. Indeed, the equations of evolution corresponds to a dynamic system of Eigen frequencies ω'_{nut} and ω'_{prec}, excited by an entry $\xi_T e^{i\,p\,t}$ rotating at frequency p.

Resonance possibly occurs when the frequency of excitation is close to inertial Eigen frequencies (11.19):

$$\omega'_{nut} = 2p_r + \operatorname{sgn}(p)\Delta\omega$$
$$\omega'_{prec} = -\operatorname{sgn}(p)\Delta\omega$$

As precession frequency is always of sign opposed to p, resonance can only occur in the nutation mode:

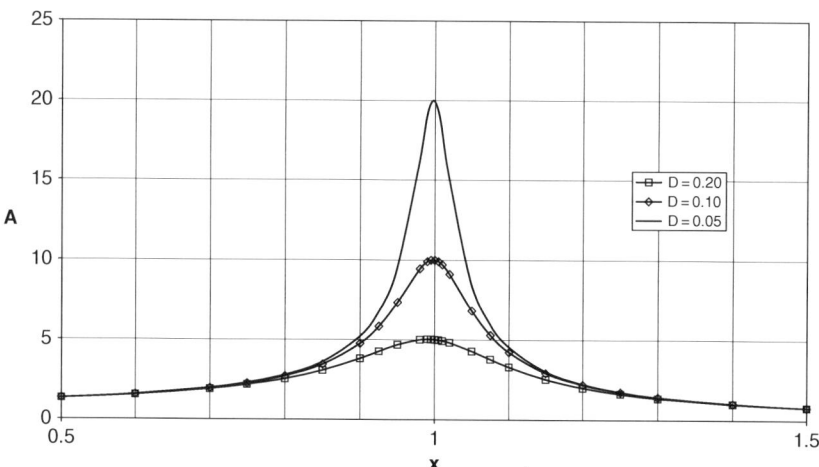

Fig. 11.4 Trim amplification factor

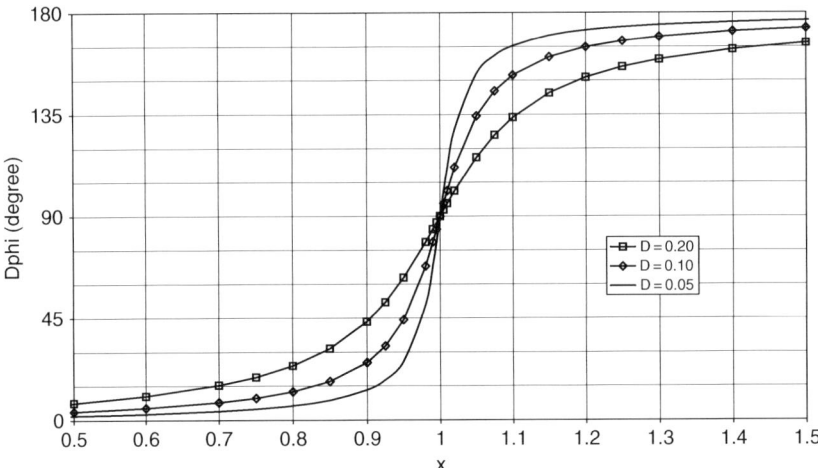

Fig. 11.5 Trim angle phase lag

$$\omega'_{nut} = \mu p + \text{sgn}(p)\Delta\omega$$

In the case of entry capsules, $\mu > 1 \rightarrow |p| < \mu |p| \leq |\omega'_{nut}|$, resonance is also impossible in the nutation mode.

We thus obtain the condition for $\mu < 1$:

$$p = \omega'_{nut} \Leftrightarrow p \approx \frac{\mu p}{2} \pm \sqrt{\omega_n^2 + \left(\frac{\mu p}{2}\right)^2} \Leftrightarrow (1-\mu)p^2 = \omega_n^2$$

The resonance occurs when the roll frequency is near the critical roll rate:

$$p \approx \pm \frac{\omega_n}{\sqrt{|1-\mu|}} = \pm p_{cr}$$

This highlights the reason why entry capsules with $\mu \geq 1$ are not subject to resonance. In fact, it is because their inertial Eigen frequencies can never be equal to roll frequency.

11.2.2.2 Lunar Motion

When the vehicle reaches its rolling trim incidence at roll rate p, we have:

$$\xi = \xi_E = \overline{\alpha}_E e^{i\phi_E} \approx cste$$
$$\Omega = \Omega_E = p\xi_E = p\overline{\alpha}_E e^{i\phi_E}$$

The relative velocity vector is motionless for the vehicle-fixed observer. From the

11.2 Instantaneous Angular Movement

point of view of an inertial observer, the CG relative velocity have a near constant direction and the axis of symmetry of the vehicle has a precession rate $\dot{\psi}$ around this relative velocity:

$$\dot{\psi}\sin\overline{\alpha}_E = |\Omega_E| = r_E = p\overline{\alpha}_E \quad \Rightarrow \quad \dot{\psi} \approx p$$

$$p = \dot{\varphi} + \dot{\psi}\cos\overline{\alpha}_E \quad \Rightarrow \quad \dot{\varphi} \approx p - p\cos\overline{\alpha}_E \approx p\frac{\overline{\alpha}_E^2}{2}$$

$$\omega_{\vec{V}} = \dot{\psi} + \dot{\varphi}\cos\overline{\alpha}_E \approx p\left(1 + \frac{\overline{\alpha}_E^2}{2}\right)$$

$$\omega_{\perp V} = \dot{\varphi}\sin\overline{\alpha}_E \approx p\frac{\overline{\alpha}_E^3}{2}$$

Rigorously, the direction of the rotation rate $\vec{\omega}$ of the vehicle is not exactly along the velocity $\vec{V}$. This vector also turns around $\vec{V}$ at the rate $\dot{\psi} = p$. However, in practice the misalignment is second order, i.e.:

$$\omega \approx \dot{\psi} \approx p$$
$$\dot{\varphi} = 0$$

Euler roll angle φ is constant: the vehicle turns on itself at roll rate p, with always the same trim meridian Z_E to the velocity vector. The motion is of the same nature as the apparent rotation motion of the moon around the earth, always presenting the same face. The lunar motion is typical of the final phase of ballistic reentry vehicles, under 15 km altitude. Let us assume that the rolling trim angle is in the meridian plane Oz_E of the vehicle, we observe the motion represented in Figure 11.6.

We will notice that the above expression of transverse angular rate at equilibrium is small compared to p. This justifies *a posteriori* the assumption to neglect q^2, r^2, and qr relative to p^2 at equilibrium to estimate the influence of the principal axis misalignment.

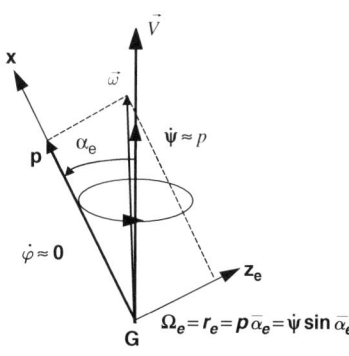

Fig. 11.6 Rolling trim motion

11.2.2.3 Representation of Angular Motion

The general motion in the presence of asymmetries is the sum of the transitory response to initial conditions and of the asymptotic trim solution. The apparent motion referenced to the aeroballistic frame is:

$$\xi' = e^{i\,p_r t}\left[\xi'_+ e^{\Lambda_+ t} e^{i\omega_a t} + \xi'_- e^{\Lambda_- t} e^{-i\omega_a t}\right] + \xi_E e^{i\,p\,t}$$

This motion of the tricycle type is shown in Fig. 11.7. The equilibrium motion corresponds to the lunar motion such that the vehicle axis turns around the relative velocity vector at the roll rate.

For an observer fixed to vehicle axis, the evolution of the complex incidence becomes,

$$\xi = e^{-i\,p(1-\frac{\mu}{2})t}\left[\xi_+ e^{\Lambda_+ t} e^{i\omega_a t} + \xi_- e^{\Lambda_- t} e^{-i\omega_a t}\right] + \xi_E.$$

This motion is of the epicycle type with origin offset ξ_E relative to G.

When $t \to \infty$, $\Lambda_+ < 0$, $\Lambda_- < 0$, the bracketed term vanishes and the vehicle reaches its rolling trim, which corresponds to a constant complex incidence ξ_E.

11.2.2.4 Critical Roll Frequency and Resonance

The critical rolling frequency is

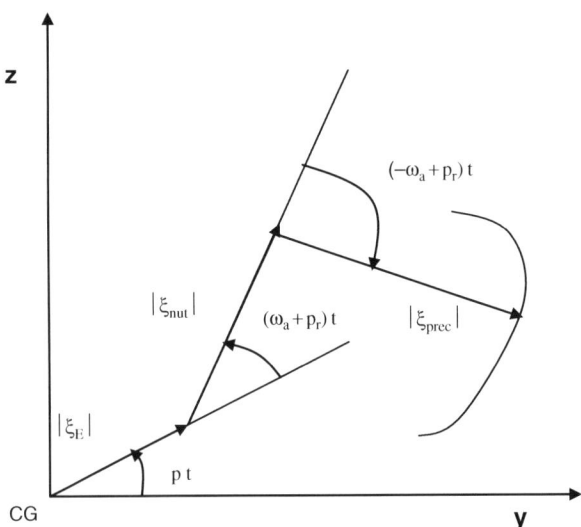

Fig. 11.7 Tricyclic motion

11.2 Instantaneous Angular Movement

$$p_{cr} = \frac{\omega_n}{\sqrt{1-\mu}}$$

where the natural pitch and yaw frequency of the vehicle is

$$\omega_n = \left(-\bar{q} S_{ref} L_{ref} \frac{C_{m\alpha/G}}{I_T}\right)^{\frac{1}{2}}$$

During the reentry, it increases with dynamic pressure and, depending on the trajectory, generally reaches a maximum before impact. Resonance conditions can be encountered more than once during the reentry. Typically, a first resonance takes place at high altitude (usually between 40 and 60 km altitude, Fig. 11.8) as the critical frequency increases to the level of the initial roll rate. A second resonance may then be crossed at low altitude (typically a few kilometers) during the decrease of the critical frequency combined with increased roll rate.

In the absence of coupling between pitch, yaw, and roll motion by the means of the CG offset, resonance crossing has only a transitory amplification of the effects of any asymmetry. The rolling trim incidence is temporarily amplified to an increase of total incidence and associated normal load factor. Figure 11.9 represents such a case of amplification of the trim associated with a 0.5° principal axis misalignment.

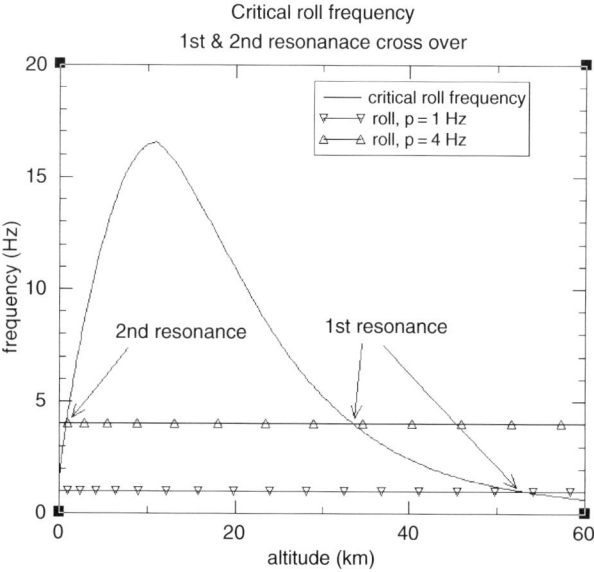

Fig. 11.8 Evolution of the critical roll frequency

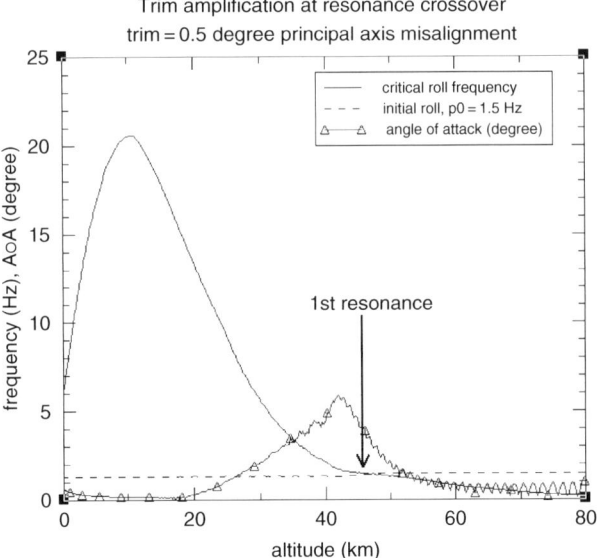

Fig. 11.9 Trim amplification at first resonance

11.3 Real Angular Motion

The preceding solutions were using two principal assumptions:

> Approximation at small angles of attack and linear dependence of the aerodynamic coefficients
> Constant velocity, dynamic pressure, and roll rate.

In reality these last parameters evolve during the reentry, as well as the coefficients of the equation of evolution. The corresponding effect during the accelerated phase could be analyzed for a vehicle with or without spin, thanks to other simplifying assumptions. In the case of an exponential evolution of the natural pitch and yaw pulsation, it is possible to obtain an "analytical" solution, analogous to that obtained in the case without spin, based on "special" functions. Rather than attempting tedious developments to finally obtain a solution with limited validity, we will restrain ambition and give a heuristic solution. This solution corrects the stationary solution of the preceding paragraph to take account of a slow evolution of these coefficients. This correction is based on the solution obtained by the method using energy of rotation in the case of the accelerated reentry and does not have rigorous theoretical justification. Thus, we obtain in the aeroballistic frame:

$$\xi' = \left(\frac{\omega_{n0}}{\omega_n}\right)^{\frac{1}{2}} \left\{ \xi_+ e^{\int_0^t \Lambda_+ d\tau} e^{i \int_0^t (p_r + \omega_a) d\tau} + \xi_- e^{\int_0^t \Lambda_- d\tau} e^{i \int_0^t (p_r - \omega_a) d\tau} \right\} + \xi_E e^{i \int_0^t p d\tau}$$

11.3 Real Angular Motion

This approximation must be handled with care and has significance only for limited durations while the vehicle is subjected to no high frequency disturbance (i.e., same order of magnitude or higher than the nutation or precession frequencies).

In this last case, only an exact numerical six degree-of-freedom solution is able to provide the realistic response of the vehicle. The effect of a disturbance at non-negligible frequency (for example, roll resonance crossing) involves in general a redistribution of energy between the nutation and precession components and thus a basic change of motion (such a phenomenon will be developed in Chap. 14). However, this approximation shows that even during the decelerated phase, the density damping plays a significant role. In fact, it is rather now a "dynamic pressure" damping since the variation of velocity is no longer negligible. As stated earlier, following maximum dynamic pressure and natural pitch frequency, their decrease can cause a destabilizing effect, compensated by the stabilizing effect of the gradient of lift force and any aerodynamic pitch damping moment of the vehicle.

Principal phenomena likely to disturb the evolution of the incidence during reentry are, in addition to the effect of external shape and mass asymmetries:

- Asymmetric progression of laminar/turbulent transition front on the reentry vehicle. It can be accompanied by destabilizing or stabilizing transverse moments.
- Effects related to the gas injection (blowing) in the boundary layer accompanying pyrolize and ablation of the heat shield. According to the thermal lag of heat shield material with respect to the heat flux and the roll rate, the corresponding moments can lead to a dynamic instability. In fact, these phenomena are generally associated with the progression of the transition and cannot be separated.
- The crossing of an unstable aerodynamic regime (e.g., sometimes in rarefied flow)
- An evolution of the center of pressure location related to a sudden change of aerodynamic configuration. The rapid evolution of the natural pitching frequency can involve a step increase or reduction in the incidence.
- The crossing of atmosphere layer with strong vertical gradient of the horizontal component of wind. Vehicles having low terminal velocity are very sensitive to this phenomenon.

This list is not exhaustive since other phenomena encountered during flight are not yet clearly explained. All the listed phenomena are likely to involve fast buildup or reductions in the angle of attack. Another potential phenomenon relates to non-linearity of the aerodynamic stability moment with respect to incidence. Indeed, for some aerodynamic shapes, one can observe a forward motion of the center of pressure when the incidence increases. When the separation from the missile induces high initial angle of attack, this may be very constraining for the center of mass location, which must guarantee a positive static margin during the reentry. For other configurations, one can observe the opposite phenomenon, i.e., an aft motion of the center of pressure with incidence at high Mach number. If packaging constraints are severe, a slight negative static margin near zero angle of attack at high Mach number can be permitted, provided it again becomes positive at lower altitudes and as the dynamic pressure increases. A more detailed analysis of some of these phenomena is done in the Chaps. 13 and 14. In Chap. 12, we will address the behavior of slightly asymmetrical vehicles when they cross roll resonance and more particularly the roll-lock-in phenomenon.

Chapter 12
Roll-lock-in Phenomenon

In Sect. 11.2, we studied the reentry of a ballistic RV with constant roll rate and highlighted the possibility, for $\mu < 1$, of a resonant amplification of the static trim angle related to asymmetries, according to the parameter $x = p/p_{cr}$.

This phenomenon is induced by the coupling between rolling motion at frequency p and pitch and yaw motion at frequency ω_{nut} and ω_{prec}. If there is a CG offset (y_G, z_G), transverse aerodynamic forces related to incidence and sideslip create a rolling moment. This applies a coupling of pitch and yaw with roll and thus closing a reaction loop. When adequate conditions are met by the parameters of the loop, a possible consequence of the crossing of resonance is a true control of roll rate on the critical frequency. This phenomenon, named "roll-lock-in" since it results in permanent resonance conditions, is likely to maintain a high rolling trim incidence at altitudes of high dynamic pressure.

This incidence may cause demise of the vehicle under very high lateral loads and heat flux. To study this phenomenon, we will use the equations of evolution established in the general case for the complex incidence and angular rates in the preceding chapter. This time, we will let the dynamic pressure evolve, as well as the roll rate according to the rolling moment.

In addition, we assume that the RV is initially at rolling trim, and its response time in pitch and yaw is negligible compared to variation of coefficients of the differential system (i.e., the dynamic pressure, the roll rate, etc). We will see later that this classical quasi-static modeling may result in poor prediction. Hence, we must remind the results are only qualitative. Under these conditions, we assume the vehicle remains close to rolling trim, and it results $\xi \approx \xi_E$, $\Omega \approx \Omega_E$:

$$\xi_E = \beta_E + i\alpha_E \approx \frac{\xi_T}{1 - x^2 + iDx}$$
$$\Omega_E = p\xi_E$$

where the complex static trim angle is $\xi_T = \xi_e + \xi_G + \xi_\theta$.

While $\dot{q} = \dot{r} = 0$, $\alpha = \alpha_E$, $\beta = \beta_E$, $q_E = p\beta_E$ and $r_E = p\alpha_E$, (11.4) gives the quasi-static roll acceleration:

$$\dot{p} \approx \dot{p}_0 - l_p p + \frac{N_\alpha}{I_x}\left[y_G \alpha_E - z_G \beta_E\right] + \frac{(1-\mu)}{\mu} p^2 \left(\theta_z \beta_E - \theta_y \alpha_E\right) \qquad (12.1)$$

From (11.22) to (11.28), we have:

$$\xi_E = A^2 \left[\left(1 - x^2\right) - i Dx \right] \xi_T$$

with

$$A^2 = \frac{1}{\left(1 - x^2\right)^2 + (Dx)^2}. \quad (12.2)$$

The incidence and sideslip angles at pitch equilibrium with spin are:

$$\alpha_E = A^2 \left[(1 - x^2)\alpha_T - D x\beta_T \right]; \beta_E = A^2 \left[D x\alpha_T + \left(1 - x^2\right) \beta_T \right] \quad (12.3)$$

with $\alpha_T = \alpha_e + \alpha_G + \alpha_\theta$; $\beta_T = \beta_e + \beta_G + \beta_\theta$.

Now, we will examine the different cases of combined asymmetries.

We will use the roll (12.1) assuming pure rolling moment $\dot{p}_0$ and damping l_p are negligible:

$$\dot{p} \approx \frac{N_\alpha}{I_x} \left[y_G \alpha_E - z_G \beta_E \right] + \frac{(1 - \mu)}{\mu} p^2 \left(\theta_z \beta_E - \theta_y \alpha_E \right) \quad (12.4)$$

12.1 Association of Aerodynamic Asymmetry and CG Offset

We choose Oz axis along the CG offset and define Φ angles as shown in Fig. 12.1.

Asymmetries are small by hypothesis, hence zero lift angles are assumed negligible compared with trim angles, $|\xi| \gg |\xi_0|$, and we neglect ξ_0 in the equations.

The simplified rolling equation becomes:

$$\dot{p} = -\frac{z_G N_\alpha \beta}{I_X} = \frac{z_G F_Y}{I_X} = -\frac{z_G |F| \sin \Phi}{I_X}$$

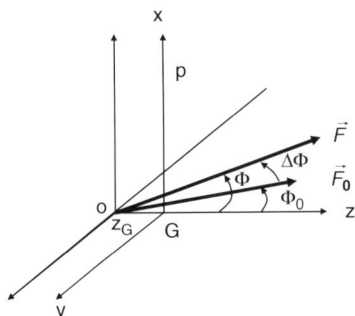

Fig. 12.1 Definition of phase lag angle for lateral force

12.1 Association of Aerodynamic Asymmetry and CG Offset

where Φ is angle of the lateral rolling trim force F at roll rate p. While $F = -N_\alpha \xi_E$, $\Phi = \Phi_0 + \Delta\Phi$, and $\xi_T = \xi_e + \xi_G$, (11.22) gives,

$$\frac{F}{F_0} = \frac{\xi_E}{\xi_e + \xi_G} = A e^{i\Delta\Phi}$$

with

$$A = \frac{1}{\sqrt{(1-x^2)^2 + (Dx)^2}}$$

$$\sin \Delta\Phi = -A\,D\,x; \quad \cos \Delta\Phi = A(1-x^2); \quad x = \frac{p}{p_{cr}}$$

$$F_0 = |F_0|\, e^{i(\frac{\pi}{2}+\Phi_0)} = -N_\alpha \left[|\xi_e|\, e^{i(\frac{\pi}{2}+\Phi_e)} + |\xi_G|\, e^{i\frac{\pi}{2}}\right]$$

$$|\xi_G| = -\frac{z_G C_A}{L_{ref} C_{m\alpha/G}} = \frac{z_G C_A}{(x_G - x_{cP}) C_{N\alpha}}$$

$$|F_0| = N_\alpha |\xi_e| \sqrt{1 + \left(\frac{|\xi_G|}{|\xi_e|}\right)^2 + 2\frac{|\xi_G|}{|\xi_e|} \cos \Phi_e}$$

$$\sin \Phi_0 = -\frac{\sin \Phi_e}{\sqrt{1 + \left(\frac{|\xi_G|}{|\xi_e|}\right)^2 + 2\frac{|\xi_G|}{|\xi_e|} \cos \Phi_e}}$$

$$\cos \Phi_0 = -\frac{\left[\cos \Phi_e + \frac{|\xi_G|}{|\xi_e|}\right]}{\sqrt{1 + \left(\frac{|\xi_G|}{|\xi_e|}\right)^2 + 2\frac{|\xi_G|}{|\xi_e|} \cos \Phi_e}}$$

Aerodynamic asymmetry being independent of z_G, static trim angle of the combination can have any orientation Φ_0.

The rolling equation is then:

$$\dot{p} = -\frac{z_G |F_0| A \sin(\Phi_0 + \Delta\Phi)}{I_x}$$

$$= \frac{z_G |F_0| \left[D x \cos \Phi_0 - (1-x^2) \sin \Phi_0\right]}{I_x \left((1-x^2) + D^2 x^2\right)} = \dot{p}_{0\,max} G$$

where $\dot{p}_{0\,max} = \frac{z_G |F_0|}{I_x}$ is the maximum "virtual" static roll acceleration corresponding to the CG offset associated to the static trim incidence, and $G = -A \sin \Phi$ the amplification of roll-acceleration in the vicinity of resonance.

12.1.1 Equilibrium on Critical Frequency

At resonance we have $x_{cr} = \frac{p}{p_{cr}} = \pm\sqrt{1 - \frac{D^2}{2}}$

$$G = \frac{[Dx \cos \Phi_0 - (1 - x^2) \sin \Phi_0]}{\left((1 - x^2)^2 + D^2 x^2\right)} = \frac{\left[\pm D\sqrt{1 - \frac{D^2}{2}} \cos \Phi_0 - \frac{D^2}{2} \sin \Phi_0\right]}{D^2 \left(1 - \left(\frac{D}{2}\right)^2\right)}$$

Let us first assume the critical frequency p_{cr} is constant. To maintain resonance as equilibrium, it is necessary to verify $\dot{p} = \dot{p}_{cr} = 0 \Leftrightarrow G = 0$, which gives:

$$\tan \Phi_0 = \pm \frac{2\sqrt{1 - \frac{D^2}{2}}}{D} \Leftrightarrow \Phi_0 \approx \pm \left|\frac{\pi}{2} - arctg\left(\frac{D}{2}\right)\right|$$

Thus there are two possible orientations of the aerodynamic trim force relative to CG offset, which lead to resonance equilibrium, close to $\pm\frac{\pi}{2}$. These configurations correspond to aerodynamic asymmetries such that:

$$\cot \Phi_0 = \frac{\cos \Phi_e + \frac{|\xi_G|}{|\xi_e|}}{\sin \Phi_e} \approx \pm \frac{D}{2}$$

For low offset values, they are close to normal to CG offset.

Asymmetries are named "out-of-plane" or "in-the-plane", depending on the angle with the CG offset.

There is another possible roll-trim for "in-the-plane" asymmetries, such $\sin \Phi_0 = 0 \rightarrow x = 0$, which does not correspond to resonance.

To have a stable lock-in on resonance it is not sufficient to have a zero moment at $p \approx p_{cr}$, moreover it is necessary that a moment exists which restores equilibrium when variations occur. This implies,

$$\Delta \dot{p} \approx K (p - p_{cr}), \text{ with } K = \left.\frac{\partial \dot{p}}{\partial p}\right|_{p \approx p_{cr}} < 0 \Leftrightarrow \left.\frac{\partial G}{\partial x}\right|_{x \approx x_{cr}} < 0$$

This gives the necessary condition:

$$\left.\frac{\partial G}{\partial x}\right|_{x=x_{cr}} = \left.\frac{D \cos \Phi_0 \left(1 + x^2 \left(2 - D^2\right) - 3x^4\right) + 2x \sin \Phi_0 \left(D^2 - 1 + 2x^2 - x^4\right)}{\left((1 - x^2)^2 + D^2 x^2\right)^2}\right|_{x=x_{cr}} < 0$$

$$\Leftrightarrow \left[D \cos \Phi_0 \left(1 + x^2 \left(2 - D^2\right) - 3x^4\right) + 2x \sin \Phi_0 \left(D^2 - 1 + 2x^2 - x^4\right)\right]\bigg|_{x=x_{cr}} < 0$$

12.1 Association of Aerodynamic Asymmetry and CG Offset

For out-of-plane asymmetries, the condition is:

$$\cos \Phi_0 \approx 0; \sin \Phi_0 \approx \pm 1 \Rightarrow \pm 2x \left(D^2 - 1 + 2x^2 - x^4\right)\bigg|_{x=x_{cr}} < 0$$

with

$$x_{cr} = \frac{p}{p_{cr}} = \pm\sqrt{1 - \frac{D^2}{2}}$$

That is to say,

$$2 \sin\left(\Phi_0 = \pm\frac{\pi}{2}\right) x_{cr} D^2 \left(1 - \left(\frac{D}{2}\right)^2\right) < 0$$

This is verified for:

$$\Phi_0 \approx \frac{\pi}{2}; \quad x_{cr} = \frac{p}{p_{cr}} = -\sqrt{1 - \frac{D^2}{2}} \approx -1$$

$$\Phi_0 \approx -\frac{\pi}{2}; \quad x_{cr} = \frac{p}{p_{cr}} = +\sqrt{1 - \frac{D^2}{2}} \approx +1$$

Thus, for out-of-plane asymmetries, stable roll lock-in at critical frequency is possible when:

- $\Phi_0 \approx -\frac{\pi}{2}$, on $+p_{cr}$
- $\Phi_0 \approx \frac{\pi}{2}$, on $-p_{cr}$

For in-the-plane asymmetries, stable roll equilibrium is possible only if $x = 0$ when:

$$D \cos \Phi_0 < 0 \rightarrow \Phi_0 = \pi.$$

Thus roll lock-in is not possible for this kind of asymmetries.

These results are illustrated in Fig. 12.2, which shows the rolling moment amplification factor G according to $x = \frac{p}{p_{cr}}$ for $\Phi_0 = 0, \frac{\pi}{2}, \pi, 3\frac{\pi}{2}$ and for $D = 0.2$.

For $\Phi_0 = \frac{\pi}{2}$, we can observe the roll-lock-in domain with high stability around $x = -1$, and a similar condition at $\Phi_0 = 3\frac{\pi}{2}$ around $x = +1$.

- Orientations $\Phi_0 = 0$ and $\Phi_0 = \pi$ at $x = \pm 1$ show strong amplification of rolling acceleration, but without possibility of equilibrium. Orientation $\Phi_0 = \pi$ at $x = 0$ (i.e., at p = 0) gives a low stability equilibrium.

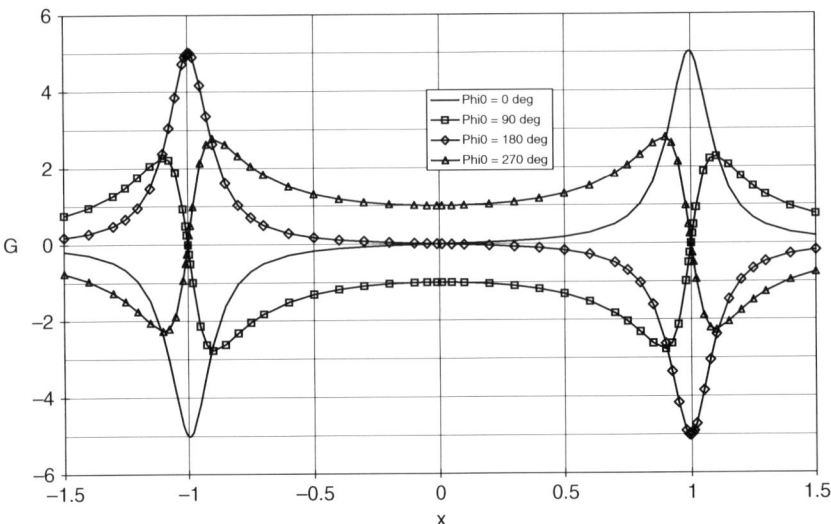

Fig. 12.2 Rolling acceleration amplification

12.1.2 Lock-in Near Resonance

We have studied the possibility of a lock-in exactly at the critical frequency; however, lock-in may also occur near the critical frequency, i.e., verifying $p = x_r p_{cr}$ with x_r constant near unity.

The necessary conditions to maintain balance around this value are:

$$G(\Phi_0, x_r) = 0; \left.\frac{\partial G}{\partial x}\right|_{x_r} < 0$$

The first condition gives:

$$\sin \Phi = 0 \Leftrightarrow \Phi_0 = -\Delta\Phi + k\pi$$
$$\Leftrightarrow \sin \Phi_0 = ADx_r; \cos \Phi_0 = A(1 - x_r^2)$$

or

$$\sin \Phi_0 = -ADx_r; \cos \Phi_0 = -A(1 - x_r^2)$$

When we introduce these values in the expression of the second condition, we obtain:

$$\pm AD\left[x_r^6 - (1 - D^2)x_r^4 - (1 - D^2)x_r^2 + 1\right] < 0$$

12.1 Association of Aerodynamic Asymmetry and CG Offset

Naming $u = x_r^2 > 0$ and factorizing, we obtain a cubic inequality:

$$\pm (u+1)\left(u^2 - \left(2 - D^2\right)u + 1\right) < 0$$

The first term has only one real root $u = -1$ and two imaginary roots. Thus, as it has a constant sign in the domain $u > 0$ and the condition is verified only for the case of negative sign, corresponding to Φ_0 values such that:

$$\sin \Phi_0 = -ADx_r; \quad \cos \Phi_0 = -A(1 - x_r^2)$$

Hence, there is a possibility of steady roll equilibrium at any arbitrary value x_r, for Φ_0 given by the preceding relation.

Figure 12.3 gives, according to Φ_0, amplification factor A and value x_r of reduced roll rate in the case of $D = 0.2$.

We note that there are two angular domains giving a rolling equilibrium with high amplification of the trim angle:

- The first is centered on 90°, which corresponds $x_r = p/p_{cr}$ near -1.
- The second is centered on 270°, which corresponds to x_r near 1.

These domains correspond to "out-of-plane" asymmetries.

For in plane asymmetries, there are two angular ranges, centered on 0° and 180°; they correspond either to the $x_r = p/p_{cr}$ value around 0 or to values appreciably higher than 1 and by consequence do not give a resonant amplification.

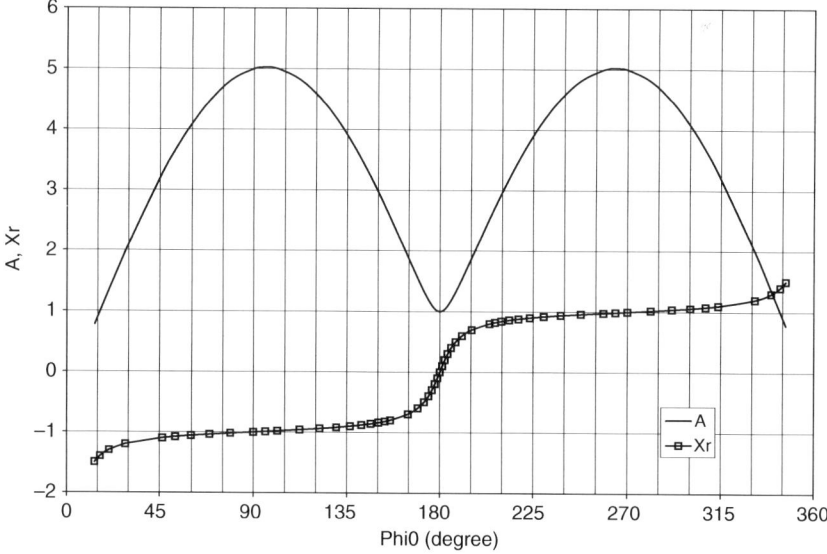

Fig. 12.3 Amplification sensitivity on Φ_0

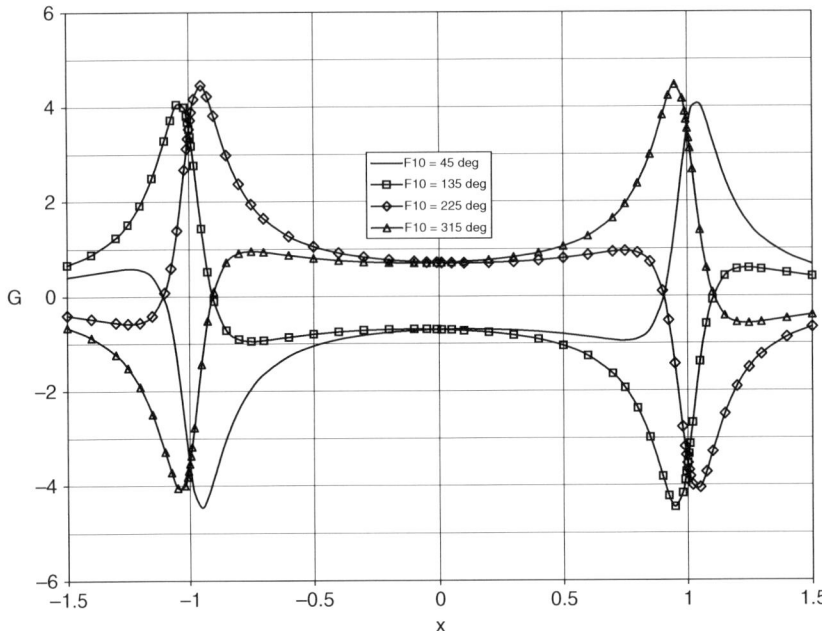

Fig. 12.4 Amplification of roll acceleration for various orientations of asymmetries

Thus, the roll-lock-in phenomenon requires "out-of-plane" asymmetries, while "in-the-plane" asymmetries are likely to give a rolling equilibrium around zero roll rate or high values of trim roll rate. Figure 12.4 shows the possibilities of roll lock-in at conditions near but distinct from resonance, for $\Phi_0 = \frac{\pi}{4}, 3\frac{\pi}{4}, 5\frac{\pi}{4}, 7\frac{\pi}{4}$ corresponding to D = 0.2.

We note that the equilibrium roll rates are $p_r \approx \pm 0.9 p_{cr}$; $p_r \approx \pm 1.1 p_{cr}$

The steady equilibriums are:

$$\Phi_0 = \frac{\pi}{4} \rightarrow p_r \approx -1.1 p_{cr}$$

$$\Phi_0 = 3\frac{\pi}{4} \rightarrow p_r \approx -0.9 p_{cr}$$

$$\Phi_0 = 5\frac{\pi}{4} \rightarrow p_r \approx +0.9 p_{cr}$$

$$\Phi_0 = 7\frac{\pi}{4} \rightarrow p_r \approx +1.1 p_{cr}$$

Corresponding amplification factors of trim angle are close to 3.2 for $p_r \approx \pm 1.1 p_{cr}$ and 3.9 for $p_r \approx \pm 0.9 p_{cr}$.

In addition, we can note that the domain of stable equilibrium in the vicinity of resonance is reduced. Equilibrium corresponding to the maximum of resonance effects are close to $p = \pm p_{cr}$.

12.1.3 Variable Critical Frequency

In the preceding analysis, we considered the possibilities of roll lock-in for the case of a constant critical roll rate. In this case, we need a zero mean rolling acceleration to maintain lock-in.

The stability condition:

$$\left.\frac{\partial \dot{p}}{\partial p}\right|_{p_r = x_r p_{cr}} = \dot{p}_{0\,max} \frac{1}{p_{cr}} \left.\frac{\partial G}{\partial x}\right|_{x_r} < 0,$$

does not imply a condition on the level of asymmetries.

When p_{cr} evolve, the stability condition remains unchanged, but to maintain equilibrium around value $p_r = x_r p_{cr}$ at constant x_r requires a nonzero mean rolling acceleration $\bar{\dot{p}} = x_r \dot{p}_{cr} \neq 0$. This mean rolling acceleration must verify:

$$\bar{\dot{p}} = \dot{p}_{0\,max} \bar{G}(x, \Phi_0) = x_r \dot{p}_{cr}$$

To maintain resonance, it is necessary that the stability domain around x_r contain a value of x such that:

$$|\dot{p}_{0\,max} G(x, \Phi_0)| \geq |x_r \dot{p}_{cr}|$$

- Moreover, around this value the sign of $\dot{p}$ must be identical to that of $x_r \dot{p}_{cr}$.
- If the maximum module of the moment generated by combination of asymmetries is insufficient, a fortiori it will always be so for nonoptimum conditions, i.e., one must satisfy the necessary conditions:

$$\max \left[|\dot{p}_{0\,max} G(x, \Phi_0)|\right] \geq |x_r \dot{p}_{cr}| \,; \, sign\left[\dot{p}_{0\,max} G_{ext}(x_{ext}, \Phi_0)\right] = sign\,(x_r \dot{p}_{cr})$$

12.1.4 Criterion for Out-of-plane Asymmetries

Optimum roll amplification factor G corresponds to:

$$\frac{\partial G}{\partial x} = 0 \Leftrightarrow \frac{D \cos \Phi_0 \left(1 + x^2 \left(2 - D^2\right) - 3x^4\right) + 2x \sin \Phi_0 \left(D^2 - 1 + 2x^2 - x^4\right)}{\left(\left(1 - x^2\right)^2 + D^2 x^2\right)^2} = 0$$

For $\cos \Phi_0 = \pm 1 \Leftrightarrow \Phi_0 = 0 \text{ or } \Phi_0 = \pi$, solutions corresponding to G optimum have analytical expressions:

$$x_{opt} \approx \sqrt{1 - \frac{D^2}{4}} \rightarrow G_{opt} \approx \frac{\cos \Phi_0}{D\sqrt{1 - \frac{3D^2}{16}}}$$

$$x_{opt} \approx -\sqrt{1 - \frac{D^2}{4}} \rightarrow G_{opt} \approx -\frac{\cos \Phi_0}{D\sqrt{1 - \frac{3D^2}{16}}}$$

For $\sin \Phi_0 = \pm 1 \Leftrightarrow \Phi_0 = \frac{\pi}{2}$ or $\Phi_0 = 3\frac{\pi}{2}$ the solutions are:

$$x_{opt} \approx \pm\sqrt{1 + D} \rightarrow G_{opt} \approx \frac{\sin \Phi_0}{D(2 + D)}$$

$$x_{opt} \approx \pm\sqrt{1 - D} \rightarrow G_{opt} \approx -\frac{\sin \Phi_0}{D(2 - D)}$$

$$x_{opt} = 0 \rightarrow G_{opt} = -\sin \Phi_0$$

For any other value of Φ_0, the solutions are accessible only numerically.

From preceding results, for out-of-plane asymmetries, stable solutions are obtained only for:

$$\Phi_0 \approx \frac{\pi}{2}; \quad x_r \approx -1$$
$$\Phi_0 \approx 3\frac{\pi}{2}; \quad x_r = +1$$

Thus optima of G in the vicinity are:

- for $\Phi_0 \approx \frac{\pi}{2}; x_r = -1$

$$x_{opt} \approx -\sqrt{1 + D} \rightarrow G_{opt} \approx \frac{1}{D(2 + D)} \approx \frac{1}{2D}$$

$$x_{opt} \approx -\sqrt{1 - D} \rightarrow G_{opt} \approx -\frac{1}{D(2 - D)} \approx -\frac{1}{2D}$$

- for $\Phi_0 \approx 3\frac{\pi}{2}; x_r = +1$,

$$x_{opt} \approx \sqrt{1 + D} \rightarrow G_{opt} \approx -\frac{1}{2D}$$

$$x_{opt} \approx \sqrt{1 - D} \rightarrow G_{opt} \approx \frac{1}{2D}$$

Hence, for out-of-plane asymmetries, the minimum size of asymmetries is given by the criterion:

$$\frac{\dot{p}_{0\,max}}{2D\,|\dot{p}_{cr}|} \geq 1$$

12.1 Association of Aerodynamic Asymmetry and CG Offset

We leave to the reader to check that, when crossing first and second resonance, the conditions of sign are met for $\Phi_0 \approx \frac{\pi}{2}$; $x_r = -1$ and for $\Phi_0 \approx 3\frac{\pi}{2}$; $x_r = +1$

The criterion can be developed using Allen's approximate expression of reentry velocity for high β vehicles [Chap. 9, (9.1)],

$$V \approx V_D e^{-K\rho}, \rho = \rho_S e^{-\frac{z}{H_R}}, K = \frac{\beta H_R}{2 \sin |\gamma|}$$

We obtain successively:

$$\dot{p}_{cr} = \sqrt{\frac{S_{ref} C_{N\alpha} \Delta X}{2(I_T - I_X)}} \frac{d}{dt}(V\sqrt{\rho}) = \sqrt{\frac{S_{ref} C_{N\alpha} \Delta X}{2(I_T - I_X)}} \left(\dot{V}\sqrt{\rho} + V \frac{d\sqrt{\rho}}{dz}(-V \sin |\gamma|) \right)$$

$$\dot{p}_{cr} = \sqrt{\frac{S_{ref} C_{N\alpha} \Delta X}{2(I_T - I_X)}} \left(-\frac{A}{m}\sqrt{\rho} + V^2 \frac{\sqrt{\rho}}{2 H_{ref}} \sin |\gamma| \right)$$

$$= \sqrt{\frac{S_{ref} C_{N\alpha} \Delta X}{2(I_T - I_X)}} V^2 \frac{\sqrt{\rho}}{2} \left(-\beta\rho + \frac{\sin |\gamma|}{H_{ref}} \right)$$

In addition,

$$D = \frac{(1-\mu)\ell_\alpha + m_q}{\omega_n \sqrt{1-\mu}} = \frac{1}{m}\sqrt{\frac{\rho S_{ref}(I_T - I_X)}{2 \Delta X C_{N\alpha}}} \left[(C_{N\alpha} - C_A) - \frac{m L_{ref}^2 C_{mq/G}}{(I_T - I_X)} \right]$$

Observing that usually $|\alpha_G| << |\alpha_e|$,

$$\dot{p}_{0\,\text{max}} \approx z_G \frac{\rho V^2}{2} S_{ref} \frac{C_{N\alpha} \alpha_e}{I_X}$$

We obtain finally the following expression of the criterion:

$$\frac{z_G \alpha_e}{[z_G \alpha_e]_1} \geq |1-\eta|$$

with,

$$\eta = \frac{H_R \rho}{\beta \sin |\gamma|}; \quad \beta = \frac{m}{S C_A}$$

$$[z_G \alpha_e]_1 = \frac{I_X}{m H_R} \sin |\gamma| \left(1 - \frac{C_A}{C_{N\alpha}} \right) \left(1 - \frac{m L_{ref}^2 C_{mq/G}}{(C_{N\alpha} - C_A)(I_T - I_X)} \right)$$

While using the usual sizing assumption $C_{mq} = 0$, we obtain:

$$[z_G\alpha_e]_1 \approx 5.7310^4 \frac{I_X}{mH_R} \sin|\gamma| \left(1 - \frac{C_A}{C_{N\alpha}}\right)_{\text{mm.degree}}$$

Let us apply these results to the conical RV (Chap. 4) of half-cone angle $8°$, $R_N = 25$ mm, $D = 0.5$ m, $m = 117$ kg, $I_x = 2\,\text{m}^2\,\text{kg}$, $I_T = 15\,\text{m}^2\,\text{kg}$, $\beta = 9170\,\text{kg/m}^2$, $\gamma = -25°$, and $x_G/Lref = 63.4\%$ from apex. At high Mach number and $1°$ angle of attack we obtain:

$$\lambda = C_{N\alpha}/C_A \approx 33.5; \quad \Delta X \approx 160\,\text{mm}$$

While $\rho_{SL} = 1.39\,\text{kg.m}^{-3}$, $H_{ref} \approx 7000$ m, and $C_{mq} \approx 0$, it results:

$$[z_G\alpha_e]_1 \approx .06\,\text{mm.degree}$$

The parameter $\eta \geq 0$ increases during reentry from zero to its maximum value at sea level $\eta_{\max} = \frac{H_{ref}\rho_0}{\beta \sin|\gamma|} = 2.51$. The critical frequency increases and passes by its maximum when $\eta = 1$, then decreases when $\eta > 1$.

Figures 12.5 and 12.6 show roll lock-in criterions according to η or altitude.

- For asymmetries whose product is lower than value $[z_G\alpha_e]_1$ lock-in may be possible in a domain of altitude around the maximum of critical roll rate (it is necessary that roll rate is reached before the critical roll value corresponding to altitude where the criterion begins to be satisfied). Roll lock-in is impossible out of this small domain. For initial roll rates slightly lower than the maximum of the

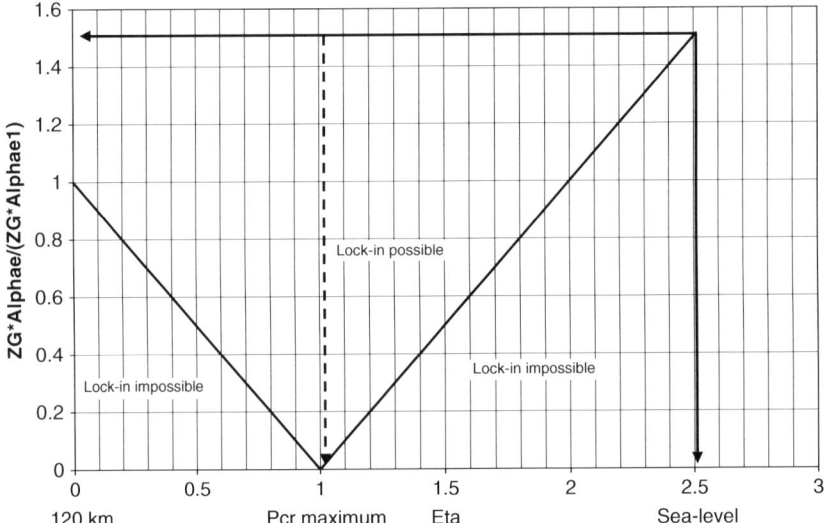

Fig. 12.5 Criterion function of η

12.1 Association of Aerodynamic Asymmetry and CG Offset

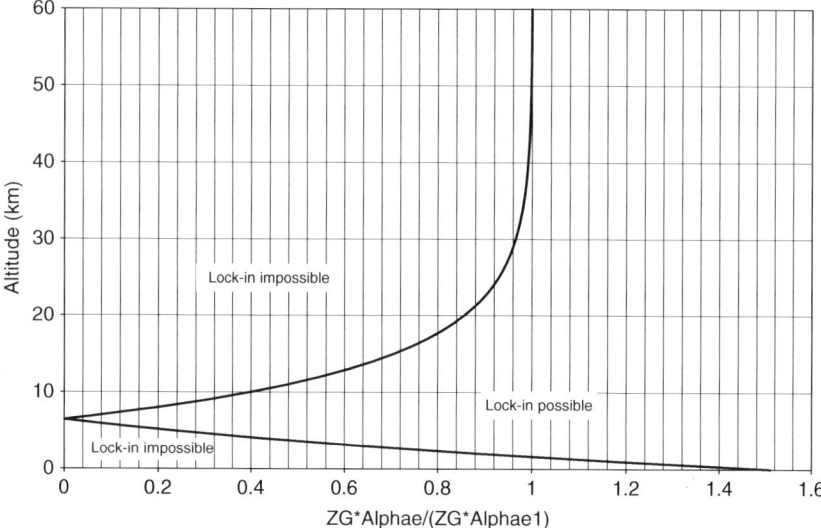

Fig. 12.6 Criterion function of altitude

critical frequency, the theoretical analysis shows the possibility of maintaining the lock-in for very low levels of asymmetries. This is due to the fact that the critical frequency is stationary in a small interval of altitude around the maximum. This does not have any practical consequences, the criterion being not satisfied out of this interval, lock-in will stop rapidly.

- For asymmetries in the range between $[z_G \alpha_e]_1$ and the sea level value $[z_G \alpha_e]_2 = |\eta_{max} - 1| [z_G \alpha_e]_1$, there may be a possibility of lock-in from the crossing of the first resonance until the altitude where the criterion is not verified, lower than the altitude of maximum critical frequency. Lock-in is not possible out of this domain. For asymmetries whose product is higher than $[z_G \alpha_e]_2$, lock-in may be possible from the crossing of the first resonance to the ground level.

Generally, this simplified theory overestimates the risks of roll lock-in compared to six degree-of-freedom calculations, particularly for the first resonance.

We did not take into account herein the dynamic pitching response of the vehicle, with characteristic time $\tau \approx 2\pi/\omega_n$. Thus, results are clearly valid only when this response time is negligible compared with the duration of resonance crossing. Another explanation could be an underestimation of damping factor D. Indeed, density damping is still effective at first resonance altitude and is not included in this theory.

Thus this criterion must only be regarded as an instructional tool to obtain orders of magnitude. A serious study requires the use of a six degree-of-freedom digital code, which yields results as shown in Fig. 12.7 corresponding to the following example:

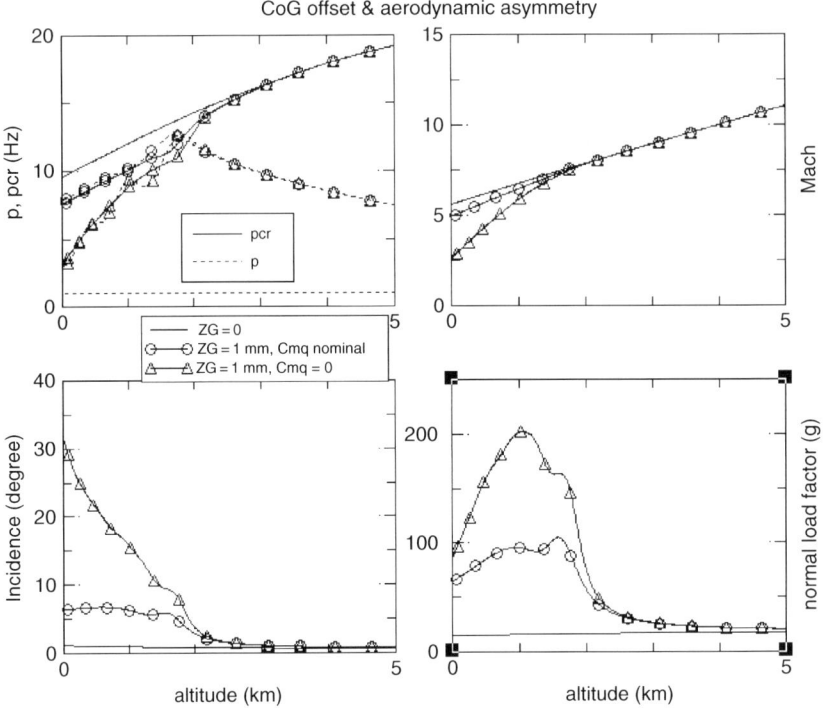

Fig. 12.7 6 DoF results for 2nd resonance lock-in with CG offset and aerodynamic asymmetry

- Same vehicle as analytic computation, CG offset $z_G = 1.\text{mm}$
- Out-of-plane aerodynamic asymmetry corresponding to a normal force at zero angle of attack exerted on the nose, directed $\Phi_e = 270°$ from CG offset:

$$F_{0n} = \bar{q}\pi R_n^2 C_{N0n}$$

C_{N0n} evolves from 0 for Z > 20 Km to 0.2 for Z < 15 km, with a linear variation. Assuming no downstream effects, from the nose force, the moment coefficient at zero angle of attack is:

$$Cm_{0G} \approx \frac{\pi R_n^2}{S_{ref}} \frac{X_G}{L_{ref}} C_{N0n}$$

- Pitch damping coefficient C_{mq}, nominal value (Chap. 4), or $C_{mq} = 0$

In the case without CG offset, the rolling moment is null and the roll rate remains constant and lower than the critical frequency. There is no crossing of the second resonance.

12.1 Association of Aerodynamic Asymmetry and CG Offset

In presence of CG offset, a positive rolling moment occurs under 20 km altitude, dependent on the development of asymmetry. The positive rolling moment increases roll rate and involves crossing of second resonance near 2 km altitude. As the orientation of asymmetry (out-of-plane), corresponds to the stable roll equilibrium condition, and available roll moment is sufficient, there is lock-in on the critical frequency $p \approx +p_{cr}$ down to the sea level for the two hypothesis of damping coefficient. Roll lock-in results in a permanent amplification and a divergence of the trim incidence compared to the undercritical reference case $p \ll p_{crit}$. This high angle of attack involves high normal loads (70–140 g) and a velocity loss (Mach difference 0.3 to 1.8 with reference case).

Design studies generally use six degree-of-freedom Monte Carlo codes as main tool for sizing RVs. From sizing hypothesis on the probability distributions of vehicle's parameters (aerodynamic coefficients, mass, inertia, axial CG location, aerodynamic, and inertial asymmetries), atmosphere, wind, and kinematics initial conditions, these codes allow to run hundreds or thousands of samples of six degree-of-freedom trajectories to assess statistics on reentry performance including load environment and dispersions.

Figure 12.8 shows distribution of lateral loads on the preceding vehicle obtained with a six degree-of-freedom Monte Carlo code, with hypothesis:

- CG offset is a random variable with uniform probability on the disk $r_G \leq R_{G,\max}$, $R_{G,\max} = 0.5, 1.0, 1.5$ mm and
- C_{N0n} is uniform in the interval [0, 0.2], ϕ is uniform in [0, 2π].

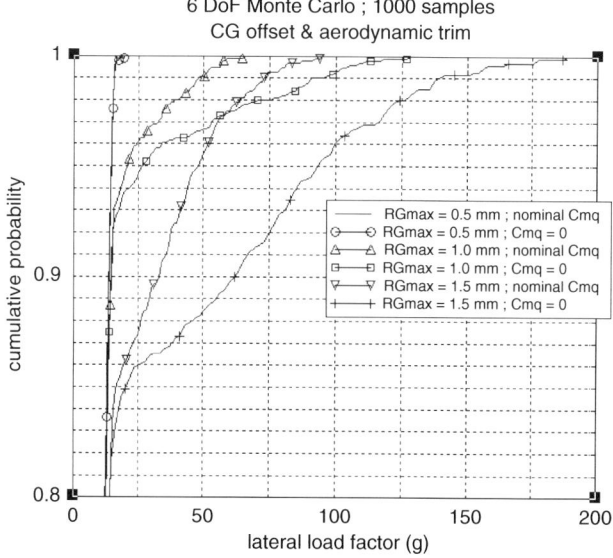

Fig. 12.8 6 DoF Monte Carlo results for 2nd resonance lock in study

While $R_{G,max} \geq 1$ mm and $C_{mq} = 0$, we obtain more than 1% probability of lateral loads in excess of 100 g.

For this vehicle, it results tolerance on the maximum CG offset must be lower than 1 mm to achieve acceptable lateral loads.

This type of results shows that Monte Carlo approach is very useful to assess requirements on stability and tolerances as well as performances of the vehicle

12.2 Isolated Center of Gravity Offset

We consider a symmetrical vehicle, except a small CG offset z_G. From previous analysis, it results static and rolling trim angles.

- At zero roll rate, the pitching moment of axial force "F_x," balanced by the pitch stability moment, results in a static trim angle $\alpha_G = \frac{z_G C_A}{C_{N\alpha}(X_F - X_G)} > 0$ in XZ plane. Figure 12.9 shows normal force $F_{ze} = -N_\alpha \alpha_G$ resulting in the opposite direction.
- At given roll rate p > 0, equilibrium results in an amplified rolling trim force $F_E = AF_{ze}$ with a shift angle $\Delta \phi \in [-\pi, 0]$ in the direction opposite to roll rate (Fig. 12.10).

The force, applied at the center of pressure, results in a rolling moment around the center of gravity. Taking into account the values of A and $\Delta \phi$ from Chap. 11, Sect. "Trim Angle of Attack," this roll moment is:

$$L_G = z_G F_E \sin \Delta \phi = -z_G N_\alpha \alpha_G A^2 Dx$$

$$L_G = -\bar{q} S C_A \frac{z_G^2}{\Delta X} \frac{Dx}{\left(1-x^2\right)^2 + (Dx)^2}$$

with $x = p/p_{cr}, \Delta X = X_G - X_{CP}$.

The rolling moment, opposed to p, cannot involve roll equilibrium except around p = 0 where it cancels. Thus, it always leads to a reduction of the amplitude of

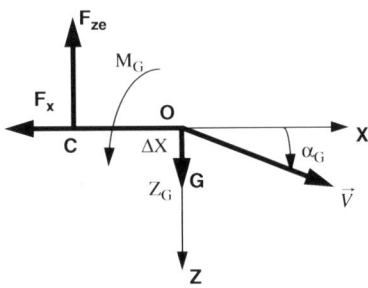

Fig. 12.9 Side view

12.3 Isolated Principal Axis Misalignment

Fig. 12.10 View from aft

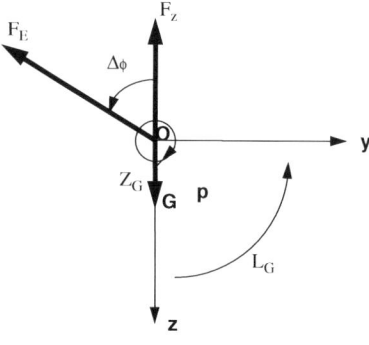

roll rate. The incidence α_G being proportional to z_G, rolling moment is proportional to the square of z_G. Taking into account the requirements on the CG offset coming from roll lock-in studies, the effect on roll rate is low, except:

- at resonance crossing where trim angle amplification is maximum,
- at the end of the trajectory when α_G increases because of the drag increase and of the possible decrease of static margin associated with lower Mach number.

While rolling moment is proportional to α and β, before decay of initial incidence, roll rate has oscillations at nutation and precession frequencies in vehicle axes. After convergence of initial angle of attack, the rolling trim angle results in small constant values of pitch and yaw angular rates (lunar motion):

$$q_E = p\beta_E \; ; \quad r_E = p\alpha_E$$

Figure 12.13 gives typical evolutions obtained from six degree-of-freedom calculations with a CG offset $Z_G = 5.\text{mm}$.

12.3 Isolated Principal Axis Misalignment

We have shown that P.A.M. results in a static trim angle proportional to the square of the roll rate.

$$\xi_\theta = \beta_\theta + i\alpha_\theta = -\text{sgn}(1-\mu)\left(\frac{p}{p_{cr}}\right)^2 (\theta_y + i\theta_z)$$

The origin of the pitching moment is the centrifugal force. While assuming a positive roll rate and a misalignment in Oz direction, this is illustrated in Fig. 12.11. We consider an equivalent system of six virtual masses having the same tensor of inertia as the vehicle. This system is composed of six point masses $m = M/6$. Two point masses are located on the rolling principal axis at a distance ℓ symmetrically

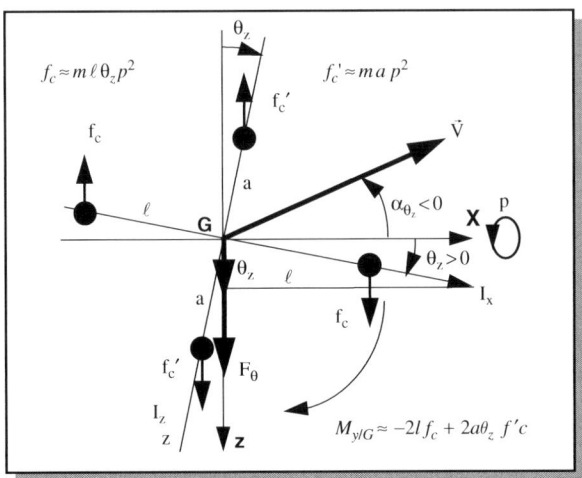

Fig. 12.11 Side view

apart from CG, and four are located symmetrically on pitch and yaw principal axes at distance "a" from center of gravity.

It is easy to show that equivalent system verifies:

$$I_{xx} = 4ma^2; \quad I_{yy} = I_{zz} = 2ma^2 + 2m\ell^2$$
$$\rightarrow 2m\ell^2 = I_{yy} - \frac{I_{xx}}{2}$$

Thus, the action of the centrifugal forces on the system of four masses located in plane Gxz results in a pitching moment around Gy:

$$M_{y/G} \approx -2m\ell^2 \theta_z p^2 + 2ma^2 \theta_z p^2 = -\left(I_{yy} - I_{xx}\right)\theta_z p^2$$

It results the static trim angle:

$$\alpha_\theta = -\frac{M_{y/G}}{\bar{q} S_{ref} L_{ref} C_{m\alpha/G}} = \frac{\left(I_{yy} - I_{xx}\right)\theta_z p^2}{\bar{q} S_{ref} L_{ref} C_{m\alpha/G}} = -\text{sgn}(1-\mu)\theta_z \left(\frac{p}{p_{cr}}\right)^2.$$

When $\mu < 1$, the masses on the roll axis which are the main contributors to transverse inertia are predominant and they create a negative pitching moment. As shown Fig. 12.11, the trim is opposed to the misalignment, applying an aerodynamic force $F_{\theta z} = -N_\alpha \alpha_\theta$ in the direction of the misalignment.

In this case α_θ represents only a convenient parameter, since this trim exists only in the presence of a roll rate. From Chap. 11, it is modified by roll rate in the same manner as other static trims. The trim force is amplified $F_E = A\,F_{\theta_z}$ and shifted an angle $\Delta\phi$ in the direction opposed to roll rate (Fig. 12.12).

12.3 Isolated Principal Axis Misalignment

Fig. 12.12 Phase lag angle of trim force

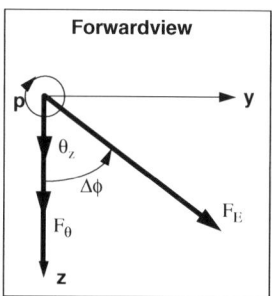

Roll acceleration is derived from (12.3) and (12.4),

$$\dot{p} \approx \frac{1-\mu}{\mu} p^2 \theta_z \beta_E$$

$$\beta_E = A^2 D x \alpha_T$$

Taking into account:

$$\alpha_T = \alpha_\theta = -\operatorname{sgn}(1-\mu) x^2 \theta_Z$$

This gives finally the roll acceleration for $\mu < 1$:

$$\dot{p} \approx -\frac{D}{\mu} \omega_n^2 \theta_z^2 \frac{x^5}{\left(1-x^2\right)^2 + (Dx)^2}$$

Thus, roll acceleration always has the sign opposed to p and cancels only for p = 0. There is no stable roll equilibrium, except possibly around p = 0. The effect of an isolated small PAM is always a reduction in the absolute value of mean roll rate (an increase for $\mu > 1$). This decrease is generally small except in the vicinity of resonance where rolling moment is amplified.

Like CG offset, before reaching rolling trim, PAM induces oscillations on roll rate at nutation and precession frequencies in vehicle axes. On the other hand, it results in first-order effects on mean values of transverse angular rates:

$$q_E = p\beta_E = -p\theta_z x^3 A^2 D = -p_{cr} D\theta_z \frac{x^4}{\left(1-x^2\right)^2 + D^2 x^2}$$

$$r_E = p\alpha_E = -p\theta_z x^2 \left(1-x^2\right) A^2 = -p_{cr}\theta_z \frac{x^3\left(1-x^2\right)}{\left(1-x^2\right)^2 + D^2 x^2}$$

This effect is maximum in the vicinity of resonance $x \approx 1 \rightarrow q_E = \frac{p_{cr}\theta_z}{D}$ and to a lesser extent at supercritical roll rates $|x| \gg 1$. It is completely negligible in the

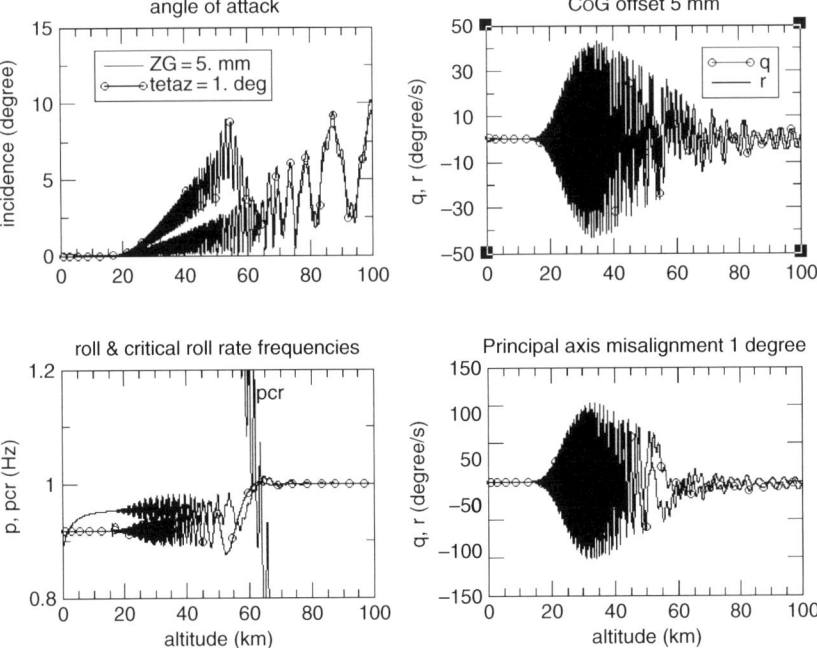

Fig. 12.13 Histories of parameters at first resonance for inertial asymmetries

case of subcritical roll rates $|x| \ll 1$ usually met at the end of the reentry (except low altitude second resonance).

In addition, we will notice that at resonance the dynamic trim angle induced by θ_z equals $\beta_E \approx -\frac{\theta_z}{D}\text{sgn}(p)$. Figure 12.13 shows typical evolutions obtained from six degree-of-freedom numerical calculation, for a misalignment $\theta_z = 1.°$.

12.4 Combined CG Offset and Principal Axis Misalignment

While we choose again z axis in the direction of CG offset (it results by definition that $z_G > 0$; $\alpha_G = \frac{z_G\, C_A}{C_{N\alpha}\Delta X} > 0$), roll acceleration (12.4) becomes:

$$\dot{p} \approx -\frac{N_\alpha}{I_x} z_G \beta_E + \frac{(1-\mu)}{\mu} p^2 \left(\theta_z \beta_E - \theta_y \alpha_E\right)$$

From (12.3) we have,

$$\beta_E = A^2 \left[\beta_\theta \left(1 - x^2\right) + (\alpha_\theta + \alpha_G)\, Dx\right]$$
$$\alpha_E = A^2 \left[(\alpha_\theta + \alpha_G)\left(1 - x^2\right) - \beta_\theta Dx\right]$$

12.4 Combined CG Offset and Principal Axis Misalignment

with $x = p/p_{cr}$; $\alpha_\theta = -x^2\theta_z$; $\beta_\theta = -x^2\theta_y$; $\alpha_G = z_G C_A / C_{N\alpha} \Delta X$

12.4.1 Out-of-plane Misalignment

In the case of a PAM normal to the plane of the CG offset $(\theta_z = 0;\ \theta_y \neq 0)$, we obtain from previous expression of rolling acceleration,

$$\dot{p} \approx -\frac{N_\alpha}{I_x} z_G \beta_E - \frac{(1-\mu)}{\mu} p^2 \theta_y \alpha_E$$

$$\alpha_E = A^2 \left[\alpha_G \left(1 - x^2\right) + \theta_y D x^3 \right]$$

$$\beta_E = A^2 \left[-x^2 \left(1 - x^2\right) \theta_y + \alpha_G D x \right]$$

While $\frac{N_\alpha Z_G}{I_x} = \frac{\lambda}{\mu} \omega_n^2 \alpha_G$, with $\lambda = \frac{C_{N\alpha}}{C_A}$, we obtain:

$$\dot{p} = \frac{A^2 \omega_n^2}{\mu} x \left[\lambda' x \left(1 - x^2\right) \alpha_G \theta_y - \lambda D \alpha_G^2 - D x^4 \theta_y^2 \right]$$

with $\lambda' = \lambda - 1 = \frac{C_{N\alpha}}{C_A} - 1$.

Then, we finally obtain,

$$\dot{p} = -\dot{p}_0 x \frac{\left[x \left(x^2 - 1\right) + \frac{D}{\lambda'} \left(v x^4 + \frac{\lambda}{v}\right) \right]}{\left(x^2 - x_r^2\right)^2 + D^2} \tag{12.5}$$

with $v = \frac{\theta_y}{\alpha_G}$, $\dot{p}_0 = \frac{\lambda' \omega_n^2}{\mu} \alpha_G \theta_y$

12.4.1.1 Rolling Equilibrium

The first solution of $\dot{p} = 0$ is trivial, i.e., $x = p = 0$. Now, we look for solutions in the vicinity of resonance $x = x_r \approx \pm\sqrt{1 - \frac{D^2}{2}}$.

For this purpose, we use the assumptions:

(i) $x_r \approx \pm 1$,
(ii) in the numerator, we assume that the terms having relatively low variation around $x_r \approx \pm 1$ are constant,

Then, using the new variable $u = x^2 - 1$, we obtain an approximate expression of rolling acceleration in the vicinity of x_r.

$$\Rightarrow \dot{p} \approx -\dot{p}_0 \text{sgn}(x_r) \frac{u \text{sgn}(x_r) + u_1}{D^2 + u^2} \tag{12.6}$$

with $u_1 = \frac{D}{\lambda'}\left(\lambda\frac{\alpha_G}{\theta_y} + \frac{\theta_y}{\alpha_G}\right)$, assuming $\frac{\lambda}{\lambda'} \approx 1$.

In the considered range, the solutions for zero rolling acceleration are obviously:

- for $x_r = 1, u = -u_1$
- for $x_r = -1, u = u_1$

i.e.:

$$x_e = \pm\sqrt{1 + u_1} \approx \pm\left[1 + \frac{u_1}{2}\right]$$

The solutions closest to resonance correspond to $\left|v + \frac{\lambda}{v}\right|$ minimum, that is to say:

$$v = \frac{\lambda}{v} \quad \Leftrightarrow \quad \theta_y = \pm\alpha_G\sqrt{\lambda}$$

They correspond to $u_e = \frac{2D\sqrt{\lambda}}{\lambda'}\text{sgn}\left(\alpha_G\theta_y\right)$. These low values justify the approximation we used.

12.4.1.2 Stability of Equilibrium

To be stable, a roll trim must check $\left.\frac{\partial\dot{p}}{\partial x}\right|_{x_e} < 0$, which is expressed according to u by,

$$\left.\frac{\partial\dot{p}}{\partial u}\frac{du}{dx}\right|_{x_e} = 2x_e \left.\frac{\partial\dot{p}}{\partial u}\right|_{u_e} < 0$$

That is to say,

- $x_e > 0 \Rightarrow \left.\frac{\partial\dot{p}}{\partial u}\right|_{u_e} < 0$
- $x_e < 0 \Rightarrow \left.\frac{\partial\dot{p}}{\partial u}\right|_{u_e} > 0$

Simple examination of roll acceleration shows that,

$$\text{sgn}\left(\left.\frac{\partial p}{\partial u}\right|_{x_e}\right) = -\text{sgn}\left(\dot{p}_0\right) = -\text{sgn}\left(\alpha_G\theta_y\right)$$

Stable roll trim corresponds, with $\alpha_G > 0$ by definition:

- For $x_e \approx +1 \rightarrow \theta_y > 0$, or, $\alpha_G\theta_y > 0$
- For $x_e \approx -1 \rightarrow \theta_y < 0$, or, $\alpha_G\theta_y < 0$

The consequence of these conditions are,

12.4 Combined CG Offset and Principal Axis Misalignment

- $x_e \approx +1 \quad \to \alpha_G \theta_y > 0 \quad \Rightarrow u_1 > 0, x_e \approx 1 - \dfrac{u_1}{2} < x_r$
- $x_e \approx -1 \quad \to \alpha_G \theta_y < 0 \quad \Rightarrow u_1 < 0, x_e \approx 1 - \left(1 + \dfrac{u_1}{2}\right) > x_r$

This means we have in every case $|x_e| < |x_r|$

12.4.1.3 Dynamic Stability

When dynamic pressure evolves, the maximum rolling acceleration $\dot{p}_{\max}$ available in the vicinity of the equilibrium point x_e must have a module at least equal to that of $x_e \dot{p}_{cr}$ and the sign required.

Thus we seek extrema of $\dot{p}$ in the vicinity of the stable trim solutions determined above.

To obtain maximum roll acceleration, we need the zero of its derivative. Examination of exact roll acceleration (12.5) shows a degree five polynomial at numerator for which analytical solutions does not exist. Hence, we will restrict our ambition to determine order of magnitude of extrema, by using the very crude approximation (12.6) of the roll acceleration:

$$\frac{\partial \dot{p}}{\partial u} \approx -\dot{p}_0 \mathrm{sgn}(x_r) \frac{(D^2 + u^2)\,\mathrm{sgn}(x_r) - 2u\,(u\,\mathrm{sgn}(x_r) - u_1)}{(D^2 + u^2)^2} = 0$$

$$\Leftrightarrow \quad (2 - \mathrm{sgn}(x_r))\,u^2 + 2u_1 u - D^2 \mathrm{sgn}(x_r) = 0$$

The extrema for $x_e \approx +1$ closest to the resonance, with $u_1 \approx \dfrac{2D}{\sqrt{\lambda}}$ correspond to:

$$u_{ext} = -u_1 \pm \sqrt{u_1^2 + D^2} = -\frac{2D}{\sqrt{\lambda}} \pm \sqrt{\frac{4D^2}{\lambda} + D^2} \approx u_e \pm D$$

Let us introduce these values into the expression (12.6) of rolling acceleration, which gives the approximate expression of corresponding extrema,

$$\dot{p}_{ext} \approx -\dot{p}_0 \frac{u_{ext} - u_e}{D^2 + u_{ext}^2} \approx \pm \dot{p}_0 \frac{D}{2D^2} = \pm \frac{\dot{p}_0}{2D}$$

To verify these results, we compare with the exact curve $G = \dot{p}/\dot{p}_0$ shown in Fig. 12.14, corresponding to critical value $v = \sqrt{\lambda}$ and $D = 0.1$. This confirms that the estimate $G_{\max} = 5$ is in qualitative agreement, although the exact results are nonsymmetric. We also note that equilibrium location at $x_e \approx 0.95$ is in agreement with $|x_e| < |x_r|$.

Hence, to follow the critical frequency, the roll rate needs an acceleration:

$$|\dot{p}_{ext}| = \frac{|\dot{p}_0|}{2D} > |\dot{p}_{cr}| \quad \Leftrightarrow \quad |\dot{p}_0| = \frac{\lambda' \omega_n^2}{\mu} |\alpha_G \theta_y| \geq 2D\,|\dot{p}_{cr}|.$$

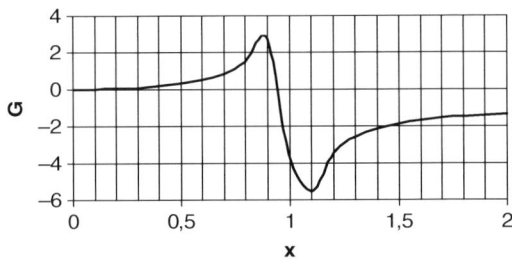

Fig. 12.14 Evolution of G for critical v

This gives the condition:

$$|\alpha_G \theta_y| \geq \frac{\mu}{\lambda'} \frac{2D|\dot{p}_{cr}|}{\omega_n^2}$$

Using the same method as Sect. 12.1.1, we obtain the expression of the expression at right member:

$$2D\left|\frac{\dot{p}_{cr}}{\omega_n^2}\right| = \frac{\ell^2 |\sin\gamma|}{\Delta X H_{ref}} \left[1 - \frac{C_A}{C_{N_\alpha}} - \frac{L_{ref}^2 C_{mq/G}}{\ell^2 C_{N_\alpha}(1-\mu)} \right] |1-\eta|$$

with:

$$\eta = 2K\rho = \frac{H_{ref}}{\beta \sin|\gamma|}\rho; \quad \ell^2 = \frac{I_T}{m}$$

This gives the expression of the lock-in criterion:

$$|\alpha_G \theta_y| \geq [\alpha_G \theta_y]_1 |1-\eta|$$

with $|\alpha_G \theta_y|_1 = \frac{1}{\lambda'} \frac{I_x |\sin\gamma|}{m\Delta X H_{ref}} \left[1 - \frac{C_A}{C_{N_\alpha}} - \frac{L_{ref}^2 C_{mq/G}}{\ell^2 C_{N_\alpha}(1-\mu)} \right]$.

Using z_G instead of α_G, we obtain:

$$|z_G \theta_y|_1 \approx 5.710^4 \frac{I_x |\sin\gamma|}{m H_{ref}} \left[1 - \frac{C_A}{C_{N_\alpha}} - \frac{L_{ref}^2 C_{mq/G}}{\ell^2 C_{N_\alpha}(1-\mu)} \right]_{mm.degree}$$

The preceding expression is identical to that obtained for the combined CG offset/aerodynamic asymmetric trim. The only difference is related to the condition on the ratio $\frac{\alpha_G}{\theta_y}$, which must equal $\frac{1}{\sqrt{\lambda}}$ to obtain the lock-in closest to the resonance, i.e.:

$$\frac{z_G}{\Delta X \theta_y} \approx \sqrt{\lambda}$$

12.4 Combined CG Offset and Principal Axis Misalignment

When we apply this criterion to the same vehicle and conditions as Sect. 12.1.4, we obtain the same results:

$$(z_G \theta_y)_1 = 0.06 \text{ mm.degree}$$

Using result shown in Fig. 12.6, this value also applies to a first resonance crossing at 45 Km. It results a PAM and a CG offset:

$$\theta_y^2 = \frac{(z_G \theta_y)}{\sqrt{\lambda} \Delta X}; \quad z_G^2 = (z_G \theta_y) \sqrt{\lambda} \Delta X$$

such that $\theta_y \approx 0.06$ degree and $z_G \approx 1$ mm.

Figure 12.15 shows the numerical results for a high roll rate ≈ 12 Hz. It results a crossing of the first resonance at 25 km and the second resonance around 1250 m, corresponding respectively to $|1 - \eta| \approx \left|1 - 2.5 e^{-\frac{Z}{7000}}\right|$ equal to 0.93 and 1.09. The roll lock-in threshold numerically determined is:

- for the second resonance, $0.16 < (z_G \theta_y)_2 \leq 0.22$ mm. degree
- for the first resonance, $0.22 < (z_G \theta_y)_1 \leq 0.37$ mm. degree

The analytical theory gives 0.057 and 0.065 mm-degree, respectively.

Thus, it underestimates the critical threshold of parameter $z_G \theta_y$ in a ratio close to 4. It corresponds to a very approximate estimate of the problem. Moreover, it does

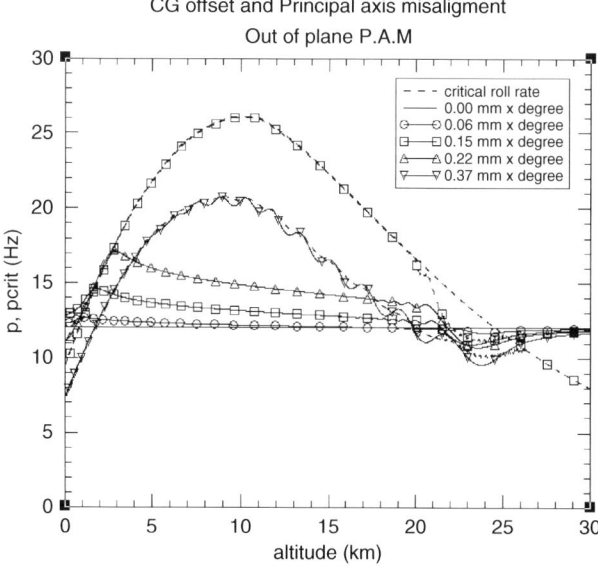

Fig. 12.15 Evolution of roll rate and critical roll rate

not allow us to determine the evolution of roll rate and thus to judge possibility of crossing the second resonance.

Figure 12.15 shows evolutions of roll rate and critical frequency for various values of the parameter $z_G \theta_y$.

12.4.2 In-plane Misalignment

In the case of a PAM in-the-plane $(\theta_y = 0; \theta_z \neq 0)$, we obtain from (12.3) and (12.4,):

$$\dot{p} = -\frac{N_\alpha}{I_x} z_G \beta_E + \frac{\omega_n^2}{\mu} x^2 \theta_Z \beta_E$$

$$\beta_E = A^2 Dx \left(\alpha_G + \alpha_\theta \right) = A^2 Dx \left(\alpha_G - x^2 \theta_z \right)$$

While $\frac{N_\alpha z_G}{I_x} = \frac{\lambda}{\mu} \omega_n^2 \alpha_G$, with $\lambda = \frac{C_{N_\alpha}}{C_A}$, we obtain:

$$\dot{p} = -\frac{\omega_n^2}{\mu} \left(\lambda \alpha_G - x^2 \theta_Z \right) \beta_E = -\frac{\omega_n^2}{\mu} A^2 Dx \left(\lambda \alpha_G - x^2 \theta_Z \right) \left(\alpha_G - x^2 \theta_Z \right)$$

Finally, the rolling acceleration is:

$$\dot{p} \approx -\dot{p}_1 \frac{\left[x^2 - a_1 \right] \left[x^2 - a_2 \right] Dx}{\left(x^2 - 1 \right)^2 + D^2 x^2}$$

with,

$$\dot{p}_1 = \frac{\omega_n^2 \theta_z^2}{\mu}; \quad a_2 = \lambda a_1; \quad a_1 = \frac{\alpha_G}{\theta_z} = \frac{z_G}{\lambda \Delta X \theta_z}$$

From this expression, we can determine the roll behavior according to parameter a_1:

- When $a_1 < 0 \Rightarrow a_2 < a_1 < 0$, the only zero of roll acceleration is for $x = 0$ and $sign(\dot{p}) = -sign(x)$. It results the same behavior than with an isolated CG offset or PAM. The amplitude of roll rate can only decrease:
- When $a_1 > 0 \Rightarrow$ the rolling moment cancels for $x = 0$ and

$$x = x_1 = \pm \sqrt{a_1}$$
$$x = x_2 = \pm \sqrt{a_2}$$

To determine if these roll trim are stable, let us study the behavior of $\dot{p}$ around $x = x_1$ and $x = x_2$.
In the domain $0 < a_1 < a_2 = \lambda a_1$ we have around x_2:

12.4 Combined CG Offset and Principal Axis Misalignment

$$\dot{p} \approx -\dot{p}_1 \frac{\left[x^2 - a_1\right][x - x_2](x + x_2)Dx}{\left(x^2 - 1\right)^2 + D^2 x^2} \approx -\dot{p}_1 \frac{[a_2 - a_1][x - x_2]2x_2 Dx_2}{\left(x^2 - 1\right)^2 + D^2 x^2}$$

As $a_2 > a_1$, $\left.\frac{\partial \dot{p}}{\partial x}\right|_{x_2} < 0$, $\left.\frac{\partial \dot{p}}{\partial x}\right|_{x_1} > 0$, and the equilibrium is stable around x_2 and unstable around x_1.

Hence, when $0 < a_1$, the only stable roll trim is $x_2 = \pm\sqrt{a_2} = \pm\sqrt{\lambda a_1}$.

However, trim near resonance can occur only when $\lambda \frac{\alpha_G}{\theta_z} \approx 1$ i.e., in a narrow domain of PAM at a high relative value $\theta_Z/\alpha_G \approx \lambda$. Clearly this event has a low probability.

Figure 12.16 shows the behavior of function $G = \dot{p}/\dot{p}_1$ in the case of stable trim near resonance, for different value of x_2 such that $x_2^2 = x_r^2 + \sigma D$.

We see that, unlike out-of-plane PAM, maximum roll amplification ($G \sim 0.5$) is low when $x_2 = x_r$. Maximum roll amplification increase when x_2 moves away, but the amplification curve becomes highly asymmetric.

12.4.2.1 Dynamic Criterion

As previously, in real condition of nonstationary flight, the available moment must be sufficient to maintain a rolling acceleration at least equal to $\dot{p}_{cr}$. On the basis of the assumption of a stable trim at initial time (stationary dynamic pressure and Mach number) at fixed values x_2 near maximum of resonance $x_2^2 = a_2 = x_r^2 + \sigma D$, we look for conditions to stay at trim when the critical frequency and dynamic pressure begin to evolve. We saw in the case of a combined CG offset and aerodynamic asymmetry that, in addition to the stable static trim conditions, which must be every time verified, it is necessary to have around the trim value extrema of rolling acceleration with amplitude at least equal to $|\dot{p}_{cr}|$, with adequate sign. While using the expression of A^2,

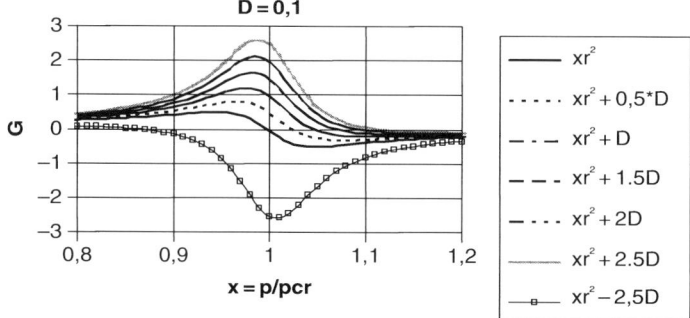

Fig. 12.16 Amplification of roll acceleration

$$A^2 = \frac{1}{D^2\left(1 - \left(\frac{D}{2}\right)^2\right) + \left(x^2 - x_r^2\right)^2} \approx \frac{1}{D^2 + \left(x^2 - x_r^2\right)^2}$$

where $x_r = \sqrt{1 - \frac{D^2}{2}}$ corresponds to resonance, the expression of rolling acceleration becomes:

$$\dot{p} = -\dot{p}_1 \frac{\left[x^2 - a_1\right]\left[x^2 - x_2^2\right] Dx}{D^2 + \left[x^2 - x_r^2\right]^2}$$

While $a_1 \approx 1/\lambda \ll 1$, $x_2 \approx x_r \approx 1$, $x^2 = x_r^2 + h$ with $h \ll 1$, we obtain:

$$\dot{p} \approx -\dot{p}_1 D \frac{(h - \sigma D)}{D^2 + h^2} = -\dot{p}_1 \frac{\frac{h}{D} - \sigma}{1 + \left(\frac{h}{D}\right)^2}$$

To obtain the extrema of $\dot{p}$ near x_1, we determine the zeros of the derivative, which gives:

$$\frac{h}{D} = \sigma \pm \sqrt{1 + \sigma^2}$$

$$\frac{\dot{p}_{\max}}{\dot{p}_1} = G_{\max \pm} = \mp \frac{1}{2} \frac{\sqrt{1 + \sigma^2}}{1 + \sigma^2 \pm \sigma\sqrt{1 + \sigma^2}}$$

Thus, as shown in Figure 12.16, two extrema of rolling moment at $(x_\pm)^2 = x_r^2 + D\left(\sigma \pm \sqrt{1 + \sigma^2}\right) = x_2^2 \pm D\sqrt{1 + \sigma^2}$ surround the roll trim value x_2. Unlike other asymmetries, maximum roll amplification is low and do not depend on damping parameter. As shown in Fig. 12.17, the maximum of amplitude on the lower side of x_2 is always positive. When $\sigma > 0$, amplitude of maximum on the left side of x_2 is higher than the amplitude on the right side, except when $x_2 = x_r$ for which $|G_{\max \pm}| = 0.5$. When $\sigma < 0$, amplitude of maximum on the right side of x_2 is the highest.

Obviously, G_- and $\sigma > 0$ apply to lock-in near first resonance (p_{cr} is increasing), G_+ and $\sigma < 0$ to the second resonance.

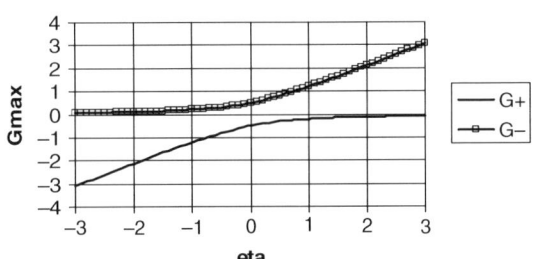

Fig. 12.17 Amplitude of maxima of roll amplification

12.4 Combined CG Offset and Principal Axis Misalignment

Hence, to prevent lock-in on x_2 such that $x_2^2 = x_r^2 + \sigma D$, we must verify:

- at first resonance, $\sigma > 0$ & $|\dot{p}_1 G_{\max -}| < |\dot{p}_{cr}|$

$$\Leftrightarrow \quad \theta_z^2 < 2\left[\sqrt{1+\sigma^2} - \sigma\right]\mu \frac{|\dot{p}_{cr}|}{\omega_n^2}$$

- at second resonance, $\sigma < 0$ & $|\dot{p}_1 G_{\max +}| < |\dot{p}_{cr}|$

$$\Leftrightarrow \quad \theta_z^2 < 2\left[\sqrt{1+\sigma^2} + \sigma\right]\mu \frac{|\dot{p}_{cr}|}{\omega_n^2}$$

Then, by using previous Allen's derivation of ω_n^2 and p_{cr}, while naming $\eta = \rho H_{ref}/\beta \sin|\gamma|$ and $I_T = m\ell^2$, we obtain,

$$\frac{|\dot{p}_{cr}|}{\omega_n^2} = \sqrt{\frac{C_A \ell^2 |\sin \gamma|}{2C_{N\alpha}\Delta X H_{ref}(1-\mu)}} \frac{|1-\eta|}{\sqrt{\eta}}$$

Hence, the condition to prevent roll equilibrium within a distance to first or second resonance defined by $\sigma = \frac{|x_2^2 - x_r^2|}{D}$ is:

$$\theta_z^2 < 2\mu\left[\sqrt{1+\sigma^2} + (-1)^i |\sigma|\right]\sqrt{\frac{C_A \ell^2 |\sin \gamma|}{2C_{N\alpha}\Delta X H_{ref}(1-\mu)}} \frac{|1-\eta|}{\sqrt{\eta}}$$

with $i = 1$ for first resonance and $i = 2$ for second resonance.

The choice of σ can be done by noting that when $x_2^2 = x_r^2 + \sigma D$, the trim angle amplification at roll equilibrium is:

$$A \approx \frac{1}{\sqrt{D^2 + (x_2^2 - x_r^2)^2}} = \frac{1}{D\sqrt{1+\sigma^2}}$$

Hence, for example, the choice $\sigma = 9$ leads to a diminution of trim amplification by the factor $1/\sqrt{1+\sigma^2} \approx 0.11$ compared to equilibrium at exact resonance condition.

For the first resonance, this choice results in the criterion:

$$\theta_z^2 < \theta_{z1}^2 \frac{|1-\eta|}{\sqrt{\eta}}$$

with

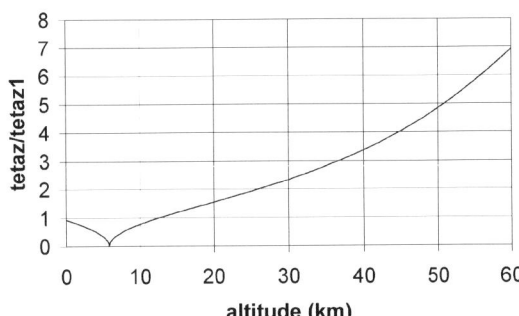

Fig. 12.18 Evolution of P.A.M. criterion with altitude

$$\theta_{z1}^2 = 0.11\mu \sqrt{\frac{C_A \ell^2 |\sin\gamma|}{2 C_{N\alpha} \Delta X H_{ref}(1-\mu)}}$$

Figure 12.18 gives the ratio $\theta_Z/\theta_{Z_1} = \left[|1-\eta|\big/\sqrt{\eta}\right]^{\frac{1}{2}}$ resulting from the criterion as a function of altitude for $\beta = 10^4$ and a flight path angle $\gamma = -25°$. For the same vehicle as in out-of-plane PAM, we obtain $\theta_{Z1} \approx 0.075°$. Using curve 12.18, we find the limiting value $\theta_Z \approx 4\theta_{Z1} = 0.3°$ for a first resonance crossing at 45 km.

In addition, we must remember that to obtain lock-in we need to have $\alpha_G \approx \frac{\theta_Z}{\lambda} \Leftrightarrow Z_G \approx \theta_Z \Delta X$, which gives an order of magnitude $Z_G \approx \theta_Z \Delta X = 0.83\,mm$. This corresponds to a product $z_G \theta_Z \approx 0.25$ mm.degree.

Hence, from this analytical theory, in-plane PAM criterion is somewhat less critical than out-of-plane PAM. However, when compared with the combination of CG offset and aerodynamic asymmetry, they both need in addition to the condition on the product of asymmetry a condition on the ratio. This makes roll lock-in a very improbable event for this type of combination.

Chapter 13
Instabilities

Preceding chapters concerning the angle-of-attack behavior used the assumption of a vehicle unconditionally stable from the static and dynamic point of view. It results that the initial incidence and/or possible disturbances of the incidence due to external factors (for example, an atmospheric gradient of wind) remain limited and dampen shortly. We deal here the intrinsic mechanisms of instability (not related to external factors), which correspond to a permanent or temporary increase in the amplitude of the angle of attack.

They are two kinds, bound respectively by the static and the dynamic moment.

13.1 Static Instabilities

We consider here the case where the vehicle is dynamically stable in the entire flight envelope.

The most radical case, that we eliminate a priori, is that of a static instability permanent and total during the whole reentry. Indeed, unless we have a huge gyroscopic moment corresponding to a very high roll rate, we will obtain a divergence of the incidence and a probable demise of the vehicle.

A more interesting case, because the diagnosis is less obvious, is that instability limited to a range of flight parameters, i.e., Knudsen number, Mach number, and incidence. We will quote two examples of this type of instability.

- Some planetary entry probes with high drag coefficient such as Viking, Mars Pathfinder, and Huygens, were unstable in the free molecular regime and at the beginning of the rarefied regime in a broad range of incidence around zero. In the absence of gyroscopic stability, i.e., for zero roll rate, these vehicles would have undergone the initiation of angle-of-attack divergence at high altitude and likely to stabilize around 180°, their back cover exposed to high pressures and heat fluxes. Needless to say that their mission would have been completely compromised. However, flight mechanics analysis shows that an adequate roll rate allowed them to stabilize during the critical phase and to limit the divergence of the incidence to a few degrees until the moment they recovered stability. Successful entries of Viking and Pathfinder in the Martian atmosphere, and Huygens in

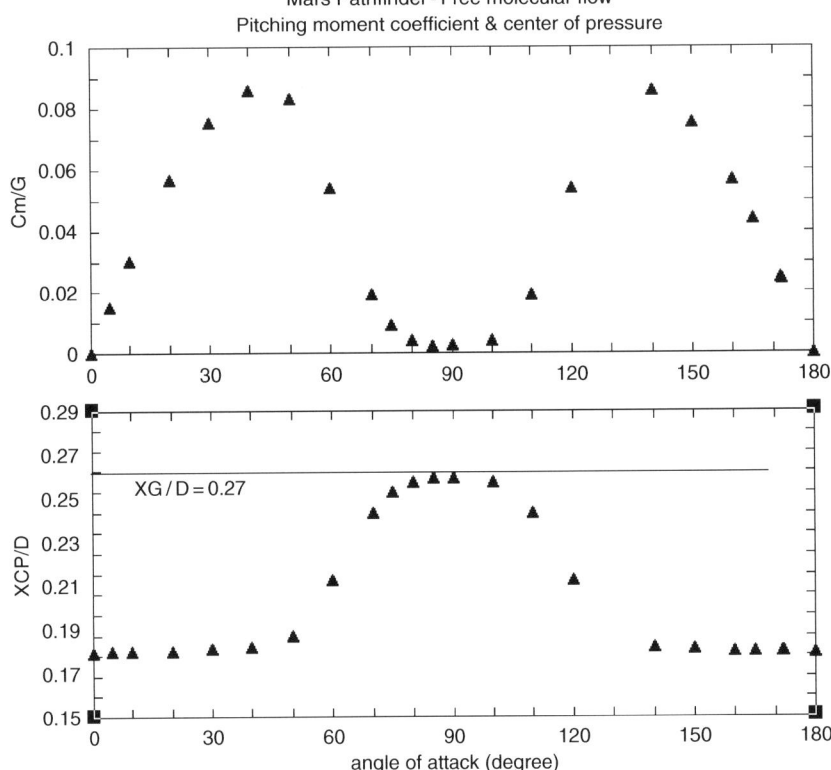

Fig. 13.1 Stability parameters in free molecular flow

the Pluto atmosphere confirmed this analysis. Figure 13.1 presents the Pathfinder configuration's evolution in incidence of center of pressure and pitching moment coefficient for $\frac{x_G}{D} = -0.27$, which were obtained using a Computational Fluid Dynamic (CFD) code in free molecular flow.

These results show that, for this CG location, the probe is unstable near zero angle of attack and stable at 180°.

Figure 13.2 [GAL] presents the evolution of the gradient of pitching moment coefficient and angle of attack for a reduced scale similar probe (diameter 1.20 m). These results were obtained from 6 DoF computations in the Martian environment. Use of a roll rate higher or equal to 5 rpm avoids any divergence of the incidence until 80 km altitude where the probe recovers its static stability. The estimate of this behavior using approximate methods of Chaps 4, 7, and 10 is proposed in exercise.

- Some reentry vehicles have a nonlinear behavior in incidence at high Mach number in continuous flow. This corresponds to a forward movement of the center of pressure when the incidence decreases, in some range $[0, \alpha_{max}]$ around zero

13.1 Static Instabilities

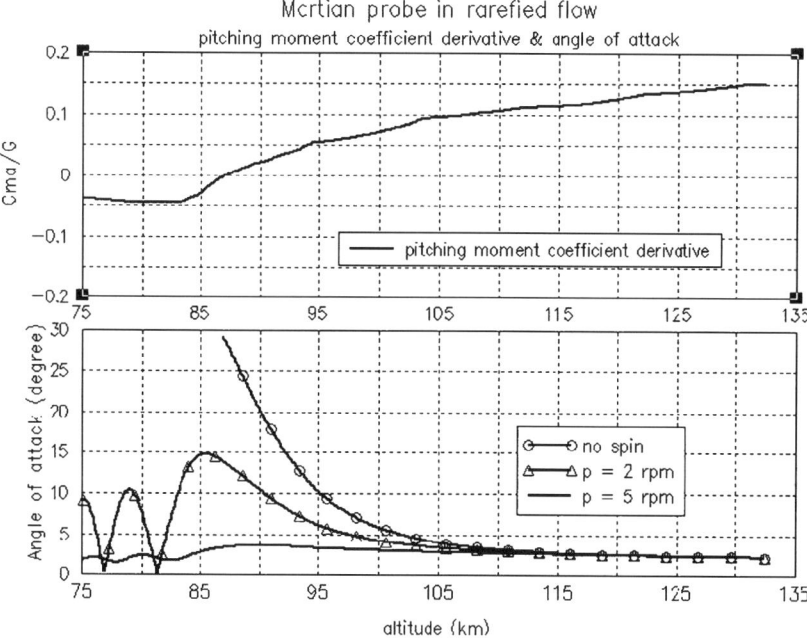

Fig. 13.2 Sensitivity of AoA divergence on spin rate

angle of attack. At higher angle of attack, the vehicle again becomes stable, i.e., the pitching moment coefficient around center of mass becomes negative. We can assess the angle of attack's behavior of this vehicle:

- If the initial amplitude of incidence is in this instability range, we will observe an increase in the incidence up to a maximum value higher than stability limits, followed by oscillations between these two values (limit cycle). When Mach number decreases because of aerodynamic deceleration, the object becomes again statically stable, and then incidence will converge quickly as in the regular case.
- If initial incidence is largely higher than the limiting value, normal convergence of angle of attack begins. When angle of attack and Mach number histories stay in the stable range, convergence continues and finishes normally.

Angle-of-attack behavior is thus either entirely normal, when the initial value is high, or a temporary increase followed by a limiting cycle, then a convergence resumes.

There are other possible nonlinear behaviors. In practice, it is often possible to cross unstable portions of the flight parameter range, in low density range with moderate dynamic pressure. Center of mass location must be validated by a 6 DoF analyses.

- Beside the predictable behavior related to vehicle geometry, Mach number, incidence, and Reynolds number, there is another case of more or less catastrophic angle-of-attack divergence met during experimental flights. They were the subject of many publications in specialized reviews since years 60 and until recently [WAT]. Two explanations most frequently quoted are respectively related to the response of heat shield material to heat flux (ablation lag phenomenon) and to asymmetrical progression of laminar-to-turbulent transition from aft to forward along the reentry vehicle. According to some authors [VFL] the effects related to transition are weak and the most probable cause would be "ablation lag" [ALG].

- Ablation lag:
Instantaneous azimuth distribution of convective heat flux around a symmetrical reentry vehicle at angle of attack is maximum on the windward meridian. According to Sect. 11.2, for an observer fixed to the vehicle, the windward meridian turns around roll axis with angular rate $\dot{\phi}_{w\pm} = -\left(1 - \frac{\mu}{2}\right) p \pm \omega_a$, with sign depending on mode of motion (nutation or precession). Heat flux affects both pyrolize and surface ablation phenomena, which result in the emanation of gas products from the surface. These products are transported downstream along the boundary layer (blowing phenomenon), and result in a thickening of the boundary layer, increasing with the mass flux of injected products. Mass flux of injected products does instantaneously respond to heat flux, but is subjected to thermal response time lag. In the case of a zero or very low incidence, the radial flow distribution has symmetry of revolution as well as the modification of displacement thickness of boundary layer and induced pressure effects. This phenomenon modifies the center of pressure, because it induces a change of the effective aerodynamic geometry of the reentry vehicle. In the case of a finite angle of attack and roll rate, the maximum of injection flux and displacement thickness on a given meridian line are affected by a time lag τ relative to time of passage of the relative velocity vector. In vehicle axes, this results in a roll angle shift $\phi_w - \phi_p = \Delta\phi = \dot{\phi}_w \cdot \tau$ of meridian line ϕ_p corresponding to the maximum of boundary layer thickness represented in Fig. 13.3 in a cross section of the vehicle.

In addition, approximate assessment of viscous effects shows that the effective shape of a reentry vehicle from the point of view of pressure distribution corresponds to the shape of the wall increased with displacement thickness of the boundary layer (this is the origin of the name of this parameter). It occurs that because of ablation lag, not only is this effective shape asymmetric, but it is not aligned to the plane of windward meridian. These results in an abnormal aerodynamic load, which corresponds in wind-fixed axes to a yaw force normal to the plane of incidence and to a yawing moment in the plane of incidence. Wind-fixed aerodynamic load directed like the regular static component (i.e., corresponding to a pitching force in

13.1 Static Instabilities

Fig. 13.3 Geometry of ablation lag

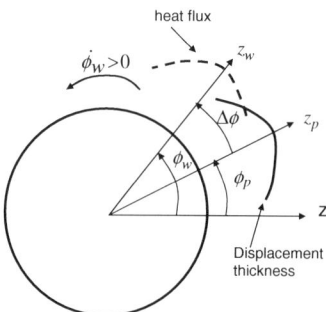

the plane of incidence and a pitching moment normal to it) is named "in-plane," and effects corresponding to yaw force and moment are named "out-of–plane."

In order to study consequences of this phenomenon, we will use approximations of Chap. 11, relating to equations of incidence motion linearized in the complex plane GYZ fixed to the vehicle. The regular aerodynamic moment is written in complex notation:

$$\bar{M} = M + iN = \bar{q} S_{ref} L_{ref} C_{m\alpha/G} \cdot e^{-i\frac{\pi}{2}} \cdot (\beta + i\alpha) = i \cdot I_T \omega_n^2 \cdot \xi$$

with

$$\xi = \beta + i\alpha = \bar{\alpha} e^{i\phi_w}$$

The origin of the abnormal effect can be modeled by a sideslip component β_p of the complex incidence relating to the meridian plane Z_p corresponding to the maximum of displacement thickness. Figure 13.4 shows the resulting moment ΔM_{zp}, in the case $\dot{\phi}_w > 0$. With these assumptions, an ablation lag have $\Delta \phi > 0$ and yaw moment along Z_p is negative or positive depending on the location of the yaw force center of pressure relating to center of mass G.

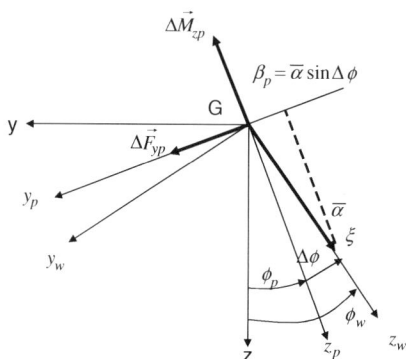

Fig. 13.4 Frontal view

The out-of-plane moment increases with the amplitude of the boundary layer displacement and with β_p angle (i.e., the time lag). At first order this moment can be put in the form:

$$\Delta M = \bar{q} S_{ref} L_{ref} K \bar{\alpha} |\sin \Delta\phi|$$

where the K term is a function of upstream flow and of thermal and ablative properties of the wall material.

Thus, the expression of the moment in vehicle axes is

$$\frac{\Delta \bar{M}}{\bar{q} S_{ref} L_{ref}} = -\varepsilon K \sin \Delta\phi \cdot \xi \cdot e^{-i\Delta\phi}$$

where $\varepsilon = \pm 1$ is positive when yaw force center of pressure is behind center of gravity, negative before. This sign depends on $\dot{\phi}_w$, vehicle shape, wall material, and upstream flow.

For small angles $\Delta\phi$, projections of this moment on wind-fixed axes are a negligible pitching moment and a yawing moment:

$$\Delta M_{yw} = -\varepsilon \left|\Delta \bar{M}\right| \sin \Delta\phi = O^2(\Delta\phi)$$
$$\Delta M_{zw} = -\varepsilon \left|\Delta \bar{M}\right| \cos \Delta\phi \approx -\varepsilon \bar{q} S_{ref} L_{ref} K \bar{\alpha} \sin \Delta\phi$$

The approximate ablation lag moment is a wind-fixed yaw moment whose expression in the vehicle frame is:

$$\Delta \bar{M} \approx \varepsilon \bar{q} S_{ref} L_{refl} K \sin \Delta\phi \cdot \xi = -\varepsilon' I_T \omega_l^2 \xi$$

with $\omega_l^2 = \frac{\bar{q} S_{ref} L_{ref} K |\sin \Delta\phi|}{I_T}$; and $\varepsilon' = \text{sgn}(\varepsilon \Delta\phi)$.

Remarks:

1) This moment has properties similar as a Magnus effects; however, its physical origin is completely different. Magnus effect is related to the modification of viscous flow by rolling motion, without any ablative phenomenon. While parameter $\frac{pR}{V}$ is small, according to experimental data and theoretical evaluations [VFL], Magnus moment is negligible for reentry vehicles. This is obviously not the case of artillery projectiles, slower and with much higher roll rate.
2) The term ω_l^2 has dimensions of the square of a pulsation.

While including the ablation lag moment, the total aerodynamic wind-fixed moment is:

$$\bar{M}_{T/G} = i I_T \left(\omega_n^2 + i\varepsilon' \omega_l^2\right) \xi$$

The influence of this moment on the instantaneous angular motion can be estimated using the results of epicycle motion of Chap. 11.2, by modifying the aerodynamic pulsation:

13.1 Static Instabilities

$$\omega_n^2 \rightarrow \omega_n'^2 = \omega_n^2 + i\varepsilon' \omega_l^2$$

By using the same approximations, Eigen frequencies $\omega_\pm$ of epicycle movement remain unchanged, and damping terms become:

$$\Lambda_\pm = -\frac{\ell_\alpha + m_q}{2} \pm \frac{1}{2\omega_a}\left[\varepsilon'\omega_l^2 + \frac{\mu p}{2}(m_q - \ell_\alpha)\right]$$

Now, let us use the hypothesis of a yaw force center of pressure aft of the center of mass, i.e., $\varepsilon > 0$. Below the first resonance altitude, depending on the sign of rotation rate of the mode, we obtain:

- for $\omega_+ = \omega_a - \left(1 - \frac{\mu}{2}\right) p > 0 \rightarrow \dot{\phi}_w = \omega_+ > 0 \rightarrow \varepsilon' > 0$

$$\Lambda_+ = -\frac{\ell_\alpha + m_q}{2} + \frac{1}{2\omega_a}\left[\omega_l^2 + \frac{\mu p}{2}(m_q - \ell_\alpha)\right]$$

- for $\omega_- = -\omega_a - \left(1 - \frac{\mu}{2}\right) p \rightarrow \dot{\phi}_w = \omega_- < 0 \rightarrow \varepsilon' < 0$

$$\Lambda_- = -\frac{\ell_\alpha + m_q}{2} - \frac{1}{2\omega_a}\left[-\omega_l^2 + \frac{\mu p}{2}(m_q - \ell_\alpha)\right]$$

We observe in both cases a positive sign of the term related to ablation lag, i.e., an unstable contribution. While neglecting the second-order term related to roll, evolution of the incidence is:

$$\frac{|\bar{\alpha}|}{|\bar{\alpha}_0|} \approx e^{-\frac{1}{2}\left(m_q + \ell_\alpha - \frac{\omega_l^2}{\omega_a}\right)t}$$

In the case where yaw force center of pressure is ahead of the mass center ($\varepsilon < 0$), we will have a stabilizing contribution.

The general expression for angle-of-attack evolution is thus:

$$\frac{|\bar{\alpha}|}{|\bar{\alpha}_0|} \approx e^{-\frac{1}{2}\left(m_q + \ell_\alpha - \varepsilon\frac{\omega_l^2}{\omega_a}\right)t}$$

We thus note that, within the framework of this approximate theory, ablation lag effects primarily modify the dynamic stability of the vehicle. The effects are similar to that of the pitch damping moment. Thus, to differentiate the origin of dynamic stability disturbances in flight observations, it is essential to be able to calculate the flow rate of ablation products injected as well as its effects on viscous flow.

Quantitative theoretical data in this phenomenon for various materials were published in [ALG].

13.2 Dynamic Instabilities

We can set aside the ablation lag phenomenon already studied, knowing that to take it into account, we can include it in the term related to traditional damping:

$$m_q \quad \to \quad m'_q = m_q - \varepsilon \frac{\omega_l^2}{\omega_a}$$

First, we will examine the regular case at low incidence and constant pitch damping moment coefficient. Then, we will give examples of nonlinear unstable behavior.

13.2.1 Approximate Study at Low Angle of Attack

We saw in Chap. 11 that while neglecting the effect of dynamic pressure variation, i.e., with respect to the short-term evolution, the expression of complex incidence in a quasi-inertial frame (aeroballistic frame) is:

$$\xi' = e^{ip_r t}\left[\xi'_+ e^{\Lambda_+ t} e^{i\omega_a t} + \xi'_- e^{\Lambda_- t} e^{-i\omega_a t}\right]$$

with $p_r = \frac{\mu p}{2}$

$$\omega_a \approx \sqrt{\omega_n^2 + p_r^2} \,;\, \Lambda_+ \approx \Lambda_- \approx \Lambda = -\frac{\ell_\alpha + m_q}{2} \,;\, m_q = -\overline{q} S_{ref} \frac{L_{ref}^2}{I_T V} C_{mq/G}$$

$$\ell_\alpha = \frac{N_\alpha - A}{mV} = \frac{\overline{q} S_{ref}}{mV}[C_{N\alpha} - CA] = \frac{\overline{q} S_{ref}}{mV} C_{L\alpha}$$

The damping pulsation can be put into the form:

$$\Lambda = -\frac{\rho V}{4}\left[\frac{1}{\beta_L} + \frac{1}{\beta_q}\right] = -\frac{\rho V}{4}\left[\frac{1}{\beta_N} - \frac{1}{\beta} + \frac{1}{\beta_q}\right]$$

with:

$$\beta = \frac{m}{S_{ref} C_A} \,;\, \beta_N = \frac{m}{S_{ref} C_{N\alpha}} \,;\, \beta_L = \frac{m}{S_{ref} C_{L\alpha}} \,;\, \beta_q = -\frac{m}{S_{ref} C_{mqG}}\left(\frac{r_G}{L_{ref}}\right)^2$$

where r_g indicates the pitching radius of gyration $r_g^2 = \frac{I_T}{m}$

Thus, we identify the dynamically stable influence of the lift gradient on the angle of attack and destabilizing influences of drag. The influence of the dynamic moment C_{mq} is either stabilizing ($C_{mq} < 0$) or destabilizing. However, to determine completely if incidence oscillations are damped or undamped, it is necessary to take account of the effect of the variation of aerodynamic pulsation on the amplitude. For this purpose, let us study in the vicinity of $t = t_1$ the behavior of:

13.2 Dynamic Instabilities

$$\eta = \frac{|\xi|}{|\xi_1|} = \left[\frac{\omega_{n1}}{\omega_n}\right]^{\frac{1}{2}} e^{\Lambda(t-t_1)}.$$

While assuming $C_{m\alpha/G}$ is constant, this expression is equivalent to:

$$\eta = \left[\frac{\rho_1 V_1^2}{\rho V^2}\right]^{\frac{1}{4}} e^{\Lambda(t-t_1)}$$

Using Allen approximation for exponential atmosphere, we obtain:

$$D = \frac{\partial Ln(\eta)}{\partial t} = \left[\Lambda + \frac{\rho V}{4\beta} - \frac{V \sin \gamma_0}{4 H_R}\right]$$

That is to say:

$$\eta \approx e^{D_1 \cdot (t-t_1)}$$

with $D_1 = -\frac{V_1}{4}\left[\rho_1\left(\frac{1}{\beta_L} + \frac{1}{\beta_q} - \frac{1}{\beta}\right) + \frac{\sin \gamma_0}{H_R}\right]$ (we must not confuse this term with the damping term in the roll resonance theory of Chaps. 11 and 12).

Thus we can define an index of dynamic stability,

$$S = -\frac{4D}{V} = \rho\left[\frac{1}{\beta_L} + \frac{1}{\beta_q} - \frac{1}{\beta}\right] + \frac{\sin \gamma_0}{H_R} > 0$$

Evolutions of the contributors to the damping factor D are shown in Figs. 13.5 and 13.6, respectively in algebraic value and in percentage of total damping.

These curves correspond to the conical vehicle of reference (Chap. 4). The hypothesis are $C_{mq/G} = -0.08$, $V_0 = 6000 \, \text{m/s}$, $\gamma_0 = -30°$, $H_R = 7000 \, \text{m}$, and $\rho_S = 1.39 \, \text{kg.m}^{-3}$.

These results reveal several observations:

1) Drag contributes twice. First, it decreases damping related to rotation of flight path vector induced by lift ($\frac{1}{\beta_L} = \frac{1}{\beta_N} - \frac{1}{\beta}$). Second it appears because of pulsation damping through variation of velocity and dynamic pressure (second term in $\frac{1}{\beta}$).
2) The expression shows the influence of the two kinds of damping factors. The group of aerodynamic factors proportional to the product ρV, and parametric damping dependent on the exponential variation of density simply proportional to V. The relative contribution of parametric damping is prevalent at high altitude ($z > 50 \, \text{Km}$) where the density is very low, then becomes negligible compared with aerodynamic damping at low altitude ($z < 20 \, \text{Km}$) when density is high. Thus dynamic instabilities are very unlikely to develop above 50 km altitude where the aerodynamic damping coefficient C_{mq} has little influence.

240   13 Instabilities

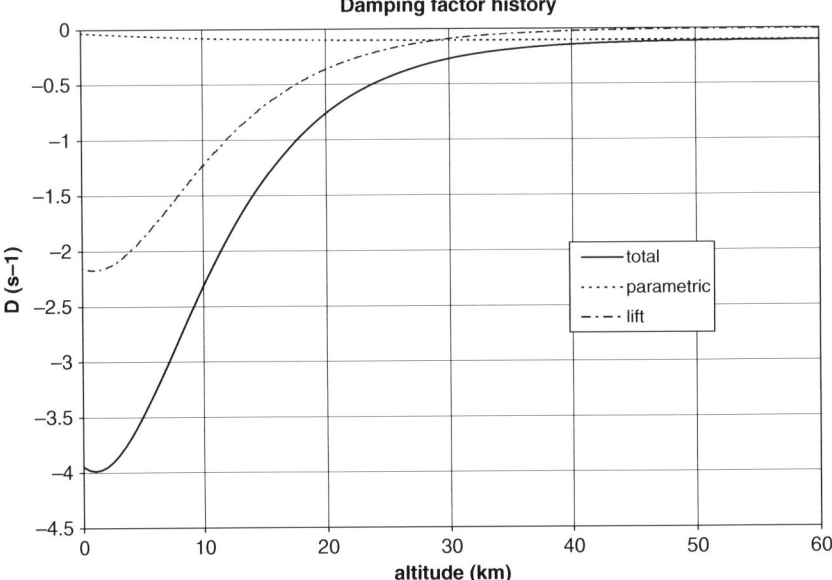

Fig. 13.5 Damping parameter

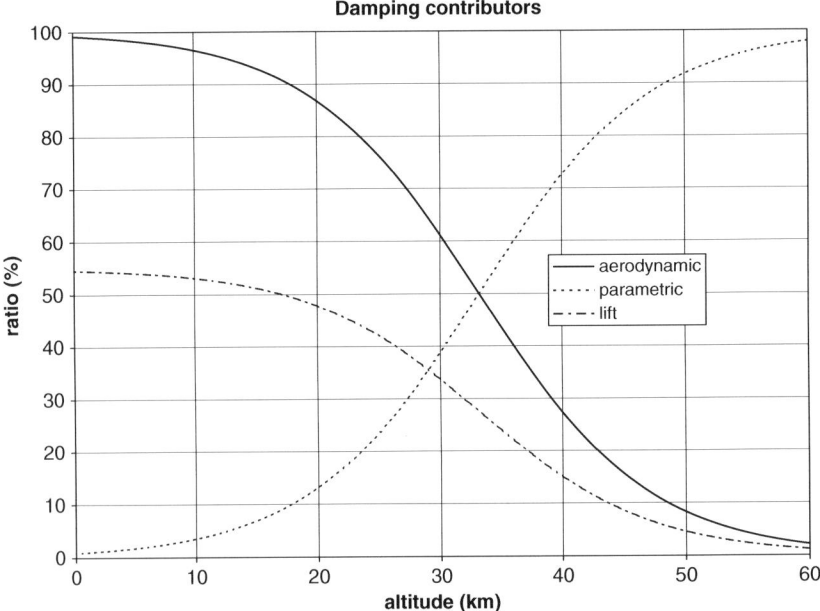

Fig. 13.6 Damping contributors

13.2 Dynamic Instabilities

3) In the group of aerodynamic factors, and for a vehicle having a high $C_{L\alpha}$ (slender vehicle), the contribution of lift to the rotation of the flight-path vector is a major contributor to damping. The vehicle can support a null or slightly negative damping coefficient $C_{mq/G}$.

13.2.2 Examples of Unstable Dynamic Behaviors

- The first case presented here was observed during flight tests of a reentry vehicle with heat shield made of resin and fiber composite material. Terms of aerodynamic damping derived from angular rate telemetries are presented Fig. 13.7. Measurements make it possible to derive the total aerodynamic damping ratio, after correction for the effect of parametric damping dependent on the variation of the aerodynamic frequency. The method uses the corrected epicycle model developed in the preceding paragraph. The contribution of lift being theoretically accessible with good accuracy, we obtain the contribution of all other factors of dynamic stability, including the pitch damping coefficient $C_{mq/G}$. We observe the zone of dynamic instability in altitude range 25–45 km.

It is interesting to note that the progression of the transition front on this material with rough ablated surface takes place inside this same range of altitude. The origin of this instability is uncertain, because it can be interpreted as well using a positive damping coefficient or an out-of-plane force and a moment related to ablation lag. Figure 13.8 gives the damping coefficient extracted from flight data,

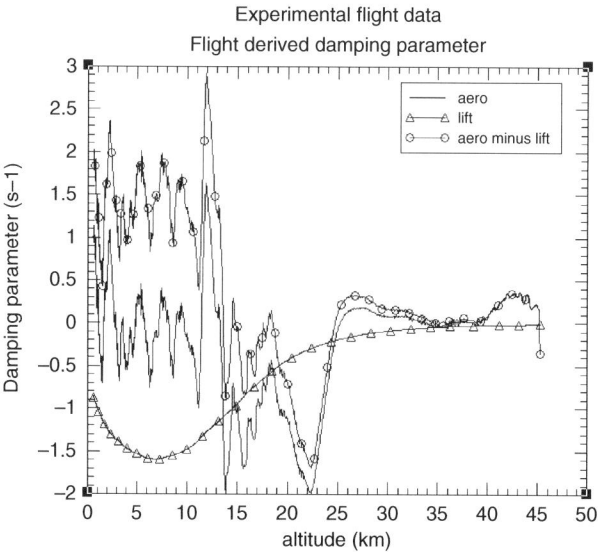

Fig. 13.7 Flight derived damping parameter

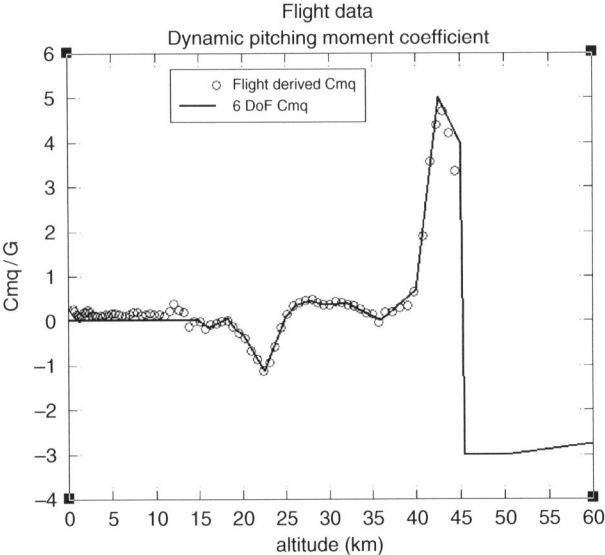

Fig. 13.8 Flight derived pitch damping moment coefficient

on the assumption of the first interpretation. The damping coefficient estimated above 45 km is highly negative, and then it becomes highly positive under this altitude. It becomes negative again below 25 km. Derived values below 15 km altitude have a strong uncertainty, because the incidence is low and damping is not easily observable. Figure 13.9 shows evolution of incidence derived from flight measurements, compared with a 6 DoF computation, using a model of pitch damping coefficient derived from flight (Fig. 13.8). The agreement between experimental and theoretical evolutions is excellent, which validate the method of identification of damping coefficient. Traditional interpretation in terms of damping coefficient, associated to boundary layer transition, assumes different locations of the transition point on windward and leeward meridians. To our knowledge, assessment of this phenomenon was not made theoretically, whereas ablation lag is now well established. Interpretation in terms of ablation lag in this particular case remains to be made.

In practice, this type of behavior was observed and studied only on flights where incidence at beginning of instability was sufficiently low. This likely corresponds to the fact that the modest incidence builtin, about 2° or 3°, is easily unperceived when initial incidence is high. This may also indicate that the phenomenon is nonlinear in incidence and that the range of dynamic instability is limited to the low incidences.

- The second example of instability relates to behavior at low Mach of a space probe of shape similar to Viking [GAL]. Figure 13.10 shows a model of dynamic stability derived from the compilation of numerous wind tunnel test results. The

13.2 Dynamic Instabilities

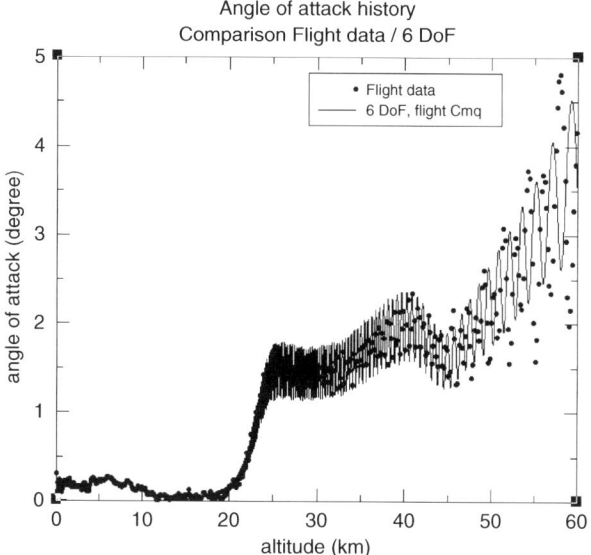

Fig. 13.9 Flight angle of attack

coefficient $C_{mq/G}$ is negative, except in the instability domain, which corresponds to Mach numbers lower than 3 and incidences lower than $3°$.

Figure 13.11 gives the evolutions of Mach number and incidence according to altitude, calculated using a 6 DoF code, for entry in the Martian atmosphere with conditions at 120 km V = 6 km/s, $\gamma = -21°$.

In the case of a probe without spin or asymmetry, incidence quickly increases in the altitude range corresponding to Mach numbers lower than 2. In the case of a probe with a spin corresponding to 5 rpm, gyroscopic effects somewhat decrease the divergence. At last, in the case with a trim angle ($4°$ at Mach > 3), bound here to a

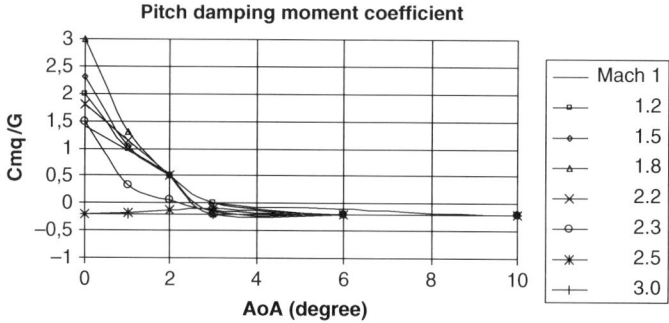

Fig. 13.10 Model of dynamic stability

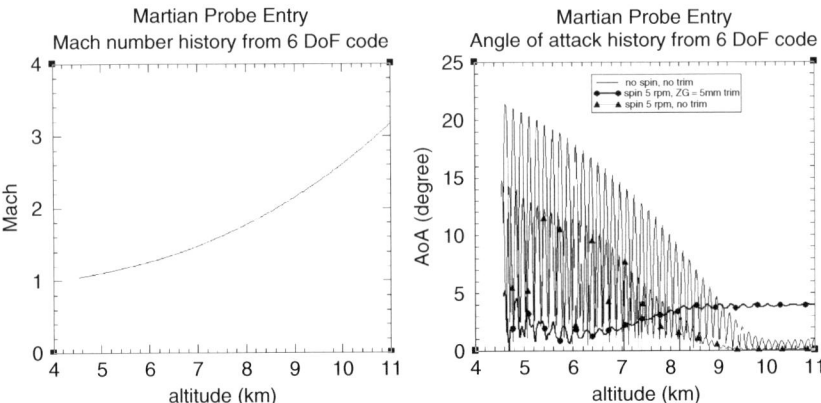

Fig. 13.11 Evolution of Mach number and angle of attack

lateral CG offset, the incidence does not diverge. Indeed, thanks to this trim angle, the time when the vehicle enters the most unstable regime near 1° of incidence is delayed until 6 km altitude. Subsequent decrease of the trim angle is related to the growth of static stability below Mach 2. Other configurations undergo a high increase in incidence below 10 km altitude. This growth of incidence corresponds to an augmentation of kinetic energy of rotation at each oscillation when crossing the unstable zone.

Chapter 14
Reentry Errors

In this chapter, we deal with the sensitivity of trajectory to various sources of dispersions. Aside from dispersions resulting from kinematics initial conditions, we can classify dispersion factors in two families, one affecting drag and the other affecting lift.

Dispersions of drag primarily affect relative velocity and chronology of the trajectory. Thus, except for a weak indirect effect on range, owing to the influence of velocity on curvature of the trajectory by gravity, drag does not contribute, strictly speaking, to trajectory deviations. On the other hand, occurrence of episodes with an average lift in a fixed inertial direction may involve large deviations of trajectory, downrange, and/or cross range errors.

We will analyze these various effects, first using a zero-incidence hypothesis, then by taking account of the angle of attack.

14.1 Zero Angle-of-attack Dispersions

14.1.1 Initial Conditions

In reentry on earth or another planet, we define initial conditions at a constant altitude corresponding to an arbitrary interface with the atmosphere. For earth, in the case of six degree-of-freedom (6 DoF) applications $H_I = 120$ km above the reference ellipsoid is used herein for most of vehicles (index I indicates the initial reentry point). For zero angle-of-attack reentry, altitude $H_I = 60$ km could be used as well. Eventually, a vehicle with very low ballistic coefficient will need a higher altitude interface. We will assume here that the planet is spherical and not revolving. Indeed, although rotation of the planet induces also non-negligible effects on a relative reentry trajectory, it has negligible effects with respect to the sensitivity to dispersions.

Initial conditions at point "I" of fixed altitude noted H_I or z_I are entirely defined by:

- The relative velocity vector V_I
- The flight path angle γ_I of this vector relating to the horizontal plane normal to the initial radius vector ($\vec{r}_I$) from the center of the planet

- Its azimuth Az$_I$ defined in the horizontal plane

For the planet having symmetry of revolution, the origin of azimuth angles is arbitrary, and we choose the origin in the direction of the nominal trajectory. In the same way, the initial latitude is indifferent in the analysis, and we can limit the study to the range from the nominal reentry point. Thus, the variations of current location and velocity compared to the nominal trajectory are:

- Δx or ΔP, downrange error
- ΔH or Δz, altitude deviation
- Δy, cross or side range error
- ΔV, deviation in relative velocity
- $\Delta \gamma$, deviation of flight path angle
- ΔAz, azimuth deviation

14.1.2 Consequences of Drag Dispersions

We analyze here the sensitivity of the trajectory to drag dispersions.

14.1.2.1 Flight Path Angle and Downrange Errors

In the case of zero angle-of-attack reentry, without wind, the drag force is:

$$C_D \equiv C_A \Rightarrow D = \frac{1}{2}\rho S_{ref} C_A V^2$$

Thus, aerodynamic dispersions of the trajectory are related to product ρC_A. While neglecting earth's rotundity, equations of motion are:

$$\frac{dV}{dt} = -\frac{D}{m} + g \sin \gamma$$
$$\frac{d\gamma}{dt} = \frac{g \cos \gamma}{V}$$

While using $dz = -V \sin \gamma \, dt$, as in Allen reentry, we obtain a differential system:

$$\frac{dV}{dz} = \frac{\rho S_{ref} C_A}{2m \sin \gamma} V - \frac{g}{V}$$
$$-\frac{\sin \gamma}{\cos \gamma}\frac{d\gamma}{dz} = \frac{d \log |\cos \gamma|}{dz} = \frac{g}{V^2}$$

14.1 Zero Angle-of-attack Dispersions

To get first-order estimate of the effect of gravity on flight path angle, we neglect the effect of gravity on velocity variation. Thus, in the case of an exponential atmosphere $\rho = \rho_0 e^{-\frac{z}{H_R}}$, we obtain (thanks to Allen's results):

$$V \approx V_I e^{-K\rho}$$

$$\ln \left| \frac{\cos \gamma}{\cos \gamma_I} \right| = \int_{z_I}^{z} \frac{g\,dz}{V^2}$$

with $K = \frac{H_R}{2\beta \sin \gamma_I}$; $\beta = \frac{m}{S_{ref} C_A}$.

That is to say,

$$\cos \gamma = \cos \gamma_I \cdot \exp\left(-\int_{z}^{z_I} \frac{g\,dz}{V^2}\right).$$

While using an approximation of the exponential, we obtain with $x = 2K\rho$:

$$\int_{z}^{z_I} \frac{g\,dz}{V^2} = \frac{gH_R}{V_I^2} \int_{2K\rho}^{2K\rho_I} \frac{e^x dx}{x} \approx \left(\frac{gH_R}{V_I^2}\right) \left(\ln|x| + x + \frac{x^2}{4} + \frac{x^3}{18} + \frac{x^4}{96} \right) \bigg|_{2K\rho}^{2K\rho_I}.$$

By noting that $x_I = 2K\rho_I \sim 0$,

$$\cos \gamma \approx \cos \gamma_I e^{\frac{g(z-z_I)}{v_I^2}} e^{-\frac{gH_R}{v_I^2} F(2K\rho)},$$

where $F(x) \approx x \left(1 + \frac{x}{4} + \frac{x^2}{18} + \frac{x^3}{96}\right)$.

We can note that the drag appears in the exponential term to the right, which gives the variation of flight path angle related to gravity and velocity loss. The first exponential corresponds to the evolution of flight path angle at constant velocity (under gravity effect, without atmosphere). We point out that, in this approximation, we neglected the effect of gravity on the velocity, as well as the effect of earth rotundity on variation of current flight path angle.

Application of this result to the calculation of flight path at ground level with conditions $\beta = 10^4$ kg.m^{-2}, $H_I = 60$ km, $V_I = 6000$ m/s, $\gamma_I = 30°$, $H_R = 7000$ m, and $\rho_S = 1.39$ kg/m^3 gives $K = 0.7$ and $x_S = 1.946$, $F(x_S) = 3.452$. This corresponds to $2.18°$ total variation of flight path angle of which $0.64°$ are the effect of the aerodynamic drag and $1.57°$ of gravity only. Results of exact numerical calculation using a three degree-of-freedom (3 DoF) code are $2.19°, 0.60°$, and $1.6°$, respectively, and are in excellent agreement with this approximate calculation.

In order to assess dispersions of current flight path angle and range, we use the assumption, valid for vehicles with high ballistic coefficient, that the relative variations of flight path angle $\frac{\delta \gamma}{\gamma} \ll 1$ are small. We obtain for flight path angle,

$$\delta\gamma(z) \approx \left(\frac{g\,H_R}{V_I^2}\right)\frac{\cos\gamma_I}{\sin\gamma_I}\left[\left(\frac{z_I - z}{H_R}\right) + F(2K\rho)\right]$$

Evolution of range variations δx with a reference trajectory with constant flight path angle is determined by the variations of flight path angle:

$$\delta x = x(\gamma_I + \delta\gamma) - x(\gamma_I)$$
$$\frac{dx}{dz} = -\frac{\cos\gamma}{\sin\gamma} \Rightarrow \frac{d(\delta x)}{dz} = \delta\left(\frac{dx}{dz}\right) \approx \frac{\delta\gamma}{\sin^2\gamma_I}$$

While using these results, we obtain an approximate expression of the reduction of range due to combined effect of gravity and drag:

$$\delta X(z) \approx \frac{g\,H_R\cos\gamma_I}{V_I^2 \sin^3\gamma_I}\int_{Z_I}^{Z}\left[\frac{z_I - z}{H_R} + F(2K\rho)\right]dz.$$

$$\delta X(z) \approx -H_R\left(\frac{g\,H_R}{V_I^2}\right)\frac{\cos\gamma_I}{\sin^3\gamma_I}\left[\frac{1}{2}\left(\frac{z - z_I}{H_R}\right)^2 + F'(2K\rho)\right]$$

with $F'(x) \approx x\left(1 + \frac{x}{8} + \frac{x^2}{54} + \frac{x^3}{384}\right)$.

This expression gives the first order of magnitude for range reduction, but may also provide the variation of range due to drag dispersions, included in the factor $x = 2\,K\rho$.

Application of this result with the preceding hypothesis gives $F'(x_S) = 2.593$ ($x_S = 1.946$ with $\rho_S = 1.39\,\text{kg/m}^3$), which corresponds to a total range reduction of 3638 m including 240 m related to aerodynamic braking and 3398 m related to FPA variation at constant velocity. These results are in agreement with 3 DOF calculations, which give 3557 m of range reduction, including 211 m dependent on aerodynamic braking.

While assuming relative variations of parameters are independent of altitude and differentiating range variation, we obtain:

$$\Delta(\delta X) = \frac{\partial(\Delta x)}{\partial K}\Delta K + \frac{\partial(\Delta x)}{\partial\rho}\Delta\rho.$$

Thus, the order of magnitude of range dispersion is:

$$\Delta(\delta X) \approx -H_R\frac{\cos\gamma_I}{\sin^4\gamma_I}\left(\frac{\rho H_R}{\beta V_I^2}\right)F''(2K\rho)\left[\frac{\Delta K}{K} + \frac{\Delta\rho}{\rho}\right]$$

with $F''(x) \approx 1 + \frac{x}{4} + \frac{x^2}{18} + \frac{x^3}{96}$ and $\frac{\Delta K}{K} = -\frac{\Delta\beta}{\beta}$.

Application of this result under the same conditions for calculation of FPA gives $F''(x_S) = 1.77$, that is to say a sensitivity of ground level range equal to 3.34 m per

14.1 Zero Angle-of-attack Dispersions

percent of variation on density or ballistic coefficient. This order of magnitude is confirmed by 3 DoF numerical calculations, which give ±3 m for ±1% variation on ballistic coefficient β.

This result justifies a posteriori the approximations. In agreement with common sense, an increase in density involves a reduction of range and an increase in ballistic coefficient, an increase in range.

It is thus confirmed that the influence on the range of dispersions of atmosphere density or ballistic coefficient is very low in the case of a vehicle with high initial velocity and high ballistic coefficient. Indeed, dividing by the variation of range, we obtain the relative value of dispersion,

$$\frac{\Delta(\delta X)}{\delta X} \approx \frac{1}{\sin \gamma_I} \frac{F''(2K\rho)\left(\frac{p}{\beta g}\right)\left[\frac{\Delta \beta}{\beta} - \frac{\Delta \rho}{\rho}\right]}{\left[\frac{1}{2}\left(\frac{z-z_I}{H_R}\right)^2 + F'(2K\rho)\right]}.$$

We can observe that relative variability of range is proportional to $\frac{\Delta \beta}{\beta^2}$ and $\frac{\Delta \rho}{\rho \beta}$. While parameter β is large, this explains the low sensitivity of range on drag dispersions.

14.1.2.2 Dispersion on Velocity

For this purpose, we use the results of Allen adapted to the case of an atmosphere at hydrostatic equilibrium, without restriction on temperature profile. These hypotheses correspond to:

$$\frac{dp}{dz} = -\rho g$$

$$p = \rho \, r T(z)$$

We established in Chap. 9 that velocity verify:

$$V \approx V_D e^{-\frac{p}{p_C}}$$

$$p_c = 2\beta \, g \sin \gamma_I$$

Note: For an exponential atmosphere, the formulations starting from ρ or of p give identical results.

By differentiating the velocity, we obtain:

$$\frac{\Delta V}{V} \approx \frac{\Delta V_I}{V_I} + \frac{p}{p_C}\left(\frac{\Delta p_C}{p_C} - \frac{\Delta p}{p}\right)$$

$$\frac{\Delta V}{V} \approx \frac{\Delta V_I}{V_I} + \frac{p}{p_C}\left(\frac{\Delta \beta}{\beta} + \frac{\Delta \gamma_I}{tg \gamma_I} - \frac{\Delta p}{p}\right)$$

We observe that this expression corresponds to a constant ballistic coefficient and constant relative variation $\frac{\Delta\beta}{\beta}$, and that velocity variation depends on dispersion of density only through the local pressure at altitude Z. Unlike range dispersions, relative sensitivity of velocity is not negligible; the magnitude depends, at first order, on relative variations of drag parameters.

Numerical application for our reference case gives a ground velocity of 2234 m/s and a sensitivity of 23 m/s per percent of constant deviation on atmosphere pressure or on ballistic coefficient.

14.1.3 Sensitivity to Initial Conditions

We analyze here dispersions related to the variations of initial velocity and flight path angle.

14.1.3.1 Sensitivity to Initial Velocity

Analysis, in Sect. "Flight Path Angle and Downrange Errors," of the effect of aerodynamic drag on the evolution of flight path angle and range has provided the reduction of range by gravity compared to a reference trajectory with constant FPA:

$$\delta X(z) \approx -Href \left(\frac{g\,H_R}{V_I^2}\right) \frac{\cos\gamma_I}{\sin^3\gamma_I} \left[\frac{1}{2}\left(\frac{z - z_I}{H_R}\right)^2 + F'(2K\rho)\right]$$

By adding the range with constant FPA, X_I, we obtain the total range:

$$X = X_I + \delta X = \frac{Z_I}{\tan\gamma_I} + \delta X$$

From this expression, we directly obtain the sensitivity of range on initial velocity:

$$\frac{\Delta X}{X} \approx \frac{\Delta(\delta X)}{X_I} \approx -2\delta X \frac{\tan\gamma_I}{Z_I} \frac{\Delta V_I}{V_I}$$

Under the same conditions as for assessment of dispersions related to drag, we obtain a relative sensitivity of 7.10^{-4} per percent variation on initial velocity. This corresponds to the nominal range 104 km to a relatively low dispersion equal to 73 m per percent. Exact 3 DoF numerical calculations give 65 m.

The sensitivity of final velocity to initial velocity is obtained in a very simple way using the Allen expression of current velocity:

$$V \approx V_I e^{-K\rho s} \Rightarrow \Delta V \approx e^{-K\rho s} \Delta V_I.$$

For our reference case, we obtain at sea level a variation 0.372 m/s per m/s of variation on initial velocity.

14.1.3.2 Sensitivity to Initial Flight Path Angle

Utilizing the expression from the preceding paragraph, we obtain:

$$\frac{\Delta X}{X} \approx \frac{\Delta X_I}{X_I} + \frac{\delta X}{X_I}\frac{\Delta(\delta X)}{\delta X}$$

which gives with respect to initial FPA,

$$\frac{\Delta X}{X} = -\left[\frac{1}{\sin\gamma_I \cos\gamma_I} + \frac{\delta X}{Z_I}\left(3 + tg^2\gamma_I\right)\right]\delta\gamma_I .$$

The second term is much smaller than first, which means that range dispersion is very close to that of the constant FPA trajectory:

$$\Delta X \approx -\frac{Z_I}{\sin\gamma_I}\left(\frac{\Delta\gamma_I}{\sin\gamma_I}\right)$$

In the case of 30° initial FPA at $z_I = 60$ km, corresponding to a nominal range about 104 km, we obtain 419 m dispersion for 0.1° FPA variation.

Influence of *initial altitude* results using the same approximation,

$$\frac{\Delta X}{X} \approx \frac{\Delta X_I}{X_I} = \frac{\Delta z_I}{z_I} \Rightarrow \Delta X \approx \frac{X_I}{z_I}\Delta z_I = \frac{\Delta z_I}{tg\gamma_I} .$$

This evaluation does not offer anything very useful, insofar as the initial reentry point is arbitrarily selected at a fixed altitude.

Let us evaluate at last the sensitivity of final velocity on *initial FPA*:

$$V \approx V_I e^{-K\rho_S} \Rightarrow \frac{\Delta V}{V} \approx -\rho_S \Delta K = K\rho_S \frac{\Delta\gamma_I}{tg\gamma_I}$$

That is to say,

$$\Delta V \approx V_I K\rho_S e^{-K\rho_S}\frac{\Delta\gamma_I}{tg\gamma_I}$$

This corresponds for our reference case +6.7 m/s for 0.1° FPA increase.

14.2 Nonzero Angle of Attack

14.2.1 Effects of Incidence on Aerodynamic Loads

Lift and drag coefficients of a symmetrical vehicle at incidence are:

$$C_D = C_A \cos\overline{\alpha} + C_N \sin\overline{\alpha} \; ; \quad C_L = C_N \cos\overline{\alpha} - C_A \sin\overline{\alpha}$$

For a slender vehicle at low angle of attack, we have:

$$C_A \approx C_{A0} + C_{A\bar{\alpha}^2}\bar{\alpha}^2 \; ; \; C_N \approx C_{N\bar{\alpha}}\bar{\alpha}$$

$$C_L \approx (C_{N\bar{\alpha}} - C_{A0})\bar{\alpha} \; ; \; C_D \approx C_{A0} + (C_{N\alpha} + C_{A\bar{\alpha}^2} - \frac{1}{2}C_{A0})\bar{\alpha}^2$$

Drag can be expressed in the equivalent form:

$$C_D \approx C_{D0} + K C_L^2 \; ; \quad \text{with } K = \frac{C_{N\bar{\alpha}} + C_{A\bar{\alpha}^2} - \frac{1}{2}C_{A0}}{(C_{N\bar{\alpha}} - C_{A0})^2}$$

This expression highlights the quadratic behavior of lift-induced drag at low incidence. Figure 14.1 illustrates this behavior at different Mach for a 8° blunted conical reentry vehicle (RV), with $R_N/R_B = 0.1$, at 0°–8° angle of attack.

The drag coefficient builds very fast with incidence (Fig. 14.2). Thus for our 8° conical body, 8° incidence corresponds to 120% increase of drag coefficient at Mach 20 (at high Reynolds continuous flow regime).

Principal effect of this drag increase on the trajectory is identical to that of zero angle-of attack drag variations analyzed in preceding paragraphs.

However, for incidences higher than 2° or 3°, relative variations of drag and lift grow very quickly, and we could think that for higher initial incidences, the effects on reentry are very important. Fortunately, for a well-designed vehicle, these effects are not governing. Indeed, as we saw in Chap. 10, the major part of angle-of-attack convergence takes place during the accelerated phase of the reentry, i.e., in a range of altitude where the dynamic pressure is low.

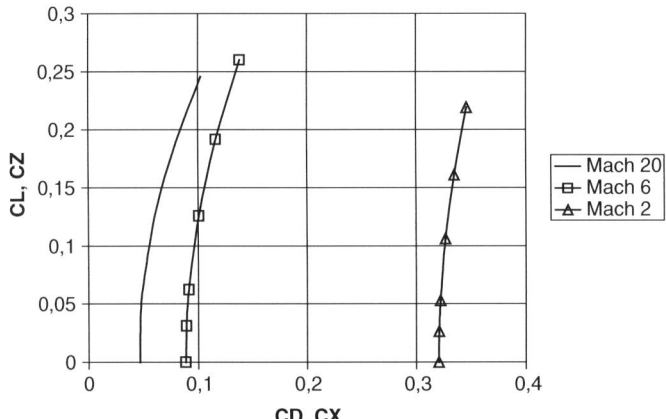

Fig. 14.1 Drag polar, 8° conical body, incidence 0° − −8°

14.2 Nonzero Angle of Attack

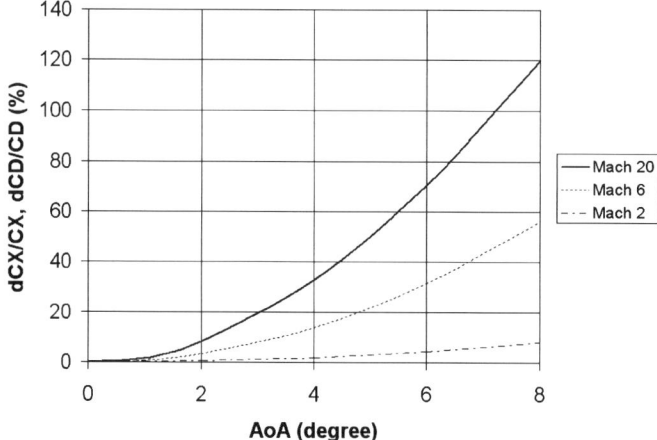

Fig. 14.2 Relative drag variation of an 8° conical

Although relative variations of drag are important, drag forces are low compared with vehicle weight and their variations have modest consequences on total acceleration as well as on the trajectory and impact point. The principal consequences are on velocity and chronology, which undergo first-order variations. Except in special cases, the most serious consequences of angle-of-attack are not related to initial reentry incidence, but to residual incidence, in high dynamic pressure zones, i.e., at low altitude. Residual angle-of-attack generally originates in aerodynamic asymmetries, possibly worsened when crossing second resonance, or static or dynamic instabilities. These modifications of trajectories are important and approximate methods of calculation are not suitable.

We will thus limit our ambitions to the analytical study of lift effects related to small disturbances in angle of attack.

14.2.2 Effects of Initial Angle-of-attack

14.2.2.1 Accelerated Phase of Center of Mass Movement

We use here hypothesis and results of Sect. 10.2, valid between 120 and 50 km altitude. We assume the relative velocity is constant module and direction, and use a quasi-inertial frame Gxyz fixed to the velocity vector (Fig. 14.3).

According to Sect. 10.2, at low angle of attack, approximate evolutions of the incidence verify:

$$\frac{\theta}{\theta_I} \approx (1+\xi)^{-\frac{1}{4}} \; ; \quad \dot{\psi}_\pm \approx p_r \left(1 \pm \sqrt{1+\xi}\right)$$

with $p_r = \frac{\mu p_0}{2}$ and $\xi = \left(\frac{\omega}{p_r}\right)^2$, $\omega^2 = -\frac{\bar{q} S_{ref} L_{ref} C_{m\alpha/G}}{I_T}$

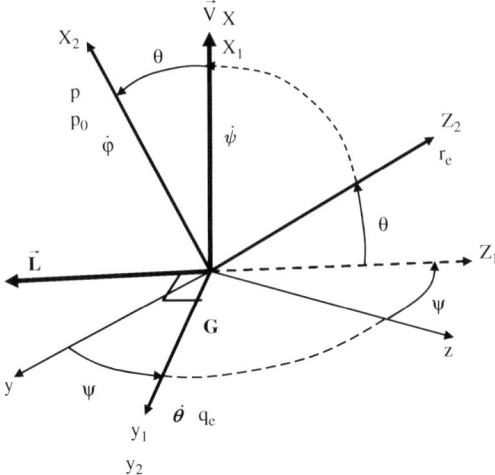

Fig. 14.3 Reference frame definition

In the plane Gyz, complex incidence and lift are written as:

$$\bar{\theta} = \theta \exp\left[i\left(\psi - \frac{\pi}{2}\right)\right]$$
$$\bar{L} = \bar{q} S_{ref} C_{L\alpha} \bar{\theta}$$

While choosing Gy directed along initial lift $\vec{L}_I$, acceleration normal to the velocity vector is:

$$\bar{A}_N = \frac{\bar{q} S_{ref} C_{L\alpha}}{m} \theta \exp[i\psi]$$

Thus, the evolution of normal velocity is:

$$\frac{d\bar{V}_N}{dt} = \bar{A}_N = \frac{V_I^2 S_{ref} C_{L\alpha}}{2m} \rho \theta \exp[i\psi]$$
$$\bar{V}_N \approx \frac{V_I^2 S_{ref} C_{L\alpha} \theta_I}{2m} \int_0^t \left\{\frac{e^{i\psi}}{[1+\xi]^{\frac{1}{4}}} \cdot \rho\right\} dt$$

with $\xi = \frac{\omega^2}{p_r^2} = a\rho$; $a = -\frac{V_I^2 S_{ref} L_{ref} C_{m\alpha/G}}{2 I_T p_r^2}$

14.2 Nonzero Angle of Attack

While using variable altitude H instead of time, we have:

$$dt = -dH/V_I \sin \gamma_I$$

$$\Rightarrow \quad \bar{V}_N \approx -\frac{V_I S_{ref} C_{L\alpha} \theta_I}{2m \sin \gamma_I} \int_{H_I}^{H} \frac{e^{i\psi}}{[1+a\rho]^{\frac{1}{4}}} \rho dH$$

where the angle ψ expression according to z is,

$$\psi_\pm = \int_0^t \dot{\psi}_\pm dt \approx -\frac{p_r}{V_I \sin \gamma_I} \int_{H_I}^{H} \left(1 \pm \sqrt{1+a\rho}\right) dH$$

Now let us analyze the behavior of ψ for the two modes of angular motion.

Figure 14.4 shows an initial free coning motion of pure precession, such that angular momentum is directed along initial velocity.

At initial altitude, $\vec{A}_N = \vec{L}/m$ rotates at finite rate $\dot{\psi}_0 = 2p_r = \mu p$ around flight path. Then $\dot{\psi}$ grows very slowly when altitude decreases and dynamic pressure increases. In this case, the normal acceleration induces oscillations of normal velocity components V_y and V_z, without significant effect on the average trajectory.

Figure 14.5 shows an initial free coning motion of pure nutation, such that initial angular momentum is along the axis of the vehicle. Initial angular rate $\dot{\psi}_-$ of the lift force is null and, when altitude decreases, grows very slowly with a sign opposed to roll rate. In this case normal acceleration, initially quasi-stationary, carries out its first half period of rotation very slowly in an altitude domain such that, because of growth of the dynamic pressure, its average long-term effect on the trajectory is not negligible.

Figures 14.6 and 14.7 show the behavior resulting from the preceding formulae, for a 8° conical vehicle, m = 117 kg, V_0 = 6000 m/s, γ_I = 30°, I_x = 2 m^2 kg, I_T = 15 m^2 kg, S_{ref} = 0.196 m^2, $x_G - x_{CP}$ = 0.162 m, $C_{N\alpha}$ = 2, $C_{L\alpha} = C_{N\alpha} - C_A$ = 1.93, p = 2π radians/s, θ_I = 20°, exponential atmosphere ρ_S = 1.39 kg/m^3, and H_R = 7000 m.

We note that in the case of pure precession, the two components of normal velocity oscillate around a zero average value. In the case of pure nutation, the component

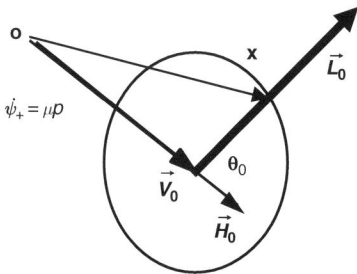

Fig. 14.4 Initial coning motion of precession

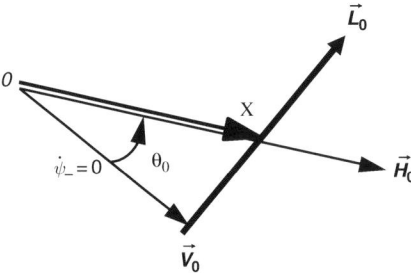

Fig. 14.5 Initial coning motion of nutation

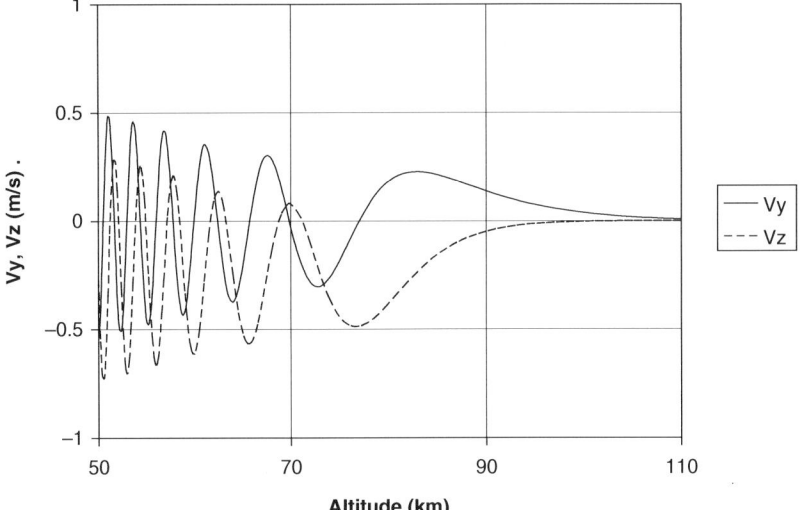

Fig. 14.6 Velocity error for nutation condition

V_y average value is zero (along the initial lift force direction) and component V_z average value is negative. This corresponds to a deviation of the mean trajectory normal to the meridian plane of initial lift, opposed to Gz in agreement with the sign of rotation rate $\dot{\psi}_- < 0$.

Figure 14.8 shows results of 6 DoF computations with the same hypothesis. One can observe, there is good agreement with the analytical solution.

Now let us develop the case that corresponds to the initial condition of pure nutation (Fig. 14.5 and 14.6). The assumption of an exponential atmosphere implies $-dH/H_R = d\rho/\rho$. While using $\xi = a\rho$ as new variable, the phase ψ_- is obtained analytically:

14.2 Nonzero Angle of Attack

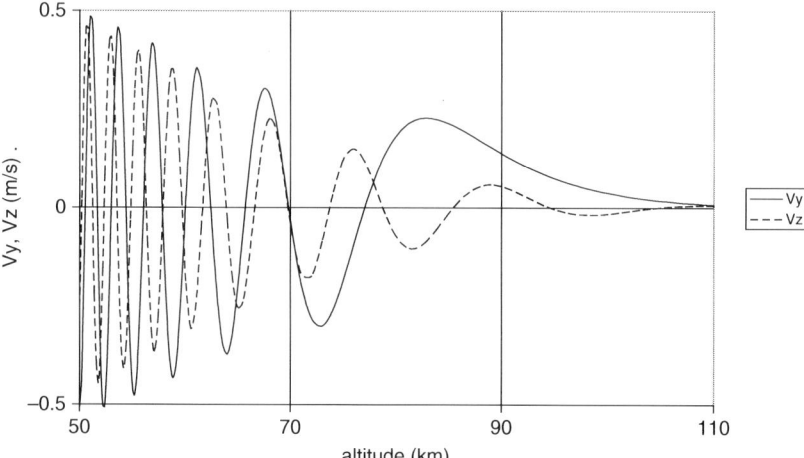

Fig. 14.7 Velocity error for precession condition

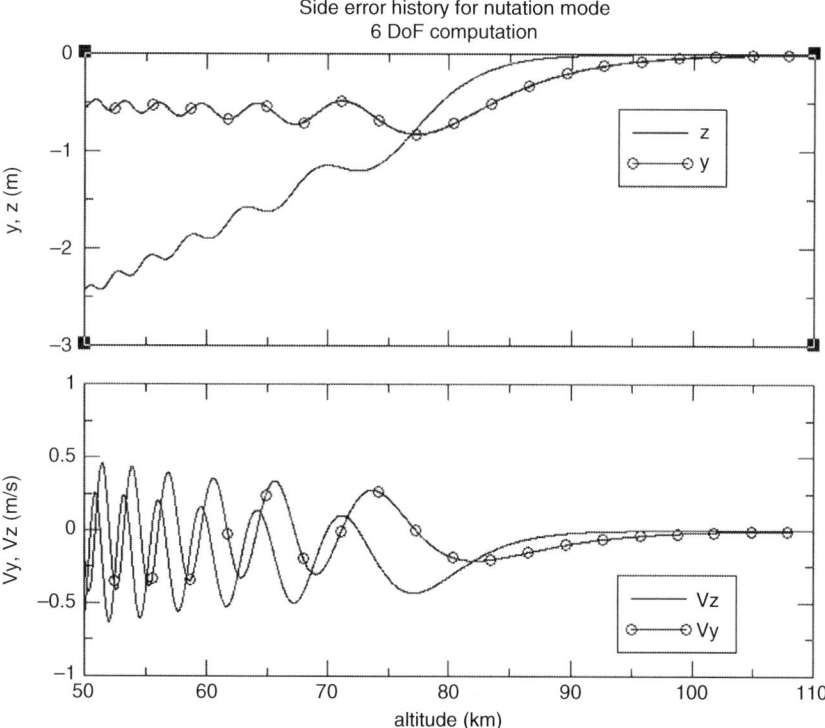

Fig. 14.8 Evolution of lateral error for nutation mode, H>50 km

$$\psi_- = -\frac{H_R p_r}{V_I \sin \gamma_I} \int_0^\xi \frac{(\sqrt{1+u}-1)}{u} du$$

$$\psi_- \approx -a' \left[2\left(\sqrt{1+\xi}-1\right) + \ln\left(4\frac{(\sqrt{1+\xi}-1)}{\xi(\sqrt{1+\xi}+1)}\right) \right]$$

with $a' = \frac{H_R p_r}{V_I \sin \gamma_I}$.

Normal velocity is expressed then as,

$$\bar{V}_N \approx \frac{H_R V_I S_{ref} C_{L\alpha}\theta_I}{2m \sin \gamma_I} \int_0^{\rho(H)} \frac{e^{i\psi_-}}{[1+a\rho]^{\frac{1}{4}}} d\rho = \frac{H_R V_I S_{ref} C_{L\alpha}\theta_I}{2maa' \sin \gamma_I} I(a',\xi)$$

with $I(a',\xi) = a' \int_0^\xi \frac{e^{i\psi_-(a',u)}}{[1+u]^{\frac{1}{4}}} du$ (dimensionless).

There is no exact analytical solution for I(a', ξ). The numerical study [PLA] of this function has shown that for large values of a', its average value is relatively independent of a':

$$I(a', x) \sim -i$$

While using more approximation, it is possible to obtain this value. First, we use a first-order approximation of phase angle, valid only during the first quarter of period:

$$\psi'_- \approx -a'\xi; \Rightarrow I(\xi, a') \approx a' \int_0^\xi \frac{e^{-ia'u}}{[1+u]^{\frac{1}{4}}} du$$

Then we neglect the variation due to the root, which gives:

$$I(\xi, a') \approx a' \int_0^\xi e^{-ia'u} du = i\left[e^{-ia'\xi}-1\right]; \Rightarrow \langle I(a',x)\rangle \approx -i$$

where operator < I > indicates the time-averaged mean value.

Thus, we obtain an approximate expression for mean normal velocity perturbation:

$$\langle V_N \rangle = -i \frac{H_{ref} V_I S_{ref} C_{L\alpha}\theta_I}{2m\, a\, a' \sin \gamma_I} = -i \frac{C_{L\alpha}\theta_I I_T\, p_r}{m(x_G - x_{CP})C_{N\alpha}} = -i \frac{C_{L\alpha}\theta_I I_x\, p}{m(x_G - x_{CP})C_{N\alpha}}.$$

$$\langle V_y \rangle = 0 \ ; \quad \langle V_z \rangle = -\frac{C_{L\alpha}\theta_I I_x\, p}{m(x_G - x_{CP})C_{N\alpha}}$$

The angular deviation of trajectory, which results, develops during the first quarter of period of angular motion (roughly between altitudes H = 90 and 70 km). Its order of magnitude is:

14.2 Nonzero Angle of Attack

$$\delta_z \approx \frac{\langle V_z \rangle}{V_I} \approx -\frac{C_{L\alpha}\theta_I I_x p}{m V_I (x_G - x_{CP}) C_{N\alpha}}.$$

This approximate formula results in $V_z = -0.22$ m/s, which agree with order of magnitude of 6 DoF numerical results in Fig. 14.8. The dependence with respect to the various parameters makes physical sense, because the error is proportional to initial angle of attack θ_I and gyroscopic angular momentum $H_x = I_x p$ (indeed, it delays the subsequent decreasing of incidence), and inversely proportional to mass and stability parameters.

The deviation of trajectory results in a lateral error $\Delta z \approx L\delta_z \approx H\delta_z/\sin \gamma_I$ at the ground impact, which gives a very small 6 m value. This estimation agrees quite well with 6 DoF computations in Fig. 14.9.

However, we point out that the purpose of these approximate developments is only to understand the origin of the induced error and to estimate its order of magnitude. It results that the effect is low for current reentry velocity and spin rate. Taking into account the coarse approximations, we needed to use, only a calculation using an exact – 6 DoF model allows a reliable result. In particular, owing to the fact that this error develops in a phase of rarefied flow, the exact lift coefficient is lower than the coefficient in continuous mode used here. In addition, we assumed constant velocity, which is roughly valid only above 60 km. It results that in this altitude range, although the incidence decreases, amplitude of oscillations of normal velocity is increasing.

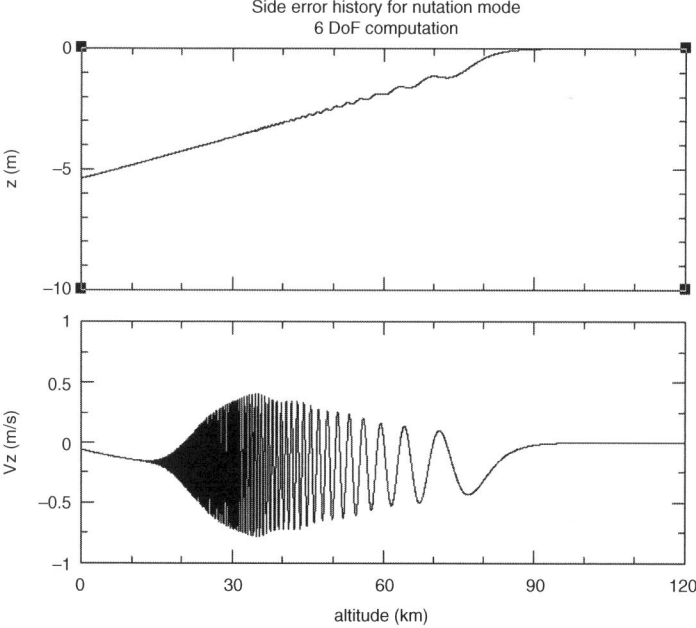

Fig. 14.9 Evolution of lateral error for nutation mode, 120 km to sea level

14.2.2.2 Decelerated Phase of Motion

Below 50 km altitude, we must take into account the increase in density and the reduction of velocity by aerodynamic drag, in addition to aerodynamic damping of incidence. The amplitudes of oscillations of normal load factor and normal velocity generally pass by a maximum before the end of the trajectory. The amplitude of these oscillations can be estimated asymptotically using present approximations and those of Chap. 11.

$$\frac{\theta}{\theta_I} \approx \xi^{-\frac{1}{4}} e^{\int_0^t \Lambda dt}$$

$$\Lambda \approx -\frac{\ell_\alpha + m_q}{2} = -\frac{1}{2}\frac{\bar{q} S_{ref}}{V}\left[\frac{C_{L\alpha}}{m} - \frac{L_{ref}^2}{I_T}C_{mq}\right]$$

$$\int_0^t \Lambda dt \approx \frac{H_R}{\sin \gamma_I} \int_0^{\rho(z)} \Lambda \frac{d\rho}{\rho V} \Rightarrow \int_0^t \Lambda dt \approx -D\rho$$

with $D = \frac{H_R S_{ref}}{4 \sin \gamma_I}\left[\frac{C_{L\alpha}}{m} - \frac{L_{ref}^2 C_{mq}}{I_T}\right]$.

By introducing the Allen approximation for velocity,

$$\xi = \frac{\omega^2}{p_r^2} = a_0 \rho \left(\frac{V}{V_I}\right)^2 = a_0 \rho e^{-2K\rho}$$

$$K = \frac{H_R}{2\beta \sin \gamma_I}$$

we obtain the amplitudes of oscillations of the incidence and normal acceleration:

$$\theta = \theta_I \xi^{-\frac{1}{4}} e^{-D\rho} = \theta_0 [a_0 \rho]^{-\frac{1}{4}} e^{-\left(D - \frac{K}{2}\right)\rho}$$

$$|A_N| = \frac{L}{m} = \frac{\bar{q} S_{ref}}{m} C_{L\alpha} \theta = \frac{S_{ref} V_I^2 C_{L\alpha} \theta_I a_0^{-\frac{1}{4}}}{2m} \rho^{\frac{3}{4}} e^{-\left(D + \frac{3}{2}K\right)\rho}$$

That is to say,

$$|A_N| = \frac{S_{ref} V_I^{\frac{3}{2}} p_r^{\frac{1}{2}} C_{L\alpha} \theta_I}{2m}\left[\frac{2I_T}{S_{ref}(x_G - x_{CP})C_{N\alpha}}\right]^{\frac{1}{4}} \rho^{\frac{3}{4}} e^{-D'\rho}$$

with $D' = D + \frac{3}{2}K = \frac{H_R S_{ref}}{m \sin \gamma_I}\left[\frac{3}{4}C_D + \frac{1}{4}\left(C_{L\alpha} - \frac{mL_{ref}^2}{I_T}C_{mq}\right)\right]$

14.2 Nonzero Angle of Attack

The maximum of normal acceleration is given by:

$$\frac{d}{d\rho}\left[\rho^{\frac{3}{4}}e^{-D'\rho}\right] = 0 \rightarrow \rho_m = \frac{3}{4D'}.$$

This corresponds to an altitude $Z_m = H_R Ln\left[\frac{\rho_S}{\rho_m}\right] = H_R Ln\left[\frac{4D'\rho_S}{3}\right]$,

and amplitude $A_{N,\max} = \frac{S_{ref} V_I^{\frac{3}{2}} p_r^{\frac{1}{2}} C_{L\alpha}\theta_I}{2m}\left[\frac{2I_T}{S_{ref}(x_G - x_{CP})C_{N\alpha}}\right]^{\frac{1}{4}}\left[\frac{3}{4eD'}\right]^{\frac{3}{4}}.$

With the same assumptions as the preceding paragraph, for $\theta_I = 20°$ initial angle of attack, we obtain $29.6\,\text{m.s}^{-2}$ for the maximum normal acceleration at 26 km altitude. The evolution of normal load is shown in Fig. 14.10, as well as the result of 6 DoF code with same hypothesis (zero gravity, exponential atmosphere). This computation provides $31.4\,\text{m.s}^{-2}$ maximum load at 26.6 km, which is in agreement with analytical result.

Now, we know the amplitude of normal acceleration, and we can calculate the lateral fluctuations of the trajectory of the center of mass. In this range of altitude, rotational frequencies of vehicle axis and of normal acceleration $\vec{A}_N$ with respect to inertial frame are, according to Chap. 11:

$$\dot{\psi}_\pm = p_r \pm \omega_a$$

The current normal acceleration is thus of the form:

$$\vec{A}_n = |A_N| e^{i\left[\psi_0 + \int_{t_0}^{t} \dot{\psi}_\pm d\tau\right]}$$

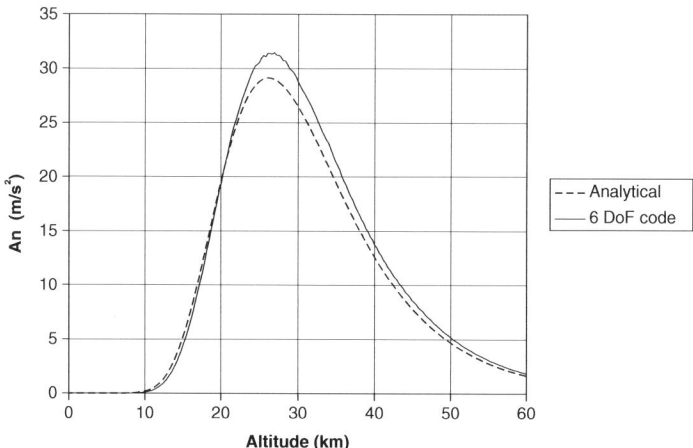

Fig. 14.10 Normal load

Thus, during an interval of time [t_0, t], evolutions of normal velocity and location are of the form:

$$\bar{V}_N = \int_{t_0}^{t} \bar{A}_N dt' + \bar{V}_{N0}$$

$$\bar{R}_N = \int_{t_0}^{t} dt' \int_{t_0}^{t'} \bar{A}_N dt'' + \bar{V}_{N0}(t - t_0) + \bar{R}_{N0}$$

We must study by interval the stability of this kind of motion, while considering on the one hand that the pulsation ω_a is slowly variable, and in addition we are in a range of altitude where nutation and precession rates are finite $\left|\dot{\psi}_\pm\right| > |p_r|$ (I.e., $\omega_a > 2|p_r|$). We must show that if at the initial time t_0 average deviations of normal velocity $\bar{V}_{N0}$ and location $\bar{R}_{N0}$ are null, the later evolution remains near zero or very low average value.

This demonstration comes within a pure mathematical field and leaves the framework of this work. Experimental observations and digital simulation show that under the conditions, which have just been defined, and for usual spin rates p_r, the lateral motion is stable and of helical nature around the mean trajectory. This phenomenon is currently named "lift averaging."

Orders of magnitude of amplitudes of oscillations of velocity and location around the mean trajectory correspond to:

$$\Delta V_N \approx \frac{|A_N|}{|\dot{\psi}|} \; ; \quad \Delta R_N \approx \frac{|A_N|}{|\dot{\psi}|^2}$$

This corresponds for a pure mode $\dot{\psi}_+$ or $\dot{\psi}_-$ with a current lateral circular motion, i.e., a circular helical trajectory (6 DoF results are shown in Fig. 14.11).

Approximate expressions of radius and pitch (wavelength) of the helix are:

$$R = \frac{|A_N|}{|\dot{\psi}|^2}; \quad \lambda = V \frac{2\pi}{|\dot{\psi}|}$$

Corresponding evolutions of radius and wave length corresponding to our reference vehicle are represented in Fig. 14.12.

We can note in Fig. 14.12 that for a vehicle of this type and in this range of altitude, amplitudes of high frequency lateral variations of trajectory generated by the evolution of the initial angle of attack is remarkably low.

For a combination of the two modes, the lateral motion is epicycle-like or in "petal plot" (to see Fig. 14.13, 6 DoF computation) already studied in Chap. 11 on the evolution of the angle of attack. The order of magnitude of variations to the average trajectory provided by preceding estimates remains valid.

14.2 Nonzero Angle of Attack

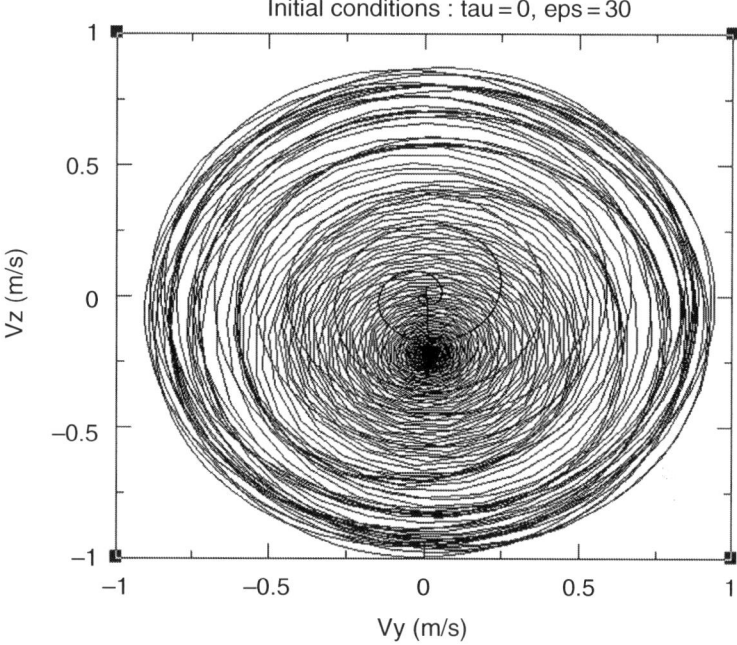

Fig. 14.11 Petal plot of lateral velocity for pure nutation

14.2.2.3 Trimmed Phase

This phase begins toward 15–20 km altitude, when incidence oscillations bound to initial incidence almost converged. In practice, residual oscillations of incidence are generally lower than $0.1°$.

We study the case of a static trim at spin rate p, assumed here as constant, far from critical roll rate.

If there is a static trim angle related to an asymmetry, the movement in incidence corresponds to lunar motion described in Chap. 11. With respect to an inertial observer, the axis of the vehicle turns around the velocity vector with an angular precession rate $\dot{\psi}_{eq} \approx p$ and maintains a constant angle equal to the trim angle $|\alpha_{eq}|$. The normal acceleration is $|A_{N,eq}| = \frac{\bar{q} S_{ref}}{m} C_{L\alpha} |\alpha_{eq}|$. This acceleration turns around the velocity vector at the same angular rate $\dot{\psi}_{eq}$.

The result is a current movement of circular helix type with pulsation p:

$$|R_N| \approx \frac{|A_{N,eq}|}{p^2} \; ; \quad \lambda = V \frac{2\pi}{p}$$

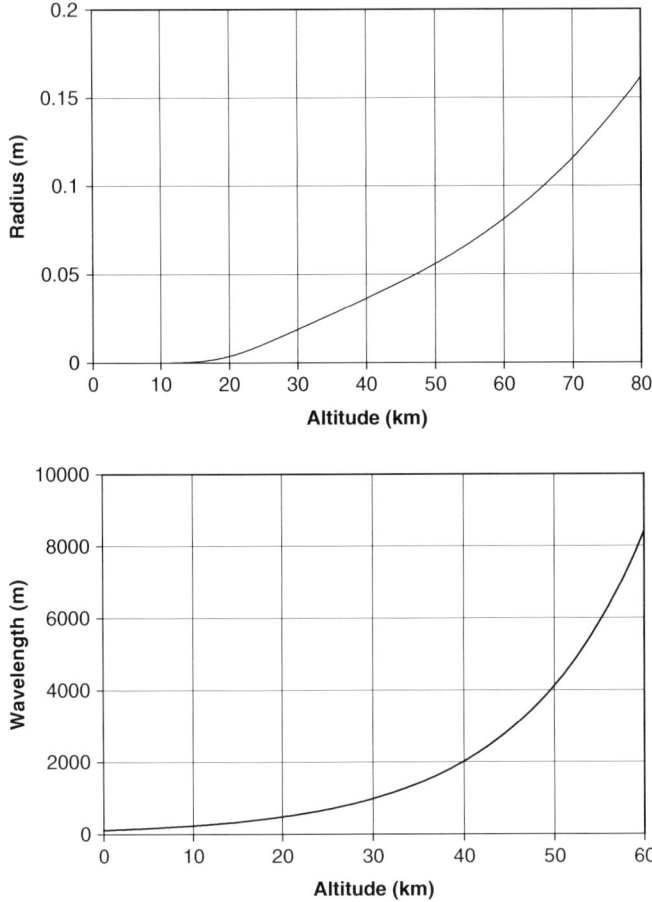

Fig. 14.12 Parameters of helix motion

The order of magnitude of the characteristics of this movement is given in Fig. 14.14 under the same conditions as the preceding paragraph, for a roll rate p = 1 rev. per second and a trim angle of attack $\alpha_{eq} = 1°$.

We can note that although relatively low, the amplitude of the fluctuations of the transverse movement corresponding to a static trim and lunar motion conditions is 10–100 times higher than fluctuations related to the initial angle of attack. This is clearly related to the precession rate $\dot{\psi}_{eq} \approx p$ of trim angle that is generally much lower than that of the oscillatory component of the incidence ($\dot{\psi}_\pm \approx p_r \pm \omega_a$). The wavelength of the normal velocity oscillations is about equal to the RV velocity. Figure 14.15 shows typical 6 DoF plots of side variation of velocity and location referenced to the zero angle-of-attack trajectory.

14.2 Nonzero Angle of Attack

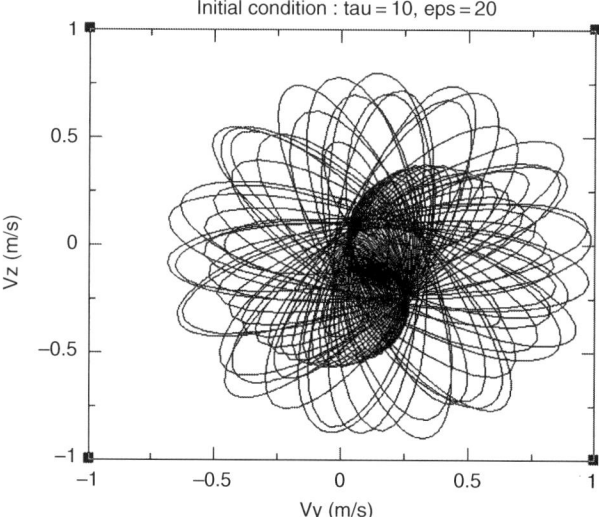

Fig. 14.13 Petal plot of lateral velocity for epicycle motion

14.2.2.4 Stability of the Lateral Motion

We admitted that when the angular precession rate $\dot{\psi}$ is slowly variable and of module $|\dot{\psi}|$ sufficient, deviations of mean trajectory related to incidence are negligible, due to lift-averaging phenomenon. The first example of instability of the mean trajectory was already studied; it corresponds to the mean normal velocity, which develops during the beginning of the accelerated reentry, associated to the nutation mode. Indeed, this mode corresponds to a zero initial rotation rate of the axis of the vehicle and lift vector around the velocity vector. Long duration of the first half period of rotation of normal acceleration involves a deviation of trajectory not compensated during the following periods, which are of much shorter duration, so lift averaging does not occur.

This phenomenon is likely to occur more generally during an inversion with passage through zero or of a temporary minimum of $|\dot{\psi}|$.

Such instability occurs in the presence of trim angle of attack and a roll moment, when roll rate crosses through zero.

Another cause of lift averaging defect is a sharp variation of some aerodynamic parameter, or its continuous variation for duration shorter than half period of precession.

Consequences of Roll Through Zero

In order to assess the side velocity disturbance, we assume positive initial roll rate and constant negative roll acceleration $\dot{p} = \dot{p}_0 < 0$, and we choose the origin of time coincident with rolling cancellation. We assume that the vehicle is at constant

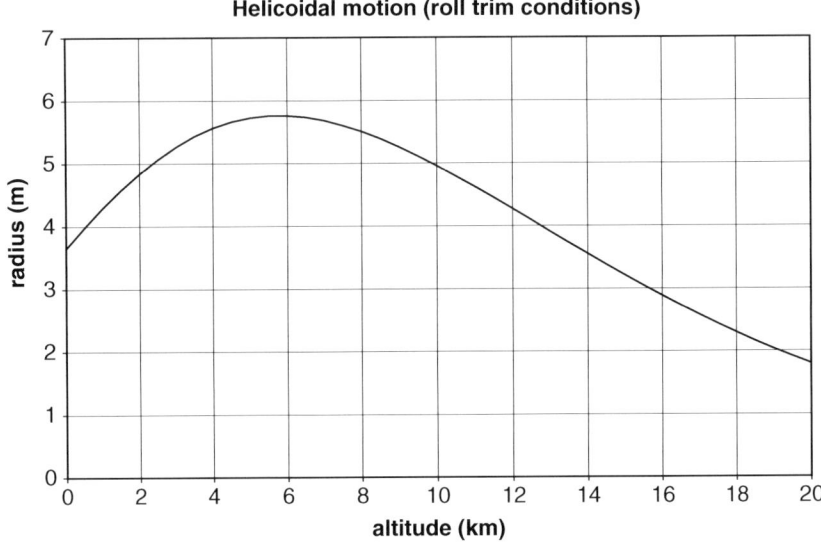

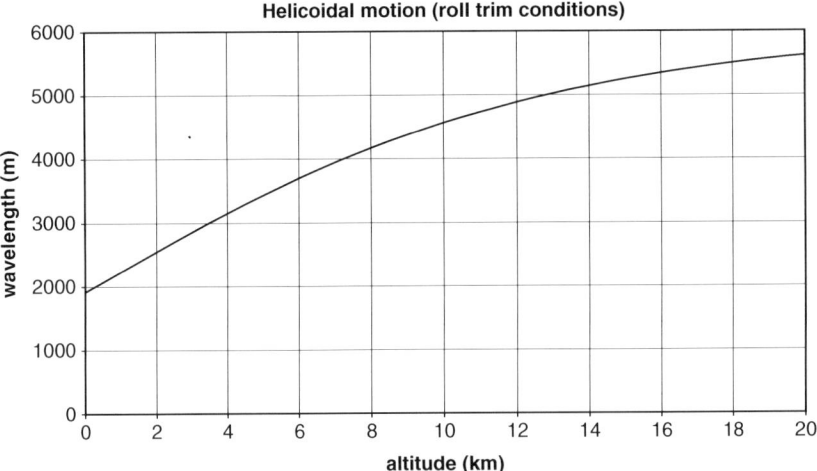

Fig. 14.14 Helix motion, roll trim conditions

14.2 Nonzero Angle of Attack

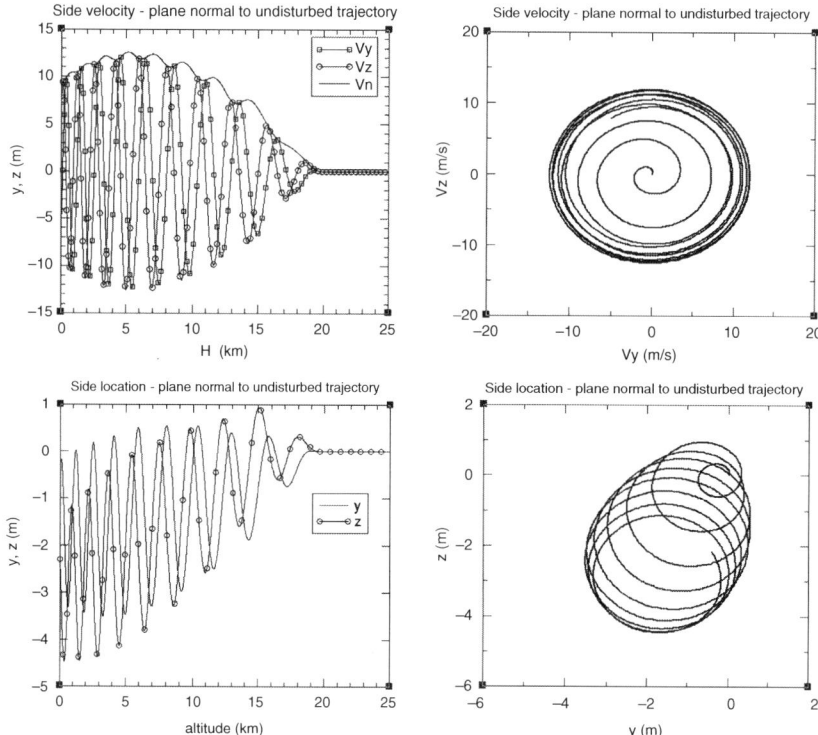

Fig. 14.15 History of motion's parameters during roll trim phase (lunar motion)

trim incidence, and we use a reference frame Gxyz such as Gx is along the nondisturbed flight path, and normal axes such as axis Gy is according to the direction of the lift force at time origin (Fig. 14.16). To finish, we assume that during duration T of normal velocity disturbance, the dynamic pressure is constant. Under these conditions, roll rate being assumed highly subcritical ($|p| < p_{cr}$), the trim incidence remains close to the static trim incidence θ_e, and amplitude of normal acceleration remains constant:

$$|A_N| = \frac{\bar{q} S_{ref}}{m} C_{L\alpha} \theta_e = |A_{N0}|$$

The vehicle is in lunar motion and according to Sect. 12.2 the phase ψ of $\bar{A}_N$ is such as:

$$\dot\psi \approx p = \dot{p}_0 t \quad \Rightarrow \quad \psi = \int_0^t \dot\psi d\tau = \dot{p}_0 \frac{t^2}{2}$$

Variation of side velocity is written as,

$$\frac{d\bar{V}_N}{dt} = \bar{A}_N = |A_{N0}|e^{i\psi} \Rightarrow \Delta\bar{V}_N = |A_{N0}|\int_{-\frac{T}{2}}^{+\frac{T}{2}} e^{i\psi}d\tau$$

While posing $x = \sqrt{\frac{|\dot{p}_0|}{2}}t$ and accounting for $\dot{p}_0 < 0 \Rightarrow \psi = -x^2$:

$$\Delta\bar{V}_N = |A_{N0}|\sqrt{\frac{2}{|\dot{p}_0|}}\int_{-x(\frac{T}{2})}^{+x(\frac{T}{2})} e^{-ix^2}dx$$

If we define T such that $x\left(\frac{T}{2}\right) = \sqrt{\frac{|\dot{p}_0|}{2}}\frac{T}{2}$ is sufficiently high, we have:

$$\Delta\bar{V}_N \approx |A_{N0}|\sqrt{\frac{2}{|\dot{p}_0|}}\int_{-\infty}^{+\infty} e^{-ix^2}dx \ .$$

We recognize the well-known Fresnel's integral (!) and we obtain finally:

$$\Delta\bar{V}_N \approx |A_{N0}|\sqrt{\frac{2}{|\dot{p}_0|}}\int_{-\infty}^{+\infty}(\cos x^2 - i\sin x^2)dx$$

$$= |A_{N0}|\sqrt{\frac{2}{|\dot{p}_0|}}\left[\sqrt{\frac{\pi}{2}} - i\sqrt{\frac{\pi}{2}}\right] = |A_{N0}|\sqrt{\frac{2\pi}{|\dot{p}_0|}}e^{-i\frac{\pi}{4}}$$

Thus, the disturbance mean velocity takes place at $\psi = -45°$ from direction of the lift force at the time of zero roll (Fig. 14.16). This disturbance is proportional to the inverse of roll acceleration. The orientation of the disturbance is simply explained in the case of a negative rolling acceleration by the fact that the phase of the lift force tends toward zero per negative value, cancel then decrease without changing sign. Direction of the lift force thus remains quasi-stationary in the fourth quadrant for the period around zero roll. In the case $\dot{p}_0 > 0$, the positive phase decreases then increases after cancellation. The disturbance will take place in the first quadrant at $\psi = +45°$, so that one can write in the general case:

$$\Delta\bar{V}_N \approx |A_{N0}|\sqrt{\frac{2\pi}{|\dot{p}_0|}}e^{i\frac{\pi}{4}\text{sgn}(\dot{p}_0)}$$

Like previously, angular deviation of trajectory and side error at the point of impact expressions are:

$$\delta_y \approx \frac{|\Delta V_N|}{V}$$

$$\Delta y \approx L\delta_y \approx \frac{H}{\sin\gamma_0}\delta_y$$

14.2 Nonzero Angle of Attack

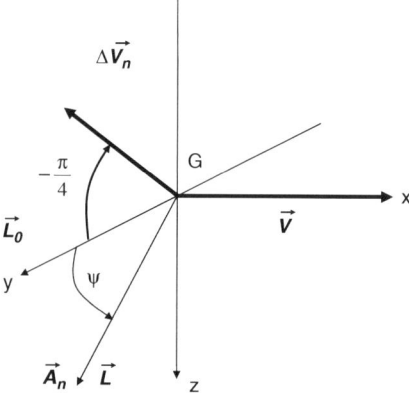

Fig. 14.16 Precession angle at roll through zero

Variations resulting from this analytical theory for our case of reference and for a roll acceleration $\dot{p}_0 = 2\pi$ radians.s^{-2} are represented in Fig. 14.17.

Figure 14.18 presents typical 6 DoF computation results of roll through zero consequences.

We can observe that the conjunction of a trim angle of attack with roll through zero is by far the largest contributor to horizontal dispersion met until now.

Consequences of the Rapid Development of a Trim Angle

Hypothesis on the movement are identical to the case of roll through zero, except

- Constant roll rate $p = p_0$
- In the vehicle frame: no trim angle for ($t < t_0$), then step static trim angle ξ_m or linear ramp $\xi_e(t) = \xi_m \frac{t-t_0}{T}$ for t_0 to $t_0 + T$

We are in lunar motion and according to Sect. 12.2, phase ψ of the vehicle verifies:

$$\dot{\psi} \approx p_0 \quad \Rightarrow \quad \psi = \int_0^t \dot{\psi} d\tau = p_0 t$$

In the aeroballistic frame, the response to a step static trim angle at time t_0 is the superposition of the trim response term turning at roll rate and the transitory terms at nutation and precession at frequencies. These last terms are of high frequency compared with p and of zero mean value. The trimmed response term at rolling frequency of p is:

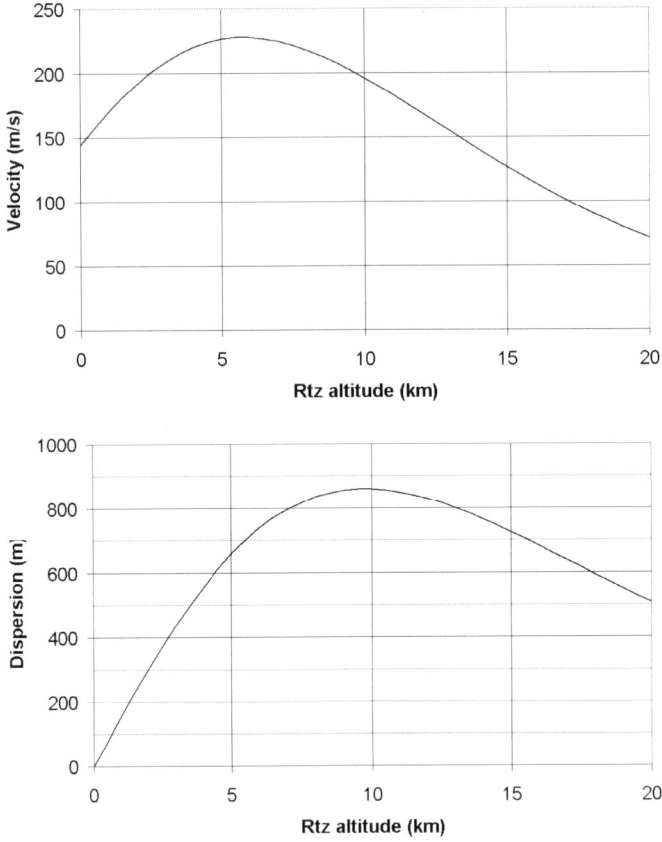

Fig. 14.17 Lateral velocity perturbation and lateral error at impact

- $t < t_0 \quad \rightarrow \quad \overline{\theta} = 0$
- $t_0 \leq t \quad \rightarrow \quad \overline{\theta} = \dfrac{\xi_m}{1-\left(\frac{p}{p_{cr}}\right)^2} e^{i\psi}$

We neglected in the preceding expression the influence of damping D on the trim incidence. This is allowable, given the assumption of a roll rate clearly subcritical.

In the ramp case, we have:

- $t < t_0 \quad \rightarrow \quad \overline{\theta} = 0$
- $t_0 \leq t \leq t_0 + T \quad \rightarrow \quad \overline{\theta} = \dfrac{\xi_m \frac{t-t_0}{T}}{1-\left(\frac{p}{p_{cr}}\right)^2} e^{i\psi}$
- $t_0 + T < t \quad \rightarrow \quad \overline{\theta} = \dfrac{\xi_m}{1-\left(\frac{p}{p_{cr}}\right)^2} e^{i\psi}$

14.2 Nonzero Angle of Attack

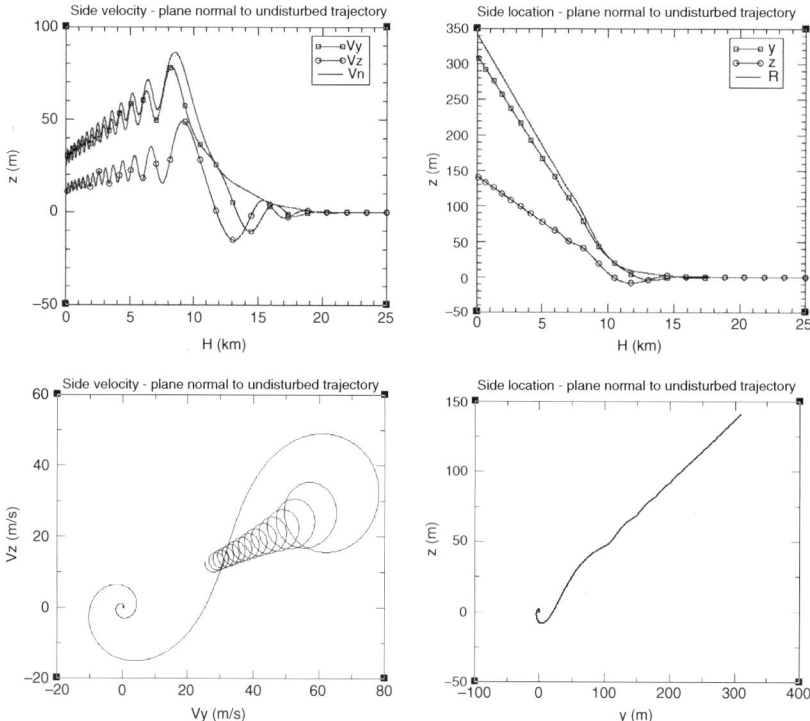

Fig. 14.18 History of motion's parameters during roll through zero

Under these conditions the aeroballistic frame is quasi-inertial and its transverse plane is very close to that of the reference frame related to the mean velocity. Complex transverse aerodynamic acceleration expression is:

$$\overline{A}_N = \frac{\overline{q} S_{ref}}{m} C_{L\alpha} \overline{\theta}$$

$$\frac{d\overline{V}_N}{dt} = \overline{A}_N \quad \Rightarrow \Delta \overline{V}_N = \int_{t_0}^{t} \overline{A}_N d\tau$$

We obtain lateral velocity disturbances by integration:

- For the step trim:

$$\Delta \overline{V}_N = i \frac{\overline{q} S_{ref} C_{L\alpha}}{mp} \frac{\xi_m}{1 - \left(\frac{p}{p_{cr}}\right)^2} \left[e^{ipt_0} - e^{ipt} \right]$$

- For the ramp and $t_0 + T < t$:

$$\Delta \overline{V}_N = i \frac{\overline{q} S_{ref} C_{L\alpha}}{mp} \frac{\xi_m}{1 - \left(\frac{p}{p_{cr}}\right)^2} \left[e^{ip\left(t_0 + \frac{T}{2}\right)} \frac{\sin\left(\frac{pT}{2}\right)}{\frac{pT}{2}} - e^{ipt} \right]$$

Terms independent of time of the two expressions represent the mean velocity disturbance.

$$|\langle \Delta \overline{V}_N \rangle|_{step} = \frac{\overline{q} S_{ref} C_{L\alpha}}{mp} \frac{\xi_m}{1 - \left(\frac{p}{p_{cr}}\right)^2}$$

$$|\langle \Delta \overline{V}_N \rangle|_{ramp} = |\langle \Delta V_N \rangle|_{step} \left| \frac{\sin\left(\frac{pT}{2}\right)}{\frac{pT}{2}} \right|$$

The curve $G = \frac{\sin u}{u}$ of Fig. 14.20 for the ramp show first that the maximum effect (equivalent to the step) takes place when the rising time T of the static trim is small compared with the roll period, in addition that this effect decreases quickly with T.

In the case of the step, the velocity perturbation has a phase lag of lag $\frac{\pi}{2}$ with respect to the direction of the lift at time t_0. In the case of the ramp, this phase lag is $\frac{\pi}{2} + \frac{pT}{2}$.

Figure 14.19 gives for 1° static trim step respectively the mean normal velocity error and the ground side variation. The assumptions are identical to preceding cases, in particular roll rate equal to 1 Hertz. It is advisable to notice that these errors are relatively high, but sudden appearance of a static trim is not very realistic, except in the event of mechanical rupture. In the case of an ablative asymmetry, the phenomenon is much more progressive and has weaker consequences, as shown in Fig. 14.20, due the influence of the rising time.

Consequences of a Static Stability Variation

Here we analyze the consequences of a fast variation of static stability. The origin can be, for example, a variation of center of mass location or of external geometry. We assume a symmetrical vehicle and consider the phase of initial convergence of the incidence. Results relating to epicycle movement of Chap. 11.2 apply. The evolution of complex incidence in the aeroballistic frame (quasi-inertial) is:

$$\xi = e^{ip_r t} \left[\xi_+ e^{i\omega_a t} + \xi'_- e^{-i\omega_a t} \right]$$

We assume that the variation of ω_a is a discontinuity, i.e., a step of amplitude $\Delta \omega = \omega_{a1} - \omega_{a0}$ at time $t = 0$. Roll rate p and spin p_r are assumed constant. We study the movement for a few periods; also we neglect the slow variation of ω_a dependent on the variation of dynamic pressure with altitude as well as the damping of the incidence.

14.2 Nonzero Angle of Attack

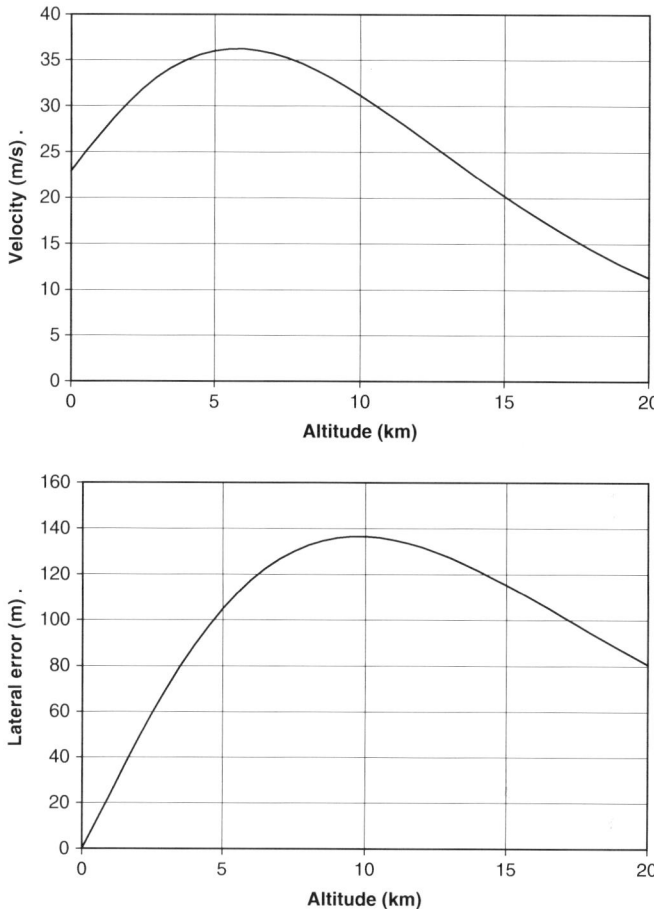

Fig. 14.19 Lateral velocity perturbation and ground error

- Before the disturbance, for $t \leq 0$ we assume that we are in the mode ξ_+ with positive pulsation, such that the complex incidence ξ is along the y axis of the aeroballistic frame at initial time:

$$\xi_0(t) = \theta_0 e^{i(p_r + \omega_{a0})t}$$

- For $0 < t$, we have again $\omega_a = \omega_{a1} = cst$ and the general evolution is

$$\xi_1(t) = e^{ip_r t}\left[\xi_{1+} e^{i\omega_{a1}t} + \xi'_{-1} e^{-i\omega_{a1}t}\right]$$

For $t \leq 0$ we have:

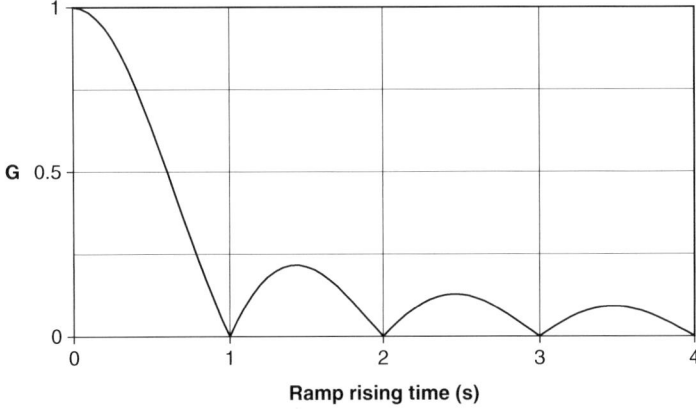

Fig. 14.20 Influence of rising time

$$\ddot{\xi}_0 - 2ip_r\dot{\xi}_0 + \omega_{n0}^2\xi_0 = 0$$

with $\omega_{n0}^2 = \omega_{a0}^2 + p_r^2$

For t > 0 we have:

$$\ddot{\xi}_1 - 2ip_r\dot{\xi}_1 + \omega_{n1}^2\xi_1 = 0$$

with $\omega_{n1}^2 = \omega_{a1}^2 + p_r^2$

The variation of pulsation is assumed to be instantaneous, and corresponds to a discontinuity of the aerodynamic moment. As the rotational movement ξ is continuous, there is no variation of ξ through discontinuity, which is by definition of null duration, thus no variation of the rotational kinetic energy E_K or the angular rate Ω (for E_K and Ω it had not been the case with respect to a disturbance like an aerodynamic moment of percussion). It results that $\dot{\xi}$ is continuous. The only discontinuous variables are the transverse angular acceleration $\dot{\Omega}$ and the second derivative of the complex incidence $\ddot{\xi}$.

$$\xi_1(0_+) = \xi_0(0).$$
$$\dot{\xi}_1(0_+) = \dot{\xi}_0(0)$$
$$\ddot{\xi}_1(0_+) - \ddot{\xi}_0(0) = -\left[\omega_{n1}^2 - \omega_{n0}^2\right]\xi_0(0) = -\left[\omega_{a1}^2 - \omega_{a0}^2\right]\xi_0(0)$$

One can show that the discontinuity corresponds to variations of potential energy and total energy of the movement:

14.2 Nonzero Angle of Attack

$$E = E_K + E_P = \frac{1}{2} I_T \left[\|\dot{\xi}\|^2 + \omega_a^2 \|\xi\|^2 \right]$$

$$\Delta E = \Delta E_P = \frac{1}{2} I_T \left(\omega_{a1}^2 - \omega_{a0}^2 \right) \|\xi(0)\|^2$$

To determine the amplitude of the two modes after the disturbance, we thus write the conditions of continuity of ξ and $\dot{\xi}$ at time zero:

$$\xi_{1+} + \xi_{1-} = \theta_0$$
$$i(p_r + \omega_{a1})\xi_{1+} + i(p_r - \omega_{a1})\xi_{1-} = i(p_r + \omega_{a0})\theta_0$$

Thus, we obtain:

$$\xi_{1+} = \frac{\omega_{a1} + \omega_{a0}}{2\omega_{a1}} \theta_0$$

$$\xi_{1-} = \frac{\omega_{a1} - \omega_{a0}}{2\omega_{a1}} \theta_0$$

We note that the disturbance involves the appearance of the mode ξ_- with negative pulsation. The evolution of the module of the incidence after the step variation is:

$$\theta_1^2 = \frac{\theta_0^2}{2} \left[1 + \left(\frac{\omega_{a0}}{\omega_{a1}} \right)^2 + \left(1 - \left(\frac{\omega_{a0}}{\omega_{a1}} \right)^2 \right) \cos(2\omega_{a1} t) \right].$$

For a decrease of static stability $\omega_{a0} > \omega_{a1}$, minimum and maximum amplitudes are:

$$\theta_{1\,\text{max}} = \frac{\omega_{a0}}{\omega_{a1}} \theta_0$$

$$\theta_{1\,\text{min}} = \theta_0$$

The initial incidence oscillation, which was circular with pulsation ω_{a0} became elliptical with pulsation ω_{a1} (see Fig. 14.21, $\theta_0 = 1°$, $\omega_{n0} = 20\pi$ radians.s^{-1}, and $\omega_{n1} = \frac{\omega_{n0}}{\sqrt{2}} \approx 14\pi$ radians.s^{-1}).

Now let us calculate the evolution of the side velocity disturbance for $t > 0$.

For this purpose, we naturally assume that side velocity before the step has zero mean value:

$$\overline{V}_{N0}(t) = \frac{\bar{q} S_{ref} C_{L_\alpha}}{m} \theta_0 \frac{e^{i(p_r + \omega_{a0})t}}{i(p_r + \omega_{a0})}$$

$$= \frac{\bar{q} S_{ref} C_{L_\alpha}}{m(p_r + \omega_{a0})} \theta_0 \left[\sin((p_r + \omega_{a0})t) - i\cos((p_r + \omega_{a0})t) \right]$$

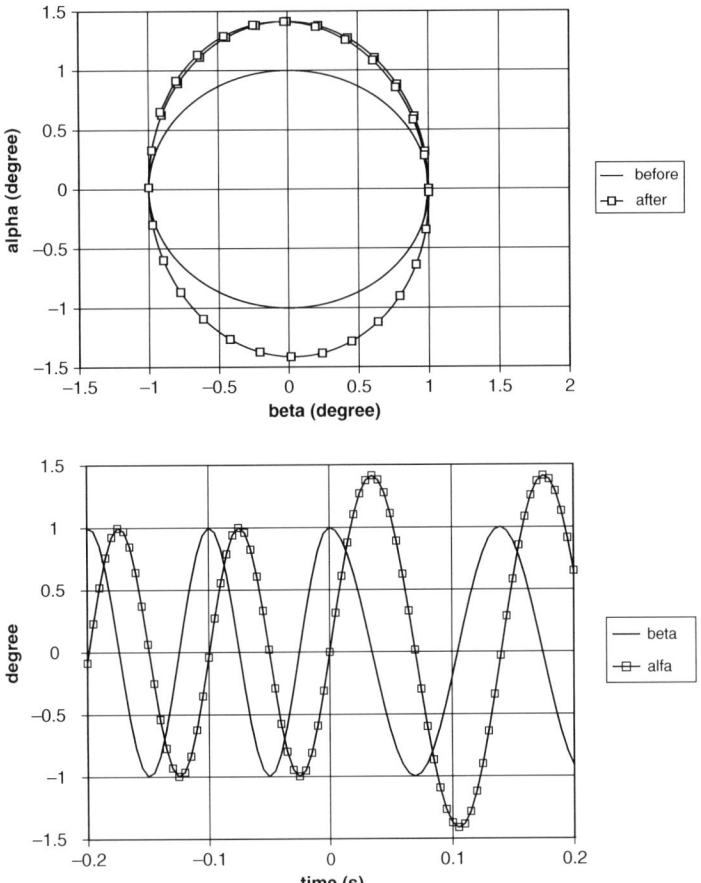

Fig. 14.21 Incidence and sideslip behavior during a step change of static stability

Lateral speed at the time t = 0 is:

$$\overline{V}_{N0}(0) = -i\frac{\overline{q}S_{ref}C_{L_\alpha}}{m}\theta_0\frac{1}{(p_r + \omega_{a0})}$$

Lateral speed at later time is:

$$\overline{V}_{N1}(t) = \overline{V}_{N0}(0) + \int_0^t \overline{A}_{N1}(\tau)\,d\tau$$

with,

14.2 Nonzero Angle of Attack

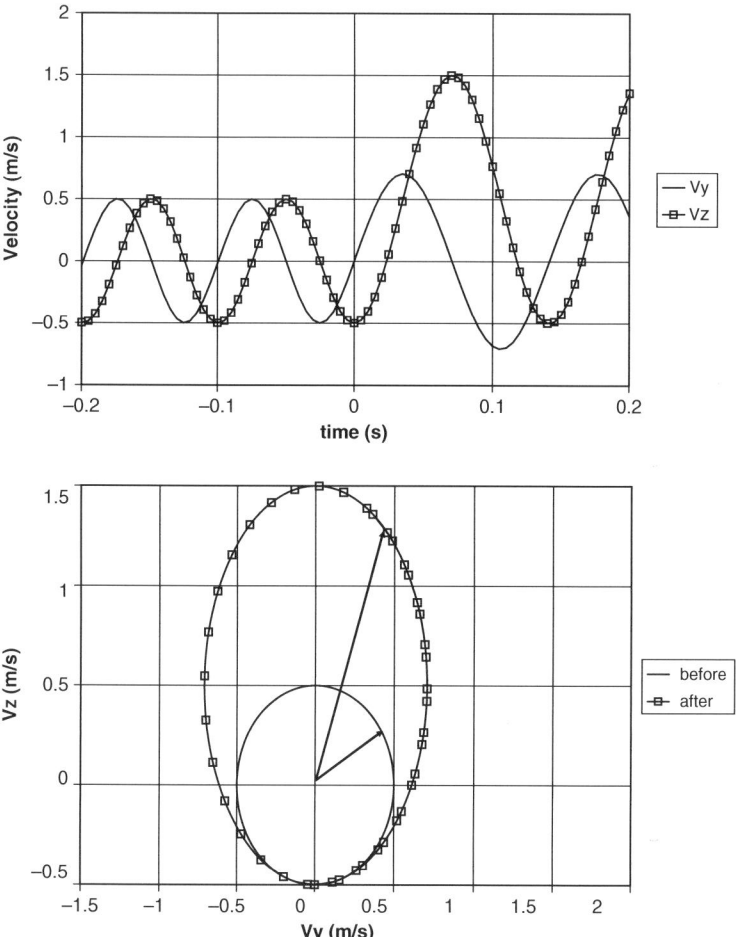

Fig. 14.22 Lateral and normal velocity behavior during a step change of static stability

$$\overline{A}_{N1}(t) = \frac{\overline{q} S_{ref} C_{L\alpha}}{m} \xi_1(t) \rightarrow \Delta \overline{V}_N(t) = \frac{\overline{q} S_{ref} C_{L\alpha}}{m} \int_0^t \xi_1(\tau) d\tau$$

We must thus evaluate,

$$\int_0^t \xi_1(\tau) d\tau = \int_0^t \left[\xi_{1+} e^{i(p_r + \omega_{a1})\tau} + \xi'_{1-} e^{i(p_r - \omega_{a1})\tau} \right] d\tau$$

We leave to the reader to check that one obtains:

$$\int_0^t \xi_1(\tau) d\tau = \frac{\theta_0}{2i\omega_{a1}} \left[\frac{\omega_{a1} + \omega_{a0}}{p_r + \omega_{a1}} \left(e^{i(p_r + \omega_{a1})t} - 1 \right) + \frac{\omega_{a1} - \omega_{a0}}{p_r - \omega_{a1}} \left(e^{i(p_r - \omega_{-a1})t} - 1 \right) \right]$$

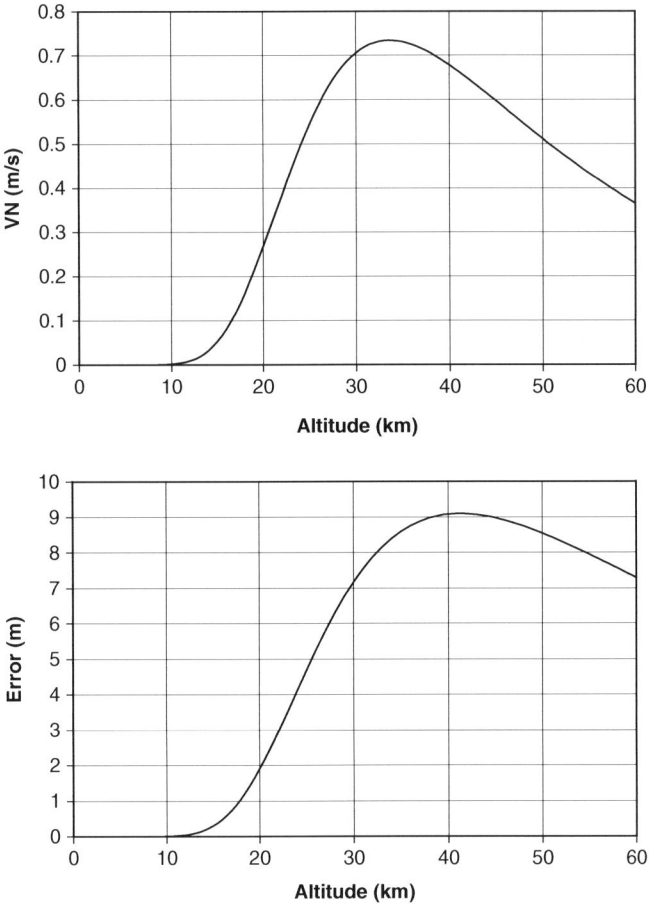

Fig. 14.23 Lateral velocity perturbation and ground error

By adding initial side velocity $\overline{V}_{N0}(0)$ to $\Delta\overline{V}_N(t)$, we obtain (while noting $\eta = \dfrac{\omega_{n0}^2}{\omega_{n1}^2} - 1$):

$$\overline{V}_{N1}(t) = \frac{\overline{q} S_{ref} C_{L\alpha}}{m(p_r + \omega_{a0})} \theta_0 \left[i\eta + e^{ip_r t} \left[\left(\frac{\omega_{a0}}{\omega_{a1}} - \eta \frac{p_r}{\omega_{a1}} \right) \sin(\omega_{a1} t) - i(\eta + 1) \cos(\omega_{a1} t) \right] \right]$$

The result is the mean lateral velocity induced by the step stability variation:

$$\langle \overline{V}_{N1}(t) \rangle = i \frac{\overline{q} S_{ref} C_{L\alpha}}{m} \frac{\theta_0}{\omega_{a0} + p_r} \eta$$

14.2 Nonzero Angle of Attack

We note that the mean velocity disturbance is proportional to the initial amplitude $\left|\vec{V}_{N0}\right|$ of lateral velocity fluctuations and, when $\omega_{n1} < \omega_{n0}$, its direction is opposed to the value of the fluctuation at the time of the disturbance:

$$\langle \vec{V}_{N1}(t) \rangle = -\eta \vec{V}_{N0}(0)$$

The multiplicative factor can be expressed from the static margin $\Delta x = x_G - x_{CP}$

$$\omega_n^2 = \frac{\bar{q} S_{ref} C_{N\alpha} \Delta x}{I_T} \quad \Rightarrow \quad \eta = \left[\frac{\omega_{n0}^2}{\omega_{n1}^2} - 1\right] = \frac{\Delta x_0}{\Delta x_1} - 1$$

Figure 14.22 gives an example of lateral velocity evolution in the transverse plane before and after the disturbance, for an initial side velocity cycle of amplitude 0.5 m/s, with the assumptions of Fig. 14.21 on aerodynamic frequency, which

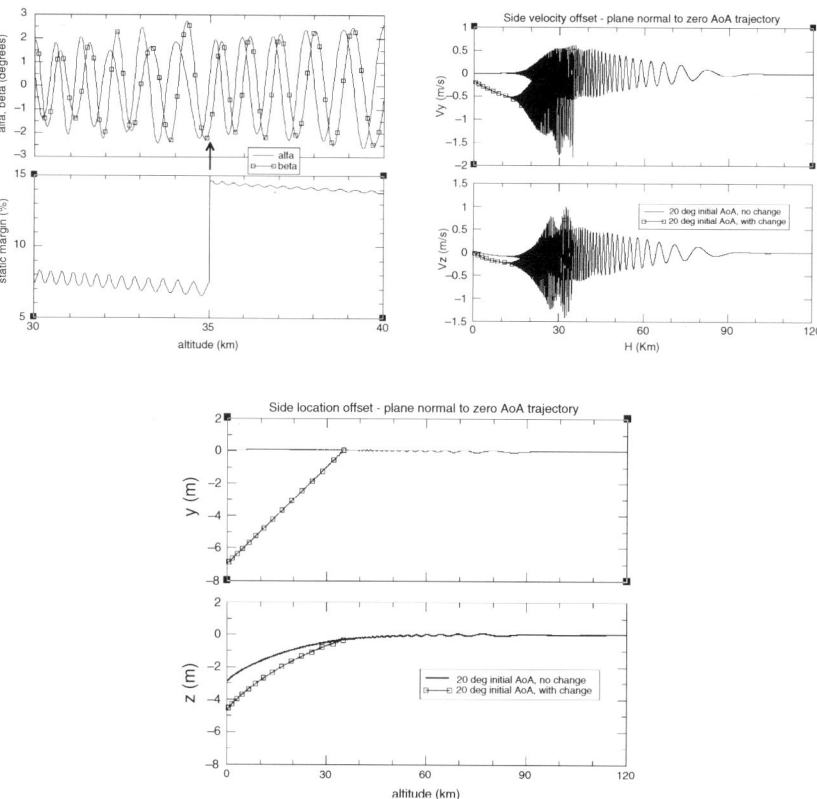

Fig. 14.24 Results of 6 DoF computation for a step change of static stability

correspond to 1/2 the static margin. In this case the induced mean velocity is equal to the amplitude of the initial velocity oscillation, i.e., 0.5 m/s.

Using assumptions and results of Sect. "Decelerated Phase of Motion" on the amplitude of side velocity oscillations at the end of the convergence of the incidence, we obtain the term $\left|\vec{V}_{N0}\right|$ versus altitude (Fig. 14.23). This term is equal to the mean side velocity disturbance for a static margin Δx divided by 2.

Figure 14.23 gives in addition the side location error at sea level under the same conditions, obtained from:

$$\Delta y_N = \eta \frac{|V_{N0}(z)|}{V_0} \frac{z}{\sin \gamma_0}$$

Figure 14.24 gives plot of typical 6 DoF computation results of this phenomenon.

This error is definitely a lower order of magnitude than that which results of a static trim angle associated with roll through zero.

Epilog

In the first five chapters, we described the basic theoretical tools necessary to analyze flight mechanics of reentry vehicles and planetary capsules. These included classical mechanics, topography and gravitational models, hypersonic aerodynamics, transform rule for change of reference frames, and inertia properties. Except hypersonic, which is specific of the subject, these topics are of general use and are provided to avoid the reader to look for results used all along the book.

The remaining is the application of these tools to ballistic phase and reentry topics. As they represent the matter of the book, it is worth to review these chapters, in order to enlighten the important results.

Chapter 6: Ballistic Trajectories

For an inertial observer, ballistic trajectories are ellipses that intersect the earth surface, of absolute range limited to less than 1/2 the earth's circumference, with velocities lower than circular orbit (about 7.9 km/s). Most interesting trajectories are of minimum energy, which minimize fuel consumption and correspond to $\gamma_0 = \frac{\pi - \alpha}{4}$ initial flight path angle, where α is angular range. At the same range, nonoptimal trajectories need higher velocities. They include lofted trajectories with $\gamma > \gamma_0$ and shallow trajectories with $\gamma < \gamma_0$. Lofted trajectories are long duration and high apogee altitude; shallow trajectories are shortest duration and low apogee altitude.

For an observer fixed to rotating earth, at given initial relative conditions, ballistic trajectories are modified depending on the initial azimuth and latitude. First the inertial velocity is the sum of the relative velocity and the local earth velocity, which modifies the absolute range; second the earth surface drift toward east, which affects the relative range. Initial eastward azimuth increases range, westward azimuth diminishes range. The highest effects are encountered at initial latitude near equator.

Chapter 7: Free Rotational Motion

Most RV and planetary probes are near axisymmetric and provided with initial roll rate p (around inertial symmetry axis) for gyroscopic stability. During the out of atmosphere motion, RV angular momentum is near constant relative to inertial frames. For symmetry reason, the roll rate remains constant. When transverse initial rotation rates are zero, the symmetry axis has a constant direction along the angular momentum. For finite initial transverse rotation rate, there is an initial angular offset $\tan\theta_0 = \frac{I_T\sqrt{q_0^2+r_0^2}}{I_X p_0}$ between axis of symmetry and angular momentum. The gyroscopic effect results in maintaining the offset θ constant (nutation angle). We observe free coning motion of the vehicle's axes around angular momentum at constant precession velocity $\dot{\psi} = \frac{I_X p_0}{I_T \cos\theta_0}$ and roll rate.

Chapter 9: Zero Angle of Attack Reentry

For high β, high initial velocity RV, effects of gravity on trajectory are second order; major effects are related to the aerodynamic drag, and flight path is very close to rectilinear. On earth, maximum deceleration (up to 100 g) occurs below 15 Km, where is 90% of the mass of the atmosphere. Most challenging effects are aerothermal effects on nosetip and heat shield of the vehicle. On nosetip, the heat flux is inversely proportional to the square root of curvature radius and can exceed 100 Mw/m^2. In the rest of the vehicle, heat flux may exceed 10 Mw/m^2. Maximum axial load and heat flux are encountered on lofted trajectories; however, because of longer reentry duration, maximum aerothermal effects (total energy, ablation mass loss, and aerodynamic asymmetries) are encountered on shallow trajectories (i.e., also on long range trajectories).

Chapter 10: Initial Reentry with Angle of Attack

At zero roll rates, angle of attack behavior depends on aerodynamic static and dynamic pitching moments, transverse inertia, and atmosphere profile. Static moment (i.e., static margin and normal force coefficient) determines only pitching frequencies and trim angle of attack. Unlike many people thinking, for a stable vehicle, angle of attack convergence is essentially related to dynamic pitching moment (aerodynamic damping) and to the rate of variation of aerodynamic frequencies (atmospheric density damping), *not directly to the static margin*. During the first part of reentry (120–60 km), CG velocity is nearly constant, dynamic pressure and aerodynamic frequencies increase exponentially; convergence of angle of attack relies mainly on density damping. Effect of the spin rate and gyroscopic moment is to slower initial convergence and to precession the angular perturbations (at $p_r = \mu p/2$) around flight path vector.

Chapter 11: End of the Convergence of the Incidence

Below 60 km, for a stable RV, the initial angle of attack has mostly converged due to density damping.

At small angle of attack, for an inertial observer, the instantaneous angular motion of a RV with spin appears tricycle. Angle of attack behavior in the incidence/sideslip plane is the sum of three terms including the static trim angle at roll frequency and two dynamic damped components, at nutation and precession frequencies. The static trim angle is due to aerodynamic and/or mass asymmetries of the vehicle. Nutation and precession frequencies are near $\pm \omega_n$ the natural aerodynamic frequency of the vehicle. During reentry, they increase with dynamic pressure, generally having a 10–30 Hz maximum before impacting. As by definition the nutation frequency has the sign of the roll rate, resonance generally occurs once at high altitude when nutation frequency equals roll frequency (first resonance condition). The roll rate at which the condition is realized is named critical roll frequency, very close to nutation frequency. A second resonance condition may also occur at low altitude depending on roll rate history. Due to resonant amplification, trim angles of attack and lateral loads are temporarily strongly increased throughout crossing, especially at second resonance where dynamic pressure is high. Except sustained resonance, final angular movement below 15 km is lunar motion, where longitudinal axis maintains a constant rolling trim angle with flight path vector, which has a constant meridian relative to the vehicle. Due to decrease of angle of attack and increase of dynamic pressure, lateral loads associated with initial angle of attack generally have a maximum around 20 km.

Chapter 12: Roll-lock-in

When any aerodynamic or inertial asymmetry is combined with a CG offset, it results a roll moment. Hence near resonance, pitch/yaw rolling trim angles and roll rate are mutually coupled. Following resonance crossing, this may result in a lock-in of the roll rate on the critical roll frequency, analog to a closed loop control, which maintains resonance for a long period with high trim angle and lateral loads. Generally, first resonance crossing occurs near 45 km and the only asymmetries existing are mass and inertia, which are not very difficult to specify to manage associated risks. After transition, high turbulent heat flux induces ablation, which may create asymmetries on nosetip and roughness on heat shield. Aerodynamics effects, including nosetip trim angles and heat shield roll moments, are difficult to control. Hence, most challenging candidates are CG offset combined with low altitude out-of-plane aerodynamic asymmetries, which may induce roll-lock-in following second resonance.

Chapter 13: Instabilities

Angle of attack divergence may result of static or dynamic instabilities. RV and probes are generally designed to provide unconditional static stability. However, if needed for packaging constraints, different levels of static instability may be admitted at high altitude and low dynamic pressure with little consequence on reentry. Dynamic instabilities are often encountered on blunted probe shapes having a positive pitch damping coefficient at low supersonic and transonic Mach number, which must be managed in the design. They are rarely a critical problem on RV at supersonic regime, except for some heat shield materials having highly blowing pyrolize or ablative properties, for which ablation lag phenomenon may have effects analog to a positive pitch damping coefficient.

Chapter 14: Dispersions

Origins of horizontal dispersions are twofold, drag and lift. Drag dispersions are the effects of variations of initial angle of attack, axial force coefficient, and atmosphere density. Lift-induced dispersions are essentially effects of asymmetries and precession rate history.

For a high performance RV, horizontal dispersions related to drag are low; effects are mainly on velocity and chronology. Lift effects associated with initial angle of attack are not negligible, but the main contributors are trim angles of attack (asymmetries). However, in both cases the key effect is that of precession rate $\dot{\psi}$ of the aerodynamic axis around the flight path vector (do not confuse $\dot{\psi}$ with precession frequency). When precession rate is a continuous function of time of minimum value, lift averaging occurs, and trajectory is helical of very small diameter around the mean flight path. Deviations of mean flight path, with large final horizontal dispersions, are consecutive to ineffective lift averaging. This occurs when precession rate becomes too small or discontinuous, combined with asymmetrical trim. This may also occur at constant precession velocity, combined with trim perturbations of duration lower than precession period. Analysis shows that the highest contributor to horizontal dispersion is low altitude trim angle, combined with roll through zero (during lunar motion, precession rate is equal to roll rate).

Exercises

- For numerical applications, use of a PC and graphical software is highly advised (or any computational software).
- All reference frames are orthogonal and right handed.

1. Noninertial reference frames (Chap. 1, Sect. 1.1.2 and Chap. 6, Sect. 6.1.1)

   Using the methodology of Sect. 6.1.1, establish the general expression of the fundamental principle of mechanics in a reference frame in nonuniform rotation and translation.

2. Accelerometers

   The principle of an accelerometer is to measure displacements of a test mass relating to the case, along its measurement axis (or measure the interior load to apply to cancel its movement relating to the case)

- To give the general expression of the interior load (accelerometer measurement)
- A triaxial accelerometer block is located at the center of mass of a missile. What do we measure in the following corresponding situations:

   - on launching area
   - during propelled phase
   - during ballistic phase out of atmosphere
   - during reentry phase

3. Vertical and apparent gravity

- Determine the equilibrium position of a line pendulum at the surface of rotational earth at given latitude.
- Give the expression of the apparent field of gravity.

4. Coriolis force

- A tourist releases a marble from the top of the Eiffel tower (300 m height approximately, latitude $45°$):

- To determine vertical and horizontal approximate motion and deviation from the vertical of the point of impact due to earth rotation,
  - Neglect aerodynamic drag and use a Z axis directed along the apparent vertical of gravity.
- Note: It is strictly forbidden to send objects from the Eiffel tower.

5. **Quaternion**

    From Sect. 6.2.3, $\underline{Q} = [q_0 \ q_1 \ q_2 \ q_3]$ with $|\underline{Q}| = 1$ represents a rotation in R^3 (like unit complex numbers in R^2).

- From q_0, q_1, q_2, q_3, determine the parameters φ and $\vec{\Delta} = [\Delta_1 \ \Delta_2 \ \Delta_3]$ of this rotation
- Determine φ and $\vec{\Delta} = [\Delta_1 \ \Delta_2 \ \Delta_3]$ for $\underline{Q} = [0 \ 1 \ 0 \ 0], \left[0 \ \frac{1}{\sqrt{3}} \ \frac{1}{\sqrt{3}} \ \frac{1}{\sqrt{3}}\right],$
  $\left[0 \ -\frac{1}{\sqrt{3}} \ -\frac{1}{\sqrt{3}} \ -\frac{1}{\sqrt{3}}\right], \left[\frac{1}{2} \ \frac{1}{2} \ \frac{1}{2} \ \frac{1}{2}\right], \left[-\frac{1}{2} \ \frac{1}{2} \ \frac{1}{2} \ \frac{1}{2}\right], \left[\frac{\sqrt{2}}{2} \ \frac{\sqrt{2}}{4} \ \frac{1}{2} \ -\frac{\sqrt{2}}{4}\right]$

6. **Euler angles and quaternions**

- Determine components of the quaternion corresponding to the operator to change of reference frame K $\to$ E through a sequence 1, 2, 1 of Euler rotations, as function of ψ, θ, φ.
- Determine quaternion corresponding to $\psi = \theta = \varphi = \pi/2; \pi/3; 2\pi/3$.
- Determine the single axis rotations corresponding to those rotations.

7. Verify equivalence of quaternion associated to change of reference frame and vector rotation.

    The quaternion associated to change of frame operator transforms components of a space fixed vector through successive rotations of reference frame (exercise 6). The rotation operator is acting on components of a vector to transform it by rotation in a unique reference frame. Determine the quaternion associated with rotation operator for a sequence 1, 2, 1 of Euler rotation. Show the equivalence with the previous one.

8. **Pendulum of Foucault**

    This experiment demonstrated the effects of the rotation of earth.

- The hinge point of a line pendulum is fixed at the top of a high tower (the line length L is such that the suspended mass is close to ground level). The pendulum is released with a low initial angle from apparent vertical ($\sim 1°$) without velocity relative to earth

General Remarks

- To determine the period of the oscillations and the evolution of the plane of the oscillations at different latitudes
- Damping related with bearing friction and aerodynamic drag will be neglected
- Use a local frame OXYZ such that O is the hinge point of the pendulum; OZ is along the ascending apparent vertical, OX eastward and in the plane normal to OZ.
- Variables are the meridian angle ψ of half plane containing the pendulum and the apparent vertical OZ (origin at the half plane containing OX) and the angle θ between the pendulum and OZ.
- Neglect second-order terms in θ.

9. Motion of the terrestrial center of mass

- Assuming earth orbit around the sun is circular of radius $R \approx 1$ astronomical unit ($UA = 1.5 \; 10^8$ km) with period 365 days, determine its velocity and centripetal acceleration. Determine the same parameters for the rotation of the moon around earth (period 27 days 8 hours, mass $7.3 \; 10^{22}$ kg). To determine the acceleration of earth's center of mass under the influence of the moon (universal constant of gravitation $G = 6.67 \; 10^{-11} \; m^3.s^{-2}.kg^{-1}$). What can we say about using a nonrotating reference frame at the earth center of mass to observe the movement of a ballistic vehicle?

10. Launch windows toward Mars

Knowing that the Martian year lasts about two terrestrial years, the most favorable launch windows from earth takes place every two years. Determine orbit (assumed circular) radius of Mars around sun as well as the minimum distance of earth/mars.

11. Energy of a solid in free rotation

Consider a rigid body of revolution around axis X, subjected only to a gravitational field (for example, a satellite). Its initial angular moment $\vec{H}_0$ and principal moments of inertia $I_x > I_y = I_z$ being fixed, determine the possible range of its kinetic energy of rotation. What can one deduce about the stability of the rotational movement around the three principal axes?

12. Angular momentum of a nonrigid satellite made of two rigid bodies

Determine the general expression of the total angular momentum of a system of two solids in rotation, of which centers of mass are in accelerated relative motion (under the effect of arbitrary interior forces). Determine first the expression of the total angular momentum and that of its time derivative in a nonrotating reference, then the expression of this derivative in a rotating reference frame fixed to one of the solids.

13. Lunar rotational motion

- It is well-known that the moon always presents the same hemisphere toward the earth.
- What is the origin of the phenomenon? What are the angular rate and the duration of the lunar day?

14. Angular stabilization of a satellite by wheels of inertia

Determine angular momentum of a system formed of a solid S in rotation, and three wheels of revolution with arbitrary rotation around their axes of symmetry. Wheel's axes are respectively directed along the three principal inertia axes of S. Center of mass of wheels are fixed to S. Give the expression of total angular momentum and its derivative in an inertial frame, then in the rotating Eulerian frame fixed to S.

15. Balancing machines

- They are used to measure centering and inertial defects of solids turning around an axis of rotation, and to determine the balancing masses to restore inertial symmetry of revolution around this axis.
- The machine consists of a structure supporting a plate turning at uniform rate around a vertical axis, balanced perfectly.
- The object to be measured is fixed rigidly to the plate, its axis in alignment with that of the plate. The method consists in measuring the effects induced on the rotation-axis's bearings by inertial asymmetries of the object.
- Using the angular momentum theorem, determine the effects induced on the axis of rotation of the plate by the defects which follow:
  - Lateral offset of the center of mass of the object relating to the axis of rotation (CG offset)
  - Angular variation between the direction of the axis of rotation and that of the corresponding principal axis of inertia (principal axis misalignment)

16. Aerodynamics of an Apollo like reentry capsule in continuous flow

We consider an earth reentry capsule with symmetry of revolution. Its forebody is a segment of sphere of radius $R = 1.2\,D$ and maximum cross section diameter $D = 3.9$ m. Its aft cover is conical with slope $-33°$. Its center of mass is located at $\frac{x_G}{D} \approx -0.26$; $y_G = 0$; $\frac{z_G}{D} = -0.035$ (the positive x axis corresponds to the front direction of heat shield).

Determine its aerodynamic coefficients in hypersonic continuous mode (Newtonian approximation) for angles of attack from $-30°$ to $+10°$:

- Coefficients of pitching moment around mass center, axial and normal force, drag and lift, with reference $S_{ref} = \pi \frac{D^2}{4}$ and $L_{ref} = D$.
- Determine its trim angle and corresponding fineness ratio.

Notice: Geometry is simplified (Apollo had a small round-off in the vicinity of the maximum diameter)

17. **Aerodynamics of an Apollo-like reentry capsule in free molecular flow regime**

- Same assumptions and questions as exercise 16
- Use accommodation coefficients $\sigma = \sigma' = 1$

18. **Aerodynamics of a planetary entry capsule of Viking type, in hypersonic continuous flow regime**

    Body of revolution with heat shield of maximum diameter D = 3.50 m, composed of a segment of sphere of radius R = 0.25 D followed by a tangent conical shroud of semiapex angle 70°. Its aft cover is a biconical shape of slope −40° and −62°. Its mass center is located in $\frac{x_G}{D} \approx -0.23$; $y_G = 0$; $\frac{z_G}{D} = -0.02$ (positive x axis is in front shield direction). Determine aerodynamic coefficients in hypersonic continuous flow for incidence from −30° to +10° (Newtonian approximation):

    - Pitching moment coefficient around center of mass, axial and normal force, lift and drag force, center of pressure, and static margin with reference $S_{ref.} = \pi \frac{D^2}{4}$ and $L_{ref} = D$.
    - Determine trim angle and corresponding fineness ratio.

    Notice: Geometry is simplified (Viking had a small round-off in the vicinity of the maximum diameter)

19. **Aerodynamics of a Viking-like planetary entry capsule in free molecular flow regime.**

- Same assumptions and questions as 18
- Use assumption $\frac{y_G}{D} = \frac{z_G}{D} = 0$
- Use accommodation coefficients $\sigma = \sigma' = 1$

20. **Aerodynamics of pathfinder planetary entry capsule in intermediate flow regime.**

    We will use for pathfinder in front of maximum cross section the geometry of Viking (18 and 19), with a maximum diameter D = 2.65 m and a center of mass in $\frac{x_G}{D} = -0.27$, $\frac{y_G}{D} = \frac{z_G}{D} = 0$.
    The coefficients C_A, C_N, and $\left.\frac{\partial C_{mG}}{\partial \alpha}\right|_{\bar{\alpha}=0}$ will be evaluated in intermediate mode, according to Knudsen for $\bar{\alpha} = 2°$, $S_{ref.} = \pi \frac{D}{4}$ and $L_{ref} = D$.

- Use Erf-Log bridging function and fitting method with one point of DSMC calculation (Sect. 4.4.2). Boundary values in free molecular mode and continuous mode result from 18 and 19. For $\frac{\partial C_{mG}}{\partial \alpha}\big|_{\overline{\alpha}=0}$, use -0.0315 instead of the Newtonian boundary value, corresponding to $C_{mG}(2°) = 0.0011$ drawn from [PAR], resulting from a NASA Navier–Stokes computation.
- Determine for each coefficient the values Kn_{mi}, Kn_m, and Kn_c knowing that $\Delta Kn = Ln(Kn_m/Kn_c) = Ln(500) = 6.2146$ and that according to the results of DSMC calculation [PAR] from NASA for $\overline{\alpha} = 5°$ and $Kn_1 = 0.109$, respective values of the function $\Phi(Kn_1) = \frac{C(Kn_1)-C_C(Kn_c)}{C_m(Kn_m)-C_c(Kn_c)}$ for C_A, C_N, are 0.392 and 0.467. For C_{mG} and $\frac{\partial C_{mG}}{\partial \alpha}\big|_{\overline{\alpha}=0}$, $\Phi(Kn_1) = 0.131$.

21. Aerodynamics of a biconical reentry vehicle in continuous flow mode

    A shape of revolution is considered. It is composed of a spherical blunted nose such that $e = \frac{2R_N}{D} = 0.1$, of a first cone C1 of half angle 10° and of a second cone C2 of half angle 5°. The length ratio is $\frac{L_{C1}}{L_{C2}} = \frac{1}{3}$. Maximum diameter is D = 0.50 m.

    - Determine by using the Newtonian approximation aerodynamic coefficients in hypersonic mode (coefficients of normal and axial force, of pitching moment and location of the center of pressure)
    - Determine center of mass location to ensure a static margin 3% of overall length
    - Determine the trim angle corresponding to $z_G = 10\,mm$ CG offset from symmetry axis

22. Accuracy of ballistic trajectories

- Consider a ballistic trajectory assumed elliptic from ground level to impact, of initial velocity V = 5000 m/s. Determine the maximum range and corresponding initial FPA γ_{opt}. Determine around preceding conditions and also for $\gamma = \gamma_{opt} \pm 10$ degrees, sensitivity of impact location to small variations of initial conditions, for the parameters:
- Initial altitude
- Initial azimuth
- Initial velocity
- Initial FPA
- Which kind of trajectories allow minimization of the influence of these last two parameters?

General Remarks

23. Stability of coning motion for a satellite

A satellite of revolution, with moments of inertia $I_x = 5000 \, \text{kg.m}^2$, $I_y = I_z = 3000 \, \text{kg.m}^2$ is spin stabilized with an initial rotation rate $\omega_x = 10$ rpm around its axis x. It is subjected to an initial transverse rotation rate $\omega_y = 60°/s$ around the axis y. Assume external moments are null. Determine the characteristics of its initial coning motion. Assume it is submitted to a slow dissipation of energy of rotation $\dot{E} = -10 \, \text{mW}$, assumed constant. Determine its angular movement:

- Evolution of roll rate ω_x and transverse rate $\omega_t = \sqrt{\omega_y^2 + \omega_z^2}$ of nutation angle θ
  - Determine the parameters of movement from the initial angular momentum, initial energy of rotation, and current value of this energy, while supposing that the movement is very close to free coning motion.
- Considering that the energy of rotation cannot decrease beyond a minimum value corresponding to a stable state of rotation, energy dissipation vanishes in this final state. Identify physical origin of dissipation and components of the movement at this origin?

24. Allen and Eggers reentry

- Consider a reentry vehicle with spherical nose, ballistic coefficient $\beta = 7500 \, \text{kg/m}^2$, and length $L = 1.5 \, \text{m}$. Initial conditions at 120 km are $V_0 = 6.5 \, \text{km/s}$ and FPA $\gamma_0 = -20°$, and $-60°$
- Calculate and trace the evolution of Mach number, Knudsen, and Reynolds numbers according to altitude in the case of a terrestrial reentry with isothermal atmosphere $H_R = 7000 \, \text{m}$, $\rho_s = 1.39 \, \text{kg/m}^3$. Determine the altitude when laminar/turbulent transition begins at the aft of the vehicle (corresponding roughly to a Reynolds number based on the overall length equal to 10^6).
- Determine the evolution of the Reynolds number and the altitude of transition on a nosetip with radius $R_N = 50 \, \text{mm}$ by using the same criterion and the nose radius as reference length.
- Determine the maximum heat flux at stagnation point for spherical noses of radius $R_N = 0.025, 0.05, 0.075, 0.100 \, m$.

25. Entries of NASA probes Viking and Mars pathfinder allowed improving Martian atmosphere models. Method of restitution of the density was based on the design model of drag coefficient and deceleration measurement during the entry with an inboard axial accelerometer.

- Give the relations allowing to estimate the density
- Determine using Allen approximation the evolution of accelerometer measurement with altitude for pathfinder (initial velocity and FPA at 125 km altitude were respectively 7470 m/s and $-13.6°$, the ballistic coefficient $\beta \sim 65$ kg.m^{-2}. Use an exponential atmosphere model with parameter values from Chap. 3).

26. Impact of meteorites

> A spherical homogeneous cosmic object is considered, of density $\rho_b = 2500$ kg/m^3 and radius R, entering with an initial velocity at 120 Km V = 12 km/s and FPA $\gamma = -30°$ in an exponential isothermal earth's atmosphere.

- Determine the expression of its hypersonic ballistic coefficient β in continuous flow according to the diameter, for diameters values D = 0.03, 3, 30, 300, and 3000 m
- Determine according to the diameter:
  ○ Ratio of the initial velocity lost while arriving at sea level
  ○ maximum convective heat fluxes and corresponding altitudes
  ○ Flow of thermal energy per unit area received in the vicinity of stagnation point and total energy received while assuming heat flux constant on frontal surface.
  ○ The temperature of the body presumed isothermal for a specific heat $C_v = 4000$ J.K^{-1} kg^{-1}
  ○ Temperature of vaporization of its material is 2500° C, what can we deduce for its survival according to its diameter?
- For the objects still quasi-intact at sea level consider their kinetic energy at impact in kilo tons of TNT (1 gram of TNT is roughly equivalent to 4000 Joules)

27. Normal load factor related to trim angle

- A 8° conical ballistic reentry vehicle is considered, of maximum diameter D = 0.5 m, mass m = 120 kg, with conditions at 120 km V = 5.5 km/s, FPA $\gamma = -20°$ and $-60°$.
- Determine the normal load factor according to altitude for constant angle of attack 1, 3, and 5°. Determine the altitude of the maximum load factor
- Use Allen approximation and neglect angle of attack effect on center of mass trajectory. Use a drag coefficient $C_D \approx 0.05$ and the Newtonian value of the coefficient of normal force slope.

28. Computer codes artifacts

- Time or altitude variation of some reentry parameters (for example, axial deceleration or heat flux) calculated using computer codes exhibit oscillations separated by points of discontinuity of first derivative relating to altitude

General Remarks

- What could be the origin of the phenomenon?
- Examine the modeling options used for the physical parameters in the expression of the aerodynamic force.

29. Gyroscopic stabilization of a planetary entry capsule.

Consider the Mars pathfinder capsule, which was unstable in rarefied flow and at the beginning of intermediate flow. Determine the evolution of the angle of attack during a Martian reentry, from 130 km to 70 km altitude, with the following hypotheses

- initial incidence $1°$ ($Z = 130$ km)
- $V = 7.48$ km/s, FPA $\gamma = -13.6°$
- Diameter $D = 2.65$ m, mass 585 kg, pitching moment of inertia $I_T = 370$ kg.m^2, roll inertia $I_X = 490$ kg.m^2, and spin rate $p = 0.5, 0.6, 0.7, 0.8, 1$, and 2 revs/minute.
- Assume constant velocity, exponential atmosphere density (for $Z > 70$ km, use $H_R = 7465$ m, $\rho_S = 0.036$)
- Use the quasi-static approximation of the evolution of the incidence (Sect. 10.2). The pitching moment coefficient derivative $C = \left.\frac{\partial C_{mG}}{\partial \alpha}\right|_{\bar{\alpha}=0}$ in intermediate mode is given by (exercise 17):

$$C(Kn) = C_C + \phi(Kn) \cdot (C_M - C_C)$$

$$\phi(Kn) = \frac{1}{2}\left[1 + erf\left(\frac{\sqrt{\pi}}{\Delta Kn} \ln\left\{\frac{Kn}{Kn_{mi}}\right\}\right)\right]$$

$$C_C = C_{m\alpha/G,C} = -0.0315 \quad ; \quad C_M = C_{m\alpha/G,M} = 0.208$$

$$L_{ref} = D \quad ; \quad S_{ref} = \pi \frac{D^2}{4}$$

$$Kn_{mi} = 1.76 \quad ; \quad \Delta Kn = Ln(500)$$

- Knudsen number is given by $Kn = \frac{0.7513 \, 10^{-7}}{\rho D}$
- Use a classical approximation of Erf function (for example, as in [ABR])

- Determine the minimum roll rate allowing it to cross the zone of instability without divergence of the incidence.

30. Effect of a small trim lift on the center of gravity motion of a reentry vehicle.

Consider a $8°$ conical vehicle without spin, of ballistic coefficient $\beta = 10^4$ kg/m^2, having a constant $0.1°$ trim angle.

- Determine the coefficient of lift gradient and the trim lift coefficient. Determine lift to drag ratio for a drag coefficient $C_A = 0.05$
- Determine equations of vertical movement while assuming the lift force vertical, exponential atmosphere, flat earth with zero gravity, for an initial velocity $V_0 = 6000$ m/s, FPA $\gamma_0 = -30°$. One will determine the evolution velocity and FPA, as well as deviation from zero angle of attack trajectory (FPA variation are considered as first order small terms).
- Compare the order of magnitude of the normal variation y with that obtained in same conditions, with a roll rate 1.5 revs/second.

31. Approximate assessment of drag dispersion effects on the trajectory and conditions at ground impact. Ballistic reentry vehicle, $\beta = 5000$ kg/m², initial velocity $V_0 = 7000$ m/s FPA $\gamma_0 = -25°$ at 120 km.

- Use a zero incidence nominal trajectory and Allen approximation
- Use 5% of relative variation on drag coefficient or atmosphere density. To determine the influence on the maximum axial load factor, maximum stagnation point heat flux, impact velocity.

32. Effect of wind on trajectory and conditions at the impact on the ground

   Using Allen approximation, determine effects of a horizontal constant wind in the plane of the trajectory

- On relative velocity and impact location
- On maximum axial load and heat flux
- Assume that the vehicle remains at zero incidence
- Ballistic RV, $V_0 = 6$ km/s, $\beta = 10000$ kg/m², FPA $\gamma_0 = -30°$ and $-60°$ at $Z_0 = 120$ km. Tail wind velocity $W = +30$ m/s and head wind $W = -30$ m/s

33. Effect of a sharp variation of static stability on evolution of amplitude of incidence oscillations. Symmetrical ballistic RV, case of plane oscillation.

- Zero spin rate
- Assume constant dynamic pressure and neglect damping
- Use expressions of total energy of rotation before and after discontinuity
- Case of an initial amplitude $\theta_0 = 5°$, and a discontinuity $\omega_{a0} = 20\pi$ radian.s^{-1}, $\omega_{a1} = 10\pi$ radian.s^{-1} at t = 0. Study the influence of the phase φ_0 of initial oscillation between 0 and 2π.

34. Skip trajectories

   For planet entries using very shallow trajectories (small FPA), Allen approximation is very inaccurate. It does not predict skip out phenomenon. From

General Remarks

Sect. "Movement in a Reference Frame Fixed with the Local Vertical" [equations (7.2) to (7.5)], by including aerodynamic drag, derive equations of motion for constant ballistic coefficient, spherical nonrotating planet. Using a FORTRAN program or a numerical software (e.g., Maple, Matlab, Scilab, or any other), derive the numerical solution for altitude, range, and velocity. Apply to pathfinder entry using Mars atmosphere and gravity data from Sect. 3.4.

Compute the following entry parameters and compare to Allen's approximation.
Altitude/time; altitude/range; velocity/altitude; axial load factor/altitude.
$\beta = 58.8 \text{ kg/m}^2$; $H_0 = 130.8 \text{ km}$, $\gamma_0 = -13.71°$; $V_0 = 7479 \text{ m/s}$;
Determine the minimum FPA at which the capsule begins to rebound on atmosphere (skip out angle).

Solutions

1 Equations of Motion Relating to a Reference Frame in Rotation

Inertial velocity and absolute acceleration of a point mass are defined as first and second temporal derivative of its location relating to an inertial observation frame K centered in O.

$$\vec{X} = x_i \vec{e}_i \quad \vec{V}_A = \dot{\vec{X}} = \dot{x}_i \vec{e}_i \quad \vec{\Gamma}_A = \ddot{x}_i \vec{e}_i$$

Consider a reference frame E centered in O', turning at rate $\vec{\Omega}$ relating to K. Velocity and acceleration of O' relative to K are $\vec{V}_e$ and $\vec{\Gamma}_e$. A point M with coordinate x'_i is in accelerated motion $\ddot{x}'_i(t)$ relative to E.

We look for the expression of inertial velocity and absolute acceleration of M in E. The locations of M expressed respectively in K and in E are:

$$\vec{X} = x_i \vec{e}_i = \overrightarrow{OO'} + x'_i \vec{e}'_i$$

By derivation we obtain:

$$\dot{\vec{X}} = \dot{x}_i \vec{e}_i = \vec{V}_e + \dot{x}'_i \vec{e}'_i + x'_i \dot{\vec{e}}'_i$$

According to Sect. 6.1.1, the inertial derivative components of E's unit vectors are expressed with the skew symmetric tensor rotation rate $\vec{\vec{\Omega}}$,

$$\dot{\vec{e}}'_i = \Omega'_{ij} \vec{e}'_j$$

$$\vec{\vec{\Omega}} = [\Omega'] = \begin{bmatrix} 0 & -\omega'_3 & \omega'_2 \\ \omega'_3 & 0 & -\omega'_1 \\ -\omega'_2 & \omega'_1 & 0 \end{bmatrix}$$

While replacing and changing the summation indices:

$$\dot{\vec{X}} = \dot{x}_i \vec{e}_i = \vec{V}_e + \dot{x}'_i \vec{e}'_i + x'_i \Omega'_{ij} \vec{e}'_j = \vec{V}_e + \dot{x}'_i \vec{e}'_i + x'_j \Omega'_{ji} \vec{e}'_i$$

$$\Leftrightarrow \dot{\vec{X}} = \dot{x}_i \vec{e}_i = v_{e,i} \vec{e}_i + v'_{Ai} \vec{e}'_i = v'_{e,i} \vec{e}'_i + (\dot{x}'_i + x'_j \Omega'_{ji}) \vec{e}'_i$$

Thus, components of inertial velocity $\vec{V}_A$ in E are:

$$v'_{Ai} = v'_{ei} + \dot{x}'_i + x'_j \Omega'_{ji}$$

where v'_{ei} are components of inertial velocity $\vec{V}_e$ of O′ in E.
Developing the third term gives,

$$v'_1 = x'_j \Omega'_{j1} = x'_2 \omega'_3 - x'_3 \omega'_2$$

$$v'_2 = x'_j \Omega'_{j2} = -x'_1 \omega'_3 + x'_3 \omega'_1$$

$$v'_3 = x'_j \Omega'_{j3} = x'_1 \omega'_2 - x'_2 \omega'_1$$

Thus, inertial velocity of a point having velocity $\vec{V}_R$ relative to E can be expressed formally in E as:

$$\vec{V}_A = \vec{V}_e + \vec{V}_R + \vec{\Omega} \wedge \overrightarrow{O'M}$$

where $\vec{\Omega} = \begin{bmatrix} \omega'_1 \\ \omega'_2 \\ \omega'_3 \end{bmatrix}$ is rotation rate "vector." While deriving the expression of the inertial velocity in E, we obtain absolute acceleration:

$$\ddot{\vec{X}} = \Gamma_{e,i} \vec{e}_i + \frac{d}{dt} \left(\dot{x}'^i \vec{e}'_i + x'_j \Omega'_{ji} \vec{e}'_i \right)$$

$$= \Gamma'_{e,i} \vec{e}'_i + \ddot{x}'_i \vec{e}'_i + \dot{x}'_i \dot{\vec{e}}'_i + \dot{x}'_j \Omega'_{ji} \vec{e}'_i + x'_j \dot{\Omega}'_{ji} \vec{e}'_i + x'_j \Omega'_{ji} \dot{\vec{e}}'_i$$

$$\ddot{\vec{X}} = \Gamma'_{e,i} \vec{e}'_i + \ddot{x}'_i \vec{e}'_i + 2\dot{x}'_j \Omega'_{ji} \vec{e}'_i + x'_j \dot{\Omega}'_{ji} \vec{e}'_i + x'_j \Omega'_{ji} \Omega'_{ik} \vec{e}'_k$$

After changing summation index in the last term, we obtain:

$$\gamma'_{Ai} = \Gamma'_{e,i} + \ddot{x}'_i + 2\dot{x}'_j \Omega'_{ji} + x'_j \dot{\Omega}'_{ji} + x'_j \Omega'_{jk} \Omega'_{ki}$$

The fifth term develops as:

$$x'_j \Omega'_{jk} \Omega'_{ki} = v'_k \Omega'_{ki}$$

which is formally equivalent to $\vec{\Omega} \wedge \vec{v}$, with $\vec{v} = \vec{\Omega} \wedge \overrightarrow{O'M}$.

Thus, absolute acceleration is expressed formally in E as:

$$\vec{\Gamma}_A = \vec{\Gamma}_e + \vec{\Gamma}_R + 2\vec{\Omega} \wedge \vec{V}_R + \dot{\vec{\Omega}} \wedge \overrightarrow{O'M} + \vec{\Omega} \wedge \left(\vec{\Omega} \wedge \overrightarrow{O'M}\right) = \frac{\vec{F}}{m}$$

2 Accelerometer Measurements

- General expression of accelerometer measurement

For an observer fixed to the case in accelerated linear motion, under the effect of external forces and a gravitational field $\vec{\varphi}$, mass "m" is subjected to the interior forces applied by the case, gravitational force $m\vec{\varphi}$, and inertia force associated with acceleration of the case. The apparent motion is written as:

$$m\vec{\gamma}_r = -m\vec{\gamma}_e + m\vec{\varphi} + \vec{F}_{int}$$

At equilibrium, $\vec{\gamma}_r = 0$, the measurement of the interior force is:

$$\vec{A} = \frac{\vec{F}_{int}}{m} = \vec{\gamma}_e - \vec{\varphi}$$

Thus, an accelerometer is sensitive only to nongravitational accelerations of the case, i.e., related to the sum of the external forces applied to the vehicle, less than the gravitational force.

$$\vec{A} = \vec{\gamma}_e - \vec{\varphi} = \frac{\Sigma_i \vec{F}_{ext,i}}{M} - \vec{\varphi}$$

- On the launching site, the missile is subjected to gravitational force $M\vec{\varphi}$ and to reaction of the ground $\vec{R} = -M\vec{g}$:

  $\vec{A} = \frac{M\vec{\varphi}+\vec{R}}{M} - \vec{\varphi} = \frac{\vec{R}}{M} = -\vec{g}$, measure acceleration created by the reaction of the ground, opposed to apparent gravity.

- During the propelled phase the missile is subjected to gravitational force $M\vec{\varphi}$, motor thrust $\vec{T}$, and aerodynamic force $\vec{F}_{ae}$:

  $\vec{A} = \frac{M\vec{\varphi}+\vec{T}+\vec{F}_{ae}}{M} - \vec{\varphi} = \frac{\vec{T}+\vec{F}_{ae}}{M}$, measure acceleration created by motor thrust and aerodynamic force.

- During the ballistic phase out of the atmosphere, it is subjected only to gravitational force: $\vec{A} = \frac{M\vec{\varphi}}{M} - \vec{\varphi} = 0$, accelerometer measurement is null and apparent gravity is zero.
- During the reentry phase:

  $\vec{A} = \frac{M\vec{\varphi}+\vec{F}_{ae}}{M} - \vec{\varphi} = \frac{\vec{F}_{ae}}{M}$, we measure acceleration created by aerodynamic force.

3 Vertical and Apparent Gravity

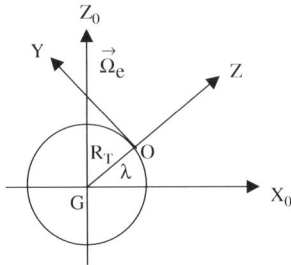

We consider a quasi-inertial frame GX_0Z_0 centered at earth's center of mass G, not rotating, such that GX_0 is in the equatorial plane and GZ_0 along earth's axis of rotation, toward north. OYZ is a rotating frame fixed to earth, centered at sea level such that OZ is along the geocentric vertical, and OY is normal to OZ, in the plane GZZ_0.

Let us consider a rotating observer fixed to OYZ, and a line pendulum centered on OZ. From the point of view of an inertial observer, the mass of the pendulum is subjected only to real forces $\vec{T}$, traction of the line, and $\vec{P'} = m\vec{\varphi}$, gravitational attraction. When the pendulum is at equilibrium, its mass is fixed to the rotating reference frame, and its absolute acceleration is equal to centripetal acceleration $m\vec{\gamma}_e \approx m\vec{\Omega}_e \wedge \left(\vec{\Omega}_e \wedge \overrightarrow{GO}\right)$ dependent on rotation around GZ_0. For the inertial observer, static trim condition is $m\overleftarrow{\gamma}_a = m\vec{\gamma}_e = \vec{T} + \vec{P'}$.

For the rotating observer, the mass is subjected to $\vec{T}$, $\vec{P'}$, and to inertia force $\vec{F}_i = -m\vec{\gamma}_e$ dependent on the motion of observation axes. The equilibrium equation is $m\vec{\gamma}_r = \vec{T} + \vec{P'} - m\vec{\gamma}_e = 0$, which leads to the same result:

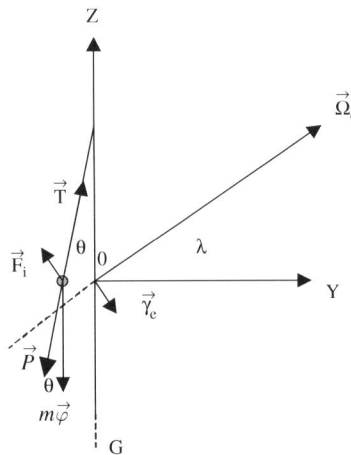

In the rotating reference frame, located at latitude λ, we have $\vec{T} = \begin{bmatrix} T\sin\theta \\ T\cos\theta \end{bmatrix}$, $\vec{P'} = \begin{bmatrix} 0 \\ -m\varphi \end{bmatrix}$, and $\vec{F}_i = -m\vec{\gamma}_e = -m\Omega_e^2 R_T \cos\lambda \begin{bmatrix} \sin\lambda \\ -\cos\lambda \end{bmatrix}$
which gives at equilibrium:

$$T\sin\theta - m\Omega_e^2 R_T \cos\lambda \sin\lambda = 0$$
$$T\cos\theta - m\varphi + m\Omega_e^2 R_T \cos^2\lambda = 0$$

And,

$$T = P = mg \approx m(\varphi - \Omega_e^2 R_T \cos^2\lambda)$$
$$\Rightarrow \sin\theta \approx \frac{\Omega_e^2 R_T \cos\lambda \sin\lambda}{g}$$

For a rotating observer it looks like there is a weight P associated to a "gravity" field "g" smaller than the true gravitation field φ. At geographic poles, the gravity is equal to the true gravitation field. At the equator, the difference with the true field of gravitation is maximum (−0.34%) and deviation is null. Elsewhere, the vertical of gravity is deviated by angle θ compared with the true field in the direction centrifuge of the axis of rotation. Maximum of deviation angle is at ±45° of latitude, and is approximately 0.1°. The normal to the ellipsoid, being by definition along the gravity field, $\lambda' = \lambda + \theta$ is close to geographical latitude.

4 Coriolis Force

We consider a reference frame E is fixed to earth, centered at the base of the tower, with axis OZ_a along the apparent vertical of gravity and axis OX_a normal to meridian plane toward east. $\vec{r} = \overleftarrow{OM}$, $\vec{v}_r$, $\vec{\gamma}_r$ are respectively location, velocity, and acceleration of marble relative to E.

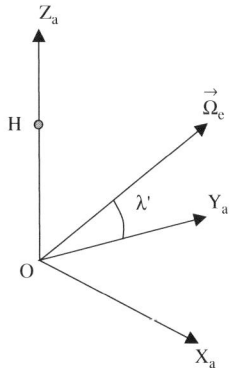

The acceleration of marble relative to the rotating frame E is:

$$\vec{\gamma}_r = \frac{\vec{P}}{m} - 2\vec{\Omega}_e \wedge \vec{v}_r - \vec{\Omega}_e \wedge \left(\vec{\Omega}_e \wedge \vec{r}\right) = \vec{g} - 2\vec{\Omega}_e \wedge \vec{v}_r + \Omega_e^2 \vec{r} - \left(\vec{\Omega}_e \cdot \vec{r}\right)\vec{\Omega}_e$$

While developing, we obtain:

$$\ddot{x} = -2\Omega_e \left[\dot{z}\cos\lambda' - \dot{y}\sin\lambda'\right] + \Omega_e^2 x$$
$$\ddot{y} = -2\Omega_e \dot{x} \sin\lambda' + \Omega_e^2 \left[y \sin^2\lambda' - z \sin\lambda'\cos\lambda'\right]$$
$$\ddot{z} = -g + 2\Omega_e \dot{x}\cos\lambda' + \Omega_e^2 \left[z\cos^2\lambda' - y\sin\lambda'\cos\lambda'\right]$$

At first order, we can neglect x and y compared with z like $\dot{x}$ and $\dot{y}$ components compared with $\dot{z}$, as well as terms including Ω_e^2:

$$\ddot{x} \approx -2\Omega_e \dot{z} \cos\lambda'$$
$$\ddot{y} \approx 0$$
$$\ddot{z} = -g$$

For $z(0) = H$, $x(0) = y(0) = \dot{x}(0) = \dot{y}(0) = \dot{z}(0) = 0$ we obtain:

$$\dot{z} \approx -gt$$
$$z \approx H - g\frac{t^2}{2} \Rightarrow z = 0 \rightarrow T = \sqrt{\frac{2H}{g}}$$
$$\Rightarrow \ddot{x} \approx 2\Omega_e g t \cos\lambda', \quad \dot{x} \approx \Omega_e g t^2 \cos\lambda', \quad x = \Omega_e g \frac{t^3}{3}\cos\lambda'$$

For $H = 300$ m, $\lambda' = 45°$, and $T = 7.8$ s, the acceleration of Coriolis moves the point of impact of the marble toward the east:

$$x(T) \approx \Omega_e g \frac{T^3}{3}\cos\lambda' \approx 0.08 \text{ m}$$

5 Quaternions

From definition (6.72),

$$\underline{Q} = \cos\frac{\phi}{2} + \sin\frac{\phi}{2}\vec{\Delta} = [q_0 \quad q_1 \quad q_2 \quad q_3]$$

While identifying corresponding terms, we obtain:

$$q_0 = \cos\frac{\phi}{2};\ q_1 = \Delta_1 \sin\frac{\phi}{2};\ q_2 = \Delta_2 \sin\frac{\phi}{2};\ q_3 = \Delta_3 \sin\frac{\phi}{2}$$

$$\left|\vec{\Delta}\right|^2 = 1 \Rightarrow \sin^2\frac{\phi}{2} = q_1^2 + q_2^2 + q_3^2$$

- If

$$-1 < q_0 < 1;\ \phi \in [0, 2\pi[\Rightarrow$$

$$q_0 > 0 \rightarrow \frac{\phi}{2} = Arc\sin\sqrt{q_1^2 + q_2^2 + q_3^2}$$

$$q_0 < 0 \rightarrow \frac{\phi}{2} = \pi - Arc\sin\sqrt{q_1^2 + q_2^2 + q_3^2}$$

$$q_0 = 0 \rightarrow \frac{\phi}{2} = \frac{\pi}{2}$$

$$\Delta_1 = \frac{q_1}{\sqrt{q_1^2 + q_2^2 + q_3^2}};\ \Delta_2 = \frac{q_2}{\sqrt{q_1^2 + q_2^2 + q_3^2}};\ \Delta_3 = \frac{q_3}{\sqrt{q_1^2 + q_2^2 + q_3^2}};$$

- If $q_0 = \pm 1,\ \phi = 0$

$$Q = [0\ 1\ 0\ 0],\ \left[0\ \tfrac{1}{\sqrt{3}}\ \tfrac{1}{\sqrt{3}}\ \tfrac{1}{\sqrt{3}}\right],\ \left[0\ -\tfrac{1}{\sqrt{3}}\ -\tfrac{1}{\sqrt{3}}\ -\tfrac{1}{\sqrt{3}}\right],\ \left[\tfrac{1}{2}\ \tfrac{1}{2}\ \tfrac{1}{2}\ \tfrac{1}{2}\right],$$
$$\left[-\tfrac{1}{2}\ \tfrac{1}{2}\ \tfrac{1}{2}\ \tfrac{1}{2}\right],\ \left[\tfrac{\sqrt{2}}{2}\ \tfrac{\sqrt{2}}{4}\ \tfrac{1}{2}\ -\tfrac{\sqrt{2}}{4}\right]$$

corresponds to:

$$\left[\varphi, \vec{\Delta}\right] = \left[\tfrac{\pi}{2}\ 1\ 0\ 0\right],\ \left[\pi\ \tfrac{1}{\sqrt{3}}\ \tfrac{1}{\sqrt{3}}\ \tfrac{1}{\sqrt{3}}\right],\ \left[\tfrac{3\pi}{2}\ \tfrac{1}{\sqrt{3}}\ \tfrac{1}{\sqrt{3}}\ \tfrac{1}{\sqrt{3}}\right],\ \left[\tfrac{2\pi}{3}\ \tfrac{1}{\sqrt{3}}\ \tfrac{1}{\sqrt{3}}\ \tfrac{1}{\sqrt{3}}\right],$$
$$\left[\tfrac{\pi}{3}\ \tfrac{1}{\sqrt{3}}\ \tfrac{1}{\sqrt{3}}\ \tfrac{1}{\sqrt{3}}\right],\ \left[\tfrac{\pi}{2}\ \tfrac{1}{\sqrt{3}}\ \tfrac{1}{\sqrt{3}}\ \tfrac{1}{\sqrt{3}}\right]$$

6 Quaternions and Euler Angles

$$\underline{R_3(\psi)\ R_2(\theta)\ R_1(\varphi)} = \left\{\cos\frac{\psi}{2} + \sin\frac{\psi}{2}\begin{bmatrix}1\\0\\0\end{bmatrix}\right\}\left\{\cos\frac{\theta}{2} + \sin\frac{\theta}{2}\begin{bmatrix}0\\1\\0\end{bmatrix}\right\}$$

$$\left\{\cos\frac{\varphi}{2} + \sin\frac{\varphi}{2}\begin{bmatrix}1\\0\\0\end{bmatrix}\right\}$$

While applying rules of quaternion product, we obtain successively:

$$\underline{R_3(\psi) R_2(\theta) R_1(\varphi)} = \left\{ \cos\frac{\psi}{2} + \sin\frac{\psi}{2} \begin{bmatrix} 0 \\ 0 \\ 1 \end{bmatrix} \right\} \left\{ \cos\frac{\theta}{2} \cos\frac{\varphi}{2} + \begin{bmatrix} \cos\frac{\theta}{2}\sin\frac{\varphi}{2} \\ \sin\frac{\theta}{2}\cos\frac{\varphi}{2} \\ -\sin\frac{\theta}{2}\sin\frac{\varphi}{2} \end{bmatrix} \right\}$$

$$\underline{R_3(\psi) R_2(\theta) R_1(\varphi)} = \left\{ \cos\frac{\psi}{2}\cos\frac{\theta}{2}\cos\frac{\varphi}{2} + \sin\frac{\psi}{2}\sin\frac{\theta}{2}\sin\frac{\varphi}{2} \right. $$
$$\left. + \begin{bmatrix} \cos\frac{\psi}{2}\cos\frac{\theta}{2}\sin\frac{\varphi}{2} - \sin\frac{\psi}{2}\cos\frac{\theta}{2}\cos\frac{\varphi}{2} \\ \cos\frac{\psi}{2}\sin\frac{\theta}{2}\cos\frac{\varphi}{2} + \sin\frac{\psi}{2}\sin\frac{\theta}{2}\sin\frac{\varphi}{2} \\ -\cos\frac{\psi}{2}\sin\frac{\theta}{2}\sin\frac{\varphi}{2} + \sin\frac{\psi}{2}\sin\frac{\theta}{2}\cos\frac{\varphi}{2} \end{bmatrix} \right\}$$

Thus, the components of quaternion are:

$$r_0 = \cos\frac{\psi}{2}\cos\frac{\theta}{2}\cos\frac{\varphi}{2} - \sin\frac{\psi}{2}\cos\frac{\theta}{2}\sin\frac{\varphi}{2}$$
$$r_1 = \cos\frac{\psi}{2}\cos\frac{\theta}{2}\sin\frac{\varphi}{2} + \sin\frac{\psi}{2}\cos\frac{\theta}{2}\cos\frac{\varphi}{2}$$
$$r_2 = \cos\frac{\psi}{2}\sin\frac{\theta}{2}\cos\frac{\varphi}{2} + \sin\frac{\psi}{2}\sin\frac{\theta}{2}\sin\frac{\varphi}{2}$$
$$r_3 = -\cos\frac{\psi}{2}\sin\frac{\theta}{2}\sin\frac{\varphi}{2} + \sin\frac{\psi}{2}\sin\frac{\theta}{2}\cos\frac{\varphi}{2}$$

While using above values of q_0, q_1, q_2, and q_3, and results of exercise 5, we derive parameters ϕ, and $\vec{\Delta}$ of the single axis rotation equivalent to the three Euler' rotations.

While $\psi = \theta = \varphi = \pi/2;\ \pi/3;\ 2\pi/3$, we obtain respectively:

$$\underline{Q} = \left[0, 1/\sqrt{2}, 1/\sqrt{2}, 0 \right] \rightarrow \phi = \pi, \vec{\Delta} = \left[1/\sqrt{2}\ 1/\sqrt{2}\ 0 \right]$$
$$\underline{Q} = \left[-1/4, \sqrt{3}/4, \sqrt{3}/2, 0 \right] \rightarrow \phi = 2\left(\pi - \text{Arc} \sin\left(\sqrt{15}/4\right)\right),$$
$$\vec{\Delta} = \left[1/\sqrt{5}\ 2/\sqrt{5}\ 0 \right]$$
$$\underline{Q} = \left[1/4, \sqrt{3}/4, \sqrt{3}/2, 0 \right] \rightarrow \phi = 2 \text{Arc} \sin\left(\sqrt{15}/4\right),\ \vec{\Delta} = \left[1/\sqrt{5}\ 2/\sqrt{5}\ 0 \right]$$

7 Equivalence of Changing Frame and Rotation Quaternion Operators for Euler Angle Sequence

From Sect. 6.3.2.2, rotation operator to transform vectors of K is:

$$\vec{X}' = \underline{R}\vec{X}\underline{R}^*,$$

with $\underline{R} = \cos\frac{\varphi}{2} + \sin\frac{\varphi}{2}\vec{\Delta}$.

Denoting $\vec{e}_1, \vec{e}_2$, and $\vec{e}_3$ the unit vectors of K frame, and using 1, 2, 1 Euler sequence of rotations, we first apply the rotation ψ around $\vec{e}_1$, then rotation θ around axis $\vec{e}'_2$ transformed of $\vec{e}_2$ by rotation ψ around $\vec{e}_1$, then rotation φ around axis $\vec{e}'_1$ transformed of $\vec{e}_1$ by rotation θ around $\vec{e}'_2$

$$\vec{e}_1 = \begin{bmatrix} 1 \\ 0 \\ 0 \end{bmatrix} ; \vec{e}'_2 = \begin{bmatrix} 0 \\ \cos\psi \\ \sin\psi \end{bmatrix} ; \vec{e}'_1 = \begin{bmatrix} \cos\theta \\ \sin\theta \sin\psi \\ -\sin\theta \cos\psi \end{bmatrix}$$

$$\underline{R} = R_3(\varphi) R_2(\theta) R_1(\psi) = \left\{\cos\frac{\varphi}{2} + \sin\frac{\varphi}{2}\vec{e}'_1\right\}\left\{\cos\frac{\theta}{2} + \sin\frac{\theta}{2}\vec{e}'_2\right\}$$
$$\left\{\cos\frac{\psi}{2} + \sin\frac{\psi}{2}\vec{e}_1\right\}$$

While using the rule of quaternion product and some trigonometry, we obtain:

$$\underline{R} = R_3(\psi) R_2(\theta) R_1(\varphi) = \left\{\cos\frac{\psi}{2}\cos\frac{\theta}{2}\cos\frac{\varphi}{2} - \sin\frac{\psi}{2}\cos\frac{\theta}{2}\sin\frac{\varphi}{2}\right.$$
$$\left. + \begin{bmatrix} \cos\frac{\psi}{2}\cos\frac{\theta}{2}\sin\frac{\varphi}{2} + \sin\frac{\psi}{2}\cos\frac{\theta}{2}\cos\frac{\varphi}{2} \\ \cos\frac{\psi}{2}\sin\frac{\theta}{2}\cos\frac{\varphi}{2} + \sin\frac{\psi}{2}\sin\frac{\theta}{2}\sin\frac{\varphi}{2} \\ -\cos\frac{\psi}{2}\sin\frac{\theta}{2}\sin\frac{\varphi}{2} + \sin\frac{\psi}{2}\sin\frac{\theta}{2}\cos\frac{\varphi}{2} \end{bmatrix}\right\}$$

This is exactly the same result as the changing of frame quaternion resulting from exercise 6, but former assessment was much easier. This verifies that the two points of view are equivalent.

8 Pendulum of Foucault

Kepler's theory for the motion of planets uses the hypothesis of central motion around the sun (Heliocentric). The explanation of the apparent motion of planets for earth's observers required assumption of a rotational motion of earth around its polar axis. Foucault's experiment was the first to provide physical proof of earth's rotational motion, within the framework of Newtonian mechanics. For a line pendulum located at the pole, initialized for a plane oscillation, the demonstration is easy. Indeed, in an inertial reference frame centered on the pole, including earth rotational axis and two nonrevolving axes in the normal plane, the plane of oscillation of the pendulum remains fixed (moment around the hinge point O of gravity forces applied to the mass is same direction as angular momentum around O, thus the direction of the angular momentum remains constant as well as the plane of oscillation). For an observer fixed to earth, which turns at angular rate $\vec{\Omega}_e$ relative to inertial frame, the plane of oscillation has an apparent rotational motion at rate $\vec{\Omega}_r = -\vec{\Omega}_e$ around vertical, i.e., a rotation toward west of one turn per 24 hours or 15°/hour. At other latitude, the phenomenon is similar, with rotation of the plane of oscillation around apparent vertical equal to $\Omega_r = -\Omega_e \sin \lambda$; however, the rigorous demonstration is definitely more difficult.

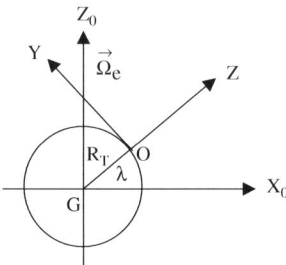

First, to simplify the equations, we will use a reference frame fixed to earth centered at hinge point O of the pendulum, with axis OZ_a along vertical of apparent gravity and axis OX_a normal to the meridian plane, toward east.

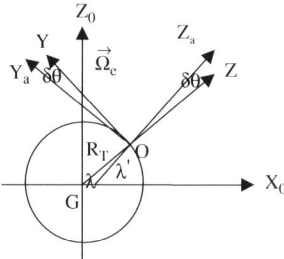

In this frame, the resultant of the gravitation force and inertia force associated with acceleration of O by terrestrial rotation results in a "gravity" force along OZ_a.

$$g \approx \varphi - \Omega_e^2 R_T \cos^2 \lambda$$

The geographical latitude λ', defined as the angle of the vertical of gravity with respect to the equatorial plane, is connected to the geocentric latitude by the relation:

$$\lambda' = \lambda + \delta\theta$$

$$\delta\theta \approx \frac{\Omega_e^2 R_T \cos \lambda \sin \lambda}{g}$$

8 Pendulum of Foucault

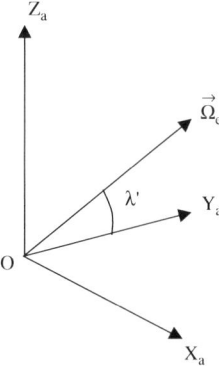

In this frame, equilibrium of the pendulum is along OZ_a.

Now let us associate frame $OX_pY_pZ_p$ with the instantaneous position of the pendulum such that the position of the mass is located in $-\ell$ along OZ_p and such that we transform $OX_aY_aZ_a$ in $OX_pY_pZ_p$ by two successive rotations:

- ψ/OZ_a, which transforms $OX_aY_aZ_a$ in $OX_1Y_1Z_1$
- θ/OX_1, which transforms $OX_1Y_1Z_1$ in $OX_pY_pZ_p$

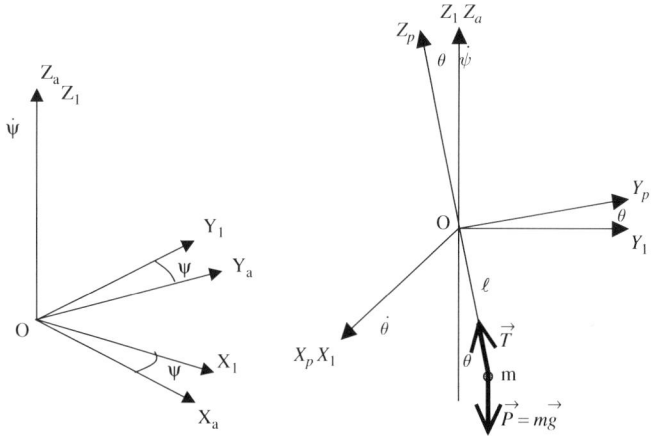

We obtain the components in "p" of a vector defined in "a" by the transformations:

$$\begin{bmatrix} V_{X1} \\ V_{Y1} \\ V_{Z1} \end{bmatrix} = \begin{bmatrix} \cos\psi & \sin\psi & 0 \\ -\sin\psi & \cos\psi & 0 \\ 0 & 0 & 1 \end{bmatrix} \begin{bmatrix} V_{Xa} \\ V_{Ya} \\ V_{Za} \end{bmatrix} = [A_\psi] \begin{bmatrix} V_{Xa} \\ V_{Ya} \\ V_{Za} \end{bmatrix}$$

and

$$\begin{bmatrix} V_{Xp} \\ V_{Yp} \\ V_{Zp} \end{bmatrix} = \begin{bmatrix} 1 & 0 & 0 \\ 0 & \cos\theta & \sin\theta \\ 0 & -\sin\theta & \cos\theta \end{bmatrix} \begin{bmatrix} V_{X1} \\ V_{Y1} \\ V_{Z1} \end{bmatrix} = [A_\theta] \begin{bmatrix} V_{X1} \\ V_{Y1} \\ V_{Z1} \end{bmatrix}$$

Angle ψ corresponds to the orientation of the plane of oscillation $OZ_a Y_p$ around OZ_a with origin in Oy_a, and the angle θ is the deflection of the pendulum from the apparent vertical OZ_a.

Rotation rate of the pendulum frame relative to observation frame "a" is:

$$\vec{\omega}_r = \dot{\psi}\ \overrightarrow{OZ_a} + \dot{\theta}\ \overrightarrow{OX_1} = \dot{\psi}\ \overrightarrow{OZ_1} + \dot{\theta}\ \overrightarrow{OX_p}$$

Rotation rate $\vec{\Omega}_e$ of "a" relative to inertial frames is in "a":

$$\vec{\Omega}_e = \Omega_e \begin{bmatrix} 0 \\ \cos\lambda' \\ \sin\lambda' \end{bmatrix}$$

Inertial rotation rate $\vec{\omega}$ of "p" is $\vec{\omega} = \vec{\Omega}_e + \vec{\omega}_r$, with components in "a":

$$\begin{bmatrix} \omega_{Xp} \\ \omega_{Yp} \\ \omega_{Zp} \end{bmatrix} = \Omega_e [A_\theta][A_\psi] \begin{bmatrix} 0 \\ \cos\lambda' \\ \sin\lambda' \end{bmatrix} + \dot{\psi}[A_\theta] \begin{bmatrix} 0 \\ 0 \\ 1 \end{bmatrix} + \dot{\theta} \begin{bmatrix} 1 \\ 0 \\ 0 \end{bmatrix}$$

$$= \begin{bmatrix} \Omega_e \sin\psi \cos\lambda' + \dot{\theta} \\ \Omega_e (\cos\theta\cos\psi\cos\lambda' + \sin\theta\sin\lambda') + \dot{\psi}\sin\theta \\ \Omega_e (-\sin\theta\cos\psi\cos\lambda' + \cos\theta\sin\lambda') + \dot{\psi}\cos\theta \end{bmatrix}$$

Let us calculate the derivative of $\vec{\omega}$ in "p":

$$\begin{bmatrix} \dot{\omega}_{Xp} \\ \dot{\omega}_{Yp} \\ \dot{\omega}_{Zp} \end{bmatrix} = \begin{bmatrix} \ddot{\theta} + \Omega_e \dot{\psi}\cos\psi\cos\lambda' \\ \ddot{\psi}\sin\theta + \dot{\psi}\dot{\theta}\cos\theta + \Omega_e \left((-\dot{\theta}\sin\theta\cos\psi - \dot{\psi}\cos\theta\sin\psi)\cos\lambda' + \dot{\theta}\cos\theta\sin\lambda'\right) \\ \ddot{\psi}\cos\theta - \dot{\psi}\dot{\theta}\sin\theta + \Omega_e \left((-\dot{\theta}\cos\theta\cos\psi + \dot{\psi}\sin\theta\sin\psi)\cos\lambda' - \dot{\theta}\sin\theta\sin\lambda'\right) \end{bmatrix}$$

The weight of the mass has components in "p":

$$\vec{P} = -mg[A_\theta]\begin{bmatrix} 0 \\ 0 \\ 1 \end{bmatrix} = -mg \begin{bmatrix} 0 \\ \sin\theta \\ \cos\theta \end{bmatrix}$$

The tension of the wire and the relative position of the mass in "p" are:

8 Pendulum of Foucault

$$\vec{T} = \begin{bmatrix} 0 \\ 0 \\ t \end{bmatrix} \quad \vec{r} = \begin{bmatrix} 0 \\ 0 \\ -\ell \end{bmatrix}$$

Velocity and acceleration of the mass relative to "p" are null by definition. The equation of motion of the mass in the rotating frame fixed to the pendulum "p" is thus, according to (1.4):

$$\vec{\gamma}_r = 0 = \frac{\vec{P}+\vec{T}}{m} - \dot{\vec{\omega}} \wedge \vec{r} - \vec{\omega} \wedge (\vec{\omega} \wedge \vec{r}) = \frac{\vec{P}+\vec{T}}{m} - \dot{\vec{\omega}} \wedge \vec{r} + \omega^2 \vec{r} - \left(\vec{\omega} \cdot \vec{r}\right) \vec{\omega}$$

We will notice that the inertial term of acceleration centrifuges related to the acceleration of O is already included in the definition of weight $\vec{P}$.

This equation develops in:

$$\dot{\omega}_{Yp} + \omega_{Zp}\omega_{Xp} = 0$$

$$\frac{g}{\ell}\sin\theta + \dot{\omega}_{Xp} - \omega_{Zp}\omega_{Yp} = 0$$

$$-g\cos\theta + \frac{t}{m} - \left(\omega_{Xp}^2 + \omega_{Yp}^2\right)\ell = 0$$

The third equation gives the expression of the tension of the wire. By using preceding expressions of components of $\vec{\omega}$ and $\dot{\vec{\omega}}$ and first-order approximations in θ, $\theta^2 \approx 0$, $\sin\theta \approx \theta$, $\cos\theta \approx 1$, we obtain for the two others equations:

$$\ddot{\psi} + \dot{\psi}\dot{\theta} + \Omega_e\left[-\left(\theta\dot{\theta}\cos\psi + \dot{\psi}\sin\psi\right)\cos\lambda' + \dot{\theta}\sin\lambda'\right]$$
$$+ \left[\Omega_e\left(-\theta\cos\psi\cos\lambda' + \sin\lambda'\right) + \dot{\psi}\right]\left(\Omega_e\sin\psi\cos\lambda' + \dot{\theta}\right) = 0$$

$$\frac{g}{\ell}\theta + \ddot{\theta} + \Omega_e\dot{\psi}\cos\lambda'\cos\psi - \left[\Omega_e\left(-\theta\cos\psi\cos\lambda' + \sin\lambda'\right) + \dot{\psi}\right]$$
$$\left[\Omega_e\left(\cos\psi\cos\lambda' + \theta\sin\lambda'\right) + \dot{\psi}\theta\right] = 0$$

That is to say,

$$\ddot{\psi}\theta + 2\dot{\psi}\dot{\theta} = -2\Omega_e\dot{\theta}\sin\lambda' + 2\Omega_e\theta\dot{\theta}\cos\lambda'\cos\psi - \Omega_e^2\cos\lambda'\sin\lambda'\sin\psi$$
$$+ \Omega_e^2\theta\cos^2\lambda'\sin\psi\cos\psi$$

$$\ddot{\theta} + \left[\frac{g}{\ell} - \dot{\psi}^2 + \Omega_e^2\left(\cos^2\psi\cos^2\lambda' - \sin^2\lambda'\right) - 2\Omega_e\dot{\psi}\sin\lambda'\right]\theta$$
$$= \Omega_e^2\sin\lambda'\cos\lambda'\cos\psi$$

Considering that Ω_e is about 10^{-4}, we neglect terms in Ω_e^2. Because of the hypothesis of small oscillations θ, we also neglect $\Omega_e\theta\dot{\theta}$ compared with $\Omega_e\dot{\theta}$ in the first equation.

We obtain:

$$\ddot{\psi}\theta + 2\left(\dot{\psi} + \Omega_e \sin\lambda'\right)\dot{\theta} \approx 0$$

$$\ddot{\theta} + \left[\frac{g}{\ell} - \dot{\psi}^2 - 2\Omega_e\dot{\psi}\sin\lambda'\right]\theta \approx 0$$

Multiplying the first equation by θ and naming $\dot{\varphi} = \dot{\psi} + \Omega_e \sin\lambda'$, we obtain:

$$\ddot{\varphi}\theta^2 + 2\dot{\varphi}\theta\dot{\theta} \approx 0 \Leftrightarrow \frac{\mathrm{d}}{\mathrm{dt}}\left(\dot{\varphi}\theta^2\right) = 0$$

In the case of an unspecified oscillation, we obtain a law similar to the area law of elliptic motion,

$$\dot{\varphi}\theta^2 = \dot{\varphi}_0\theta_0^2 = \text{constant}$$

When we initialize the pendulum such that $\dot{\varphi}_0 = 0$ (planar oscillation in inertial frame), we must verify $\dot{\varphi}\theta^2 = 0 \quad \forall\theta \quad \Rightarrow \quad \dot{\varphi} = 0 \quad \Rightarrow \dot{\psi} = -\Omega_e \sin\lambda'$

Using this value $\dot{\psi} \sim 10^{-4}$ in the equation in θ, we can neglect corresponding terms and we obtain:

$$\ddot{\theta} + \left[\frac{g}{\ell}\right]\theta \approx 0$$

i.e., a quasi-planar oscillation with pulsation $\sqrt{\frac{g}{\ell}}$ whose plane turns slowly around the apparent vertical at angular rate $-\Omega_e \sin\lambda'$. At 45° latitude, rotation rate is 0.7 revs per 24 hours or 10.5° per hour. To the equator, the plane of oscillation does not turn.

In situations where $\dot{\varphi}_0 \neq 0$, relative rotation rate of oscillation of the plane around Z_p is:

$$\dot{\psi} = -\Omega_e \sin\lambda' + \frac{\left(\dot{\psi}_0 + \Omega_e \sin\lambda'\right)\theta_0^2}{\theta^2}$$

We can observe that the pendulum will have an elliptic-like oscillation with a fast rotation rate corresponding to the second term, and a slow rotation rate $-\Omega_e \sin\lambda'$. The major axis of trajectory turn around the apparent vertical at rate $-\Omega_e \sin\lambda'$. Deflection of the pendulum verifies by equation:

$$\ddot{\theta} + \left[\frac{g}{\ell} - \frac{K}{\theta^4}\right]\theta \approx 0$$

with $K = \dot{\psi}_0^2\theta_0^4$.

The experiment of Foucault corresponds to an initialization with $\dot{\varphi}_0 \approx 0$ and the oscillation is planar and sinusoidal.

9 Motion of the Earth's Mass Center

For a period of the terrestrial orbit T = 365 × 24 × 3600 s, a perimeter of orbit L = $2\pi R \approx 9.4 \ 10^{11}$ m, the circular velocity is V_c = L/T ≈ 30 10^3 m/s, and centripetal acceleration is $\gamma_c = \frac{V_c^2}{R} \approx 5.9 \ 10^{-3}$ m.s^{-2}.

According to the third law of Kepler, the radius of lunar orbit is such that $R^3 = a^3 = \mu \left(\frac{T}{2\pi}\right)^2$. For T = 27 days and 8 h, a = 3.83 10^8 m = 383000 km, its circular velocity is $V_c \approx 1.02$ km/s. The lunar constant of attraction is $\mu_L = Gm_L = 4.87 \ 10^{12} \ m^3 \ s^{-2}$. The acceleration applied on earth's center of mass is thus $\gamma = \frac{\mu_L}{a^2} = 3.3 \ 10^{-5}$ m.s^{-2}. This acceleration is negligible with respect to earth's gravitation. To study motion of ballistic vehicle relative to earth, we can use a nonrotating frame with origin at its center of mass like an inertial frame.

10 Launch Windows to Mars

Duration of the Martian year is close to two terrestrial years. According to the third law of Kepler, $\left(\frac{a_{Mars}}{a_{Earth}}\right)^3 = \left(\frac{T_{Mars}}{T_{Earth}}\right)^2 \Rightarrow a_{Mars} \approx 1.6 \, a_{Earth} \approx 1.6 UA$ (240 million Km).

The minimum distance is about 0.6 UA, that is to say 90 million Km when the two planets have same solar longitude. The maximum distance is 2.6 UA, which is about 4.3 higher, when the planets are in opposition of phase. As Martian period of revolution around sun is double of the earth's period, these optimum launch windows to minimize flight duration occur every two years.

11 Energy of a Solid in Free Rotation

According to (1.39), angular momentum of a solid expressed in the Eulerian frame corresponding to its principal axes of inertia is:

$$\vec{H} = \begin{bmatrix} I_x \omega_x \\ I_y \omega_y \\ I_z \omega_z \end{bmatrix}$$

For a solid in free rotational motion with given angular momentum, we have:

$$(I_x \omega_x)^2 + (I_y \omega_y)^2 + (I_z \omega_z)^2 = H^2 \tag{1}$$

Thus, rotation rate is minimum when $\vec{H}$ is along the axis of maximum moment of inertia Gx, and maximum for $\vec{H}$ along the axes of lowest moment of inertia Gy or Gz. According to (1.44), kinetic energy of rotation is:

$$E = \frac{1}{2}\vec{\omega}\cdot\vec{H} = \frac{1}{2}\left(I_x\omega_x^2 + I_y\omega_y^2 + I_z\omega_z^2\right) \quad (2)$$

Let us replace ω_x^2 by its expression from (1) while posing $I_y = I_z = I_T$ and $\omega_T^2 = \omega_y^2 + \omega_z^2$, we obtain:

$$E = \frac{1}{2}\left[\frac{H^2}{I_x} + I_T\left(1 - \frac{I_T}{I_x}\right)\omega_T^2\right]$$

By hypothesis, $I_x > I_T$, which is representative of satellites, the coefficient of the term in ω_T^2 is positive and the kinetic energy of rotation is minimum for $\omega_T^2 = 0$, i.e., $E_{\min} = \frac{1}{2}\frac{H^2}{I_x}$. In this case angular momentum is directed along Gx and rotation rate is minimum, $\omega = \frac{H}{I_x}$. The energy of rotation is maximum when ω_T is maximum, i.e., according to (1) when $\omega_x = 0$, and $\omega_T = \frac{H}{I_T}$, $E_{\max} = \frac{1}{2}\frac{H^2}{I_T}$.

We finally have the possible range for kinetic energy of rotation:

$$\frac{1}{2}\frac{H^2}{I_x} \leq E \leq \frac{1}{2}\frac{H^2}{I_T}$$

When the body is not absolutely rigid and there are dissipations of rotational energy, the final equilibrium state corresponds to minimum energy, that is to say such that axis Gx is along the angular momentum.

12 Total Angular Momentum of a Deformable System in Gravitation

For an inertial observer, the angular momentum around common center of mass "G" is

$$\vec{H} = \vec{H}_1 + \vec{H}_2 + \overrightarrow{GG_1} \wedge m_1\vec{V}_1 + \overrightarrow{GG_2} \wedge m_2\vec{V}_2$$

where $\vec{H}_1$ and $\vec{H}_2$ are angular momentum of the two solids relative to its mass center G_1 and G_2. From definition of G,

$$m_1\overrightarrow{GG_1} + m_2\overrightarrow{GG_2} = 0$$
$$\Rightarrow \overrightarrow{GG_2} = \frac{m_1}{m_1+m_2}\overrightarrow{G_1G_2}; \quad \overrightarrow{GG_1} = -\frac{m_2}{m_1+m_2}\overrightarrow{G_1G_2}$$

Hence, we obtain:

$$\vec{H} = \vec{H}_1 + \vec{H}_2 + \mu\vec{R} \wedge \dot{\vec{R}} \quad (1)$$

where $\vec{R} = \overrightarrow{G_1 G_2}$, $\dot{\vec{R}} = \vec{V}_2 - \vec{V}_1$ and $\mu = \frac{m_1 m_2}{m_1 + m_2}$ is "equivalent mass" associated with relative motion.

Deriving (1) with respect to time we obtain:

$$\dot{\vec{H}} = \dot{\vec{H}}_1 + \dot{\vec{H}}_2 + \mu \vec{R} \wedge \ddot{\vec{R}} = 0 \tag{2}$$

According to the principle of action and reaction, the resultant of interior effects is null. Using G_2 as reference point for resulting moment we obtain:

$$\vec{F}_{12} + \vec{F}_{21} = 0$$
$$\vec{M}_{12} + \vec{M}_{21} + \vec{R} \wedge \vec{F}_{12} = 0$$

where $\vec{F}_{12}$ and $\vec{M}_{12}$ are force and moment (around G_2) exerted by 1 on 2, $\vec{F}_{21}$ and $\vec{M}_{21}$ are force and moment (around G_1) exerted by 2 on 1.

Applying the fundamental principle of dynamics to solids 1 and 2 we obtain:

$$\vec{R} = \vec{V}_2 - \vec{V}_1; \ \ddot{\vec{R}} = \dot{\vec{V}}_2 - \dot{\vec{V}}_1 = \frac{\vec{F}_{12}}{m_2} - \frac{\vec{F}_{21}}{m_1} + \vec{\varphi}_2 - \vec{\varphi}_1 \approx \frac{\vec{F}_{12}}{\mu}$$

$$\Rightarrow \vec{F}_{12} = \mu \ddot{\vec{R}}$$

Taking into account this relation in (2) we obtain alternate expression (3) of the derivative of the total angular momentum in an inertial frame:

$$\dot{\vec{H}} = \dot{\vec{H}}_1 + \dot{\vec{H}}_2 + \vec{R} \wedge \vec{F}_{12} = 0 \tag{3}$$

Let us express (2) in the Eulerian frame E fixed to solid 1, while noting $\vec{\Omega}$ inertial rotation rate of 1 and $\vec{\Omega}_R$ rotation rate of 2 relative to E:

$$\ddot{\vec{R}} = \ddot{\vec{R}}_E + 2\vec{\Omega} \wedge \dot{\vec{R}}_E + \vec{\Omega} \wedge \left(\vec{\Omega} \wedge \vec{R}_E\right) + \dot{\vec{\Omega}} \wedge \vec{R}_E \tag{4}$$

$$\dot{\vec{H}}_1 = \dot{\vec{H}}_{1E} + \vec{\Omega} \wedge \vec{H}_{1E} \tag{5}$$

$$\dot{\vec{H}}_2 = \dot{\vec{H}}_{2E} + \vec{\Omega} \wedge \vec{H}_{2E} \tag{6}$$

$$\vec{H}_{1E} = [I_1]\vec{\Omega} \quad \Rightarrow \quad \dot{\vec{H}}_{1E} = [I_1]\dot{\vec{\Omega}} \tag{7}$$

Since 2 is in rotation relative to E:

$$\vec{H}_{2E} = [I_{2E}]\left[\vec{\Omega} + \vec{\Omega}_R\right] \Rightarrow \dot{\vec{H}}_{2E} = [I_{2E}]\left[\dot{\vec{\Omega}} + \dot{\vec{\Omega}}_R\right] + [\dot{I}_{2E}]\left[\vec{\Omega} + \vec{\Omega}_R\right]$$
$$+ \vec{\Omega} \wedge [I_{2E}]\left[\vec{\Omega} + \vec{\Omega}_R\right] \tag{8}$$

We consider rotation matrix A, which transform coordinates relative to E into coordinates relative to E2 associated to solid 2 and $[I_2]$ the matrix of inertia of 2 relative to E_2:

According to (1.41) we have:

$$[I_{2E}] = [A]^T [I_2][A]$$

According to 6.24, we have $[\dot{A}] = [A][\Omega_R]$ where $[\Omega_R]$ indicates the tensor rotation rate of 2 relative to 1.

Using these two equations, we obtain:

$$[\dot{I}_{2E}]\left(\vec{\Omega} + \vec{\Omega}_R\right) = [I_{2E}]\left(\vec{\Omega}_R \wedge \vec{\Omega}\right) - \vec{\Omega}_R \wedge [I_{2E}]\left(\vec{\Omega} + \vec{\Omega}_R\right) \qquad (9)$$

Using (3) to (8), conservation of total angular momentum (2) is written in E, while gathering on the left the terms in $\dot{\vec{\Omega}}$:

$$[[I_1] + [I_{2E}]]\dot{\vec{\Omega}}_E + \mu \vec{R} \wedge \left(\dot{\vec{\Omega}} \wedge \vec{R}\right)$$
$$= -[I_{2E}]\dot{\vec{\Omega}}_R - \vec{\Omega} \wedge \left\{[[I_1] + [I_{2E}]]\vec{\Omega} + [I_{2E}]\vec{\Omega}_R\right\}$$
$$- [I_{2E}]\left(\vec{\Omega}_R \wedge \vec{\Omega}\right) + \vec{\Omega}_R \wedge [I_{2E}]\left(\vec{\Omega} + \vec{\Omega}_R\right)$$
$$- \mu \vec{R} \wedge \left\{\ddot{\vec{R}}_E + 2\vec{\Omega} \wedge \dot{\vec{R}}_E + \vec{\Omega} \wedge \left(\vec{\Omega} \wedge \vec{R}\right)\right\} \qquad (10)$$

The second term of the left member is written as:

$$\mu \vec{R} \wedge \left(\dot{\vec{\Omega}} \wedge \vec{R}\right) = [I_\mu]\dot{\vec{\Omega}}$$

with

$$[I_\mu] = \mu \begin{bmatrix} y^2 + z^2 & -xy & -xz \\ -xy & x^2 + z^2 & -yz \\ -xz & -yz & x^2 + y^2 \end{bmatrix} ; \quad \vec{R} = \begin{bmatrix} x \\ y \\ z \end{bmatrix} \quad \text{in } E$$

Then we define:

$$[I] = [I_1] + [I_{2E}] + [I_\mu],$$

which represents the tensor of inertia of the deformable system around its center of mass in a frame having axes parallel to E.

Finally (10) is written as:

$$[I]\vec{\dot{\Omega}} = -[I_{2E}]\vec{\dot{\Omega}}_R - \vec{\Omega} \wedge \left\{[I]\vec{\Omega} + [I_{2E}]\vec{\Omega}_R\right\} - [I_{2E}]\left(\vec{\Omega}_R \wedge \vec{\Omega}\right)$$
$$+ \vec{\Omega}_R \wedge [I_{2E}]\left(\vec{\Omega} + \vec{\Omega}_R\right) - \mu \vec{R} \wedge \left\{\vec{\ddot{R}}_E + 2\vec{\Omega} \wedge \vec{\dot{R}}_E\right\}$$

This expression of the conservation of the total angular momentum in E determines the evolution of $\vec{\Omega}$ (i.e., of the complete system) when the relative motion of 2 relative to 1 ($\vec{R}_E, \vec{\dot{R}}_E, \vec{\ddot{R}}_E, \psi_R, \theta_R, \varphi_R, \vec{\Omega}_R, \vec{\dot{\Omega}}_R$) is given.

13 Lunar Motion

The origin of this particular motion is the moment related to the gradient of gravity (7.2). In the case of a homogeneous solid comparable to a long cylinder, this couple tends to align the axis of the cylinder along the direction of gravitation field (there are two positions of steady trim, 0° and 180°, and two unstable positions, 90° and 270°). Generally, for any solid, direction of the principal axis of highest moment of inertia is a stable trim, which is the case of the Moon. During its rotation around the common mass center of earth/moon system, the orientation of the moon is such that its principal axis of highest inertia is directed on average toward the earth mass center (there are small fluctuations around this direction). It thus always presents the same side to terrestrial observers. The motion of the moon's CG relative to a frame with origin at common mass center is thus coupled with its own rotational motion. Its angular rate Ω corresponds to the period T of its orbit around earth $\Omega = \frac{2\pi}{T}$, T = 27 days and 8 hours. The duration of the lunar day is thus equal to its period of orbit T.

14 Angular Stabilization with Inertia Wheels

Let us name $\vec{\Omega}$ satellite angular rate, $\vec{\omega}_{R1} = \begin{bmatrix} \omega_{R1} \\ 0 \\ 0 \end{bmatrix}$, $\vec{\omega}_{R2} = \begin{bmatrix} 0 \\ \omega_{R2} \\ 0 \end{bmatrix}$, and $\vec{\omega}_{R3} = \begin{bmatrix} 0 \\ 0 \\ \omega_{R3} \end{bmatrix}$ wheels angular rates relative to the satellite, I_{R1}, I_{R2}, I_{R3} moments of inertia of the wheels around their relative rotation axes. $[I]$ denotes the tensor of inertia of the complete satellite, including wheels, which is diagonal per hypothesis. Wheels are assumed symmetrical around their relative rotational axes. Thus the total tensor of inertia is independent of the relative angular position of the wheels.

Total angular momentum is the sum of the momentum corresponding to the rigid motion at rate $\vec{\Omega}$ (i.e., including the wheels at angular rate $\vec{\Omega}$) and momentums created by relative rotation of wheels.

$$\vec{H} = [I]\vec{\Omega} + I_{R1}\vec{\omega}_{R1} + I_{R2}\vec{\omega}_{R2} + I_{R3}\vec{\omega}_{R3}$$

The conservation of the total angular momentum is expressed in the Eulerian reference frame of satellite by:

$$\dot{\vec{H}} + \vec{\Omega} \wedge \vec{H} = 0$$

$$[I]\dot{\vec{\Omega}} = -I_{R1}\dot{\vec{\omega}}_{R1} - I_{R2}\dot{\vec{\omega}}_{R2} - I_{R2}\dot{\vec{\omega}}_{R3} - \vec{\Omega} \wedge [I]\vec{\Omega} - I_{R1}\vec{\Omega} \wedge \vec{\omega}_{R1}$$
$$- I_{R2}\vec{\Omega} \wedge \vec{\omega}_{R2} - I_{R3}\vec{\Omega} \wedge \vec{\omega}_{R3}$$

While projecting, we obtain:

$$I_1\dot{\Omega}_1 = -I_{R1}\dot{\omega}_{R1} - I_{R3}\Omega_2\omega_{R3} + I_{R2}\Omega_3\omega_{R2} - (I_3 - I_2)\Omega_2\Omega_3$$
$$I_2\dot{\Omega}_2 = -I_{R2}\dot{\omega}_{R2} - I_{R1}\Omega_3\omega_{R1} + I_{R3}\Omega_1\omega_{R3} - (I_1 - I_3)\Omega_3\Omega_1$$
$$I_3\dot{\Omega}_3 = -I_{R3}\dot{\omega}_{R3} - I_{R2}\Omega_1\omega_{R2} + I_{R1}\Omega_2\omega_{R1} - (I_2 - I_1)\Omega_1\Omega_2$$

While varying rotation rates of wheels we obtain inertial reaction moments. This makes it possible to control the orientation of the satellite.

15 Balancing Machines

In a Eulerian frame E with origin at the center of mass of the test vehicle, with Gx axis parallel to the axis of rotation and Gy horizontal and parallel to the plate, rotation rate $\vec{\Omega} = \begin{bmatrix} \Omega_x \\ 0 \\ 0 \end{bmatrix}$ and angular momentum $\vec{H} = [I']\vec{\Omega}$ of vehicle are constant.

According to the expression (5.6), the tensor of inertia around the center of gravity in the presence of a principal axis misalignment (θ_1, Φ_1) and of a rotation of the transverse axes of angle φ_1 is:

$$[I'] \approx \begin{bmatrix} I_x & -(I_y - I_x)\theta_1 \cos\Phi_1 & -(I_z - I_x)\theta_1 \sin\Phi_1 \\ -(I_y - I_x)\theta_1 \cos\Phi_1 & I_y + (I_z - I_y)\sin^2\varphi_1 & -(I_z - I_y)\sin\varphi_1 \cos\varphi_1 \\ -(I_z - I_x)\theta_1 \sin\Phi_1 & -(I_z - I_y)\sin\varphi_1 \cos\varphi_1 & I_z - (I_z - I_y)\sin^2\varphi_1 \end{bmatrix}$$

We deduce expression in E of the angular momentum created by constant rotation rate $\vec{\Omega}$:

$$\vec{H} \approx \begin{bmatrix} I_x\Omega_x \\ -(I_y - I_x)\theta_1 \cos\Phi_1 \Omega_x \\ -(I_z - I_x)\theta_1 \sin\Phi_1 \Omega_x \end{bmatrix}$$

15 Balancing Machines

We consider an inertial reference frame, centered at a point O on the axis of rotation. In this frame, the motion of the center of mass is a circular motion of radius $r_G = \sqrt{y_G^2 + z_G^2}$. According to the fundamental principle, the solid is subjected to a centripetal force $F = M\Omega_x^2 r_G$ applied by the axis of rotation having reaction on bearings that is measurable. The expression in E of this centripetal force is:

$$\vec{F} = \begin{bmatrix} 0 \\ -M\Omega_x^2 y_G \\ -M\Omega_x^2 z_G \end{bmatrix}$$

In this inertial frame, angular momentum is a vector rotating at rate $\vec{\Omega}$. According to angular momentum theorem, the moment applied on the solid around G is equal to the time derivative of the angular momentum in the inertial observation frame, that is to say:

$$\vec{C} = \dot{\vec{H}} = \vec{\Omega} \wedge \vec{H}$$

The expression of $\vec{C}$ in E is:

$$\vec{C} = \begin{bmatrix} 0 \\ (I_z - I_x)\Omega_x^2 \theta_I \sin \Phi_I \\ -(I_y - I_x)\Omega_x^2 \theta_I \cos \Phi_I \end{bmatrix}$$

The expression in E of total effect around center of mass G is thus:

$$\left(\vec{F}, \vec{C} \right)$$

The expression in E of effect around the reference point O is:

$$\left(\vec{F}, \quad \vec{M} = \vec{C} + \vec{OG} \wedge \vec{F} \right)$$

That is to say $\vec{M} = \begin{bmatrix} 0 \\ \left[(I_z - I_x)\theta_I \sin \Phi_I + Mx_G z_G \right] \Omega_x^2 \\ -\left[(I_y - I_x)\theta_I \cos \Phi_I + Mx_G y_G \right] \Omega_x^2 \end{bmatrix}$

Measurements of components F_y, F_z, M_y, and M_z applied by rotation axis to maintain the motion provide an estimate of CG offset and the principal axis misalignment:

$$y_G = -\frac{F_y}{M\Omega_x^2} \; ; \; z_G = -\frac{F_z}{M\Omega_x^2}$$

$$\theta_I \sin \Phi_I = \frac{M_y + x_G F_z}{(I_z - I_x)\Omega_x^2} \; ;$$

$$-\theta_I \cos \Phi_I = \frac{M_z - x_G F_y}{(I_y - I_x)\Omega_x^2}$$

Remarks:

- The applied force and the moment are constant in the reference axes of the revolving plate. They are revolving for an inertial observer. Output signals of the strain gauges fixed to bearings are thus sinusoidal. Amplitudes of signals allow assessment of asymmetries and its phases of their orientation relative to axes of revolving plate.
- This measurement must be accompanied by a measurement of the moments of inertia and of the location x_G of the center of mass. A measurement of pendulum frequency around the reference axes is generally used to measure the moments of inertia.

16 Aerodynamics of Apollo Reentry Capsule in Continuous Flow

For the incidences $\bar{\alpha} < 33°$ where back cover is not impinged by the flow, only the spherical front shield contributes to the aerodynamic effects.

According to Sect. "Pressure Coefficients" in Chap. 4, coefficients of axial and normal force of a segment of sphere in Newtonian flow are with references $S_{\text{ref}} = \pi \cdot R^2$, $L_{\text{ref}} = R$ and θ_a, semiapex angle of tangent cone at the edge of sphere (reference point for moments is at upstream pole O of sphere):

$$C_{Aw} = 1 - \sin^4 \theta_a - \frac{(1 + 3\sin^2 \theta_a)\cos^2 \theta_a}{2} \cdot \sin^2 \bar{\alpha}$$

$$C_{Nw} = \frac{2}{3} \cos^4 \theta_a \cdot \sin 2\bar{\alpha}$$

$$C_{mw} = -C_{Nw} \Leftrightarrow x_{CP} = -R$$

While $\bar{\alpha} < 33°$, $\theta_a = \frac{\pi}{2} - \theta_C$, $\cos \theta_a = \sin \theta_C = \frac{D}{2R}$; $\frac{R}{D} = 1.2$, maximum cross section and diameter D as reference, coefficients are:

16 Aerodynamics of Apollo Reentry Capsule in Continuous Flow

$$C'_A = \frac{S_{ref}}{\pi \frac{D^2}{4}} C_A = 4\left(\frac{R}{D}\right)^2 C_A = 1.826 - 1.740 \sin^2 \bar{\alpha}$$

$$C'_{Nw} = 4\left(\frac{R}{D}\right)^2 C_{Nw} = 0.0868 \sin(2\bar{\alpha})$$

$$C_D = C_X = C'_A \cos\bar{\alpha} + C'_{Nw} \sin\bar{\alpha}$$
$$C_L = C_Z = -C'_A \sin\bar{\alpha} + C'_{Nw} \cos\bar{\alpha}$$
$$f = \frac{C_Z}{C_X} = \frac{C_L}{C_D}$$

With D as reference length, pitching moment coefficient relative to O is:

$$C'_{mw/O} = -\frac{R}{D} C'_{Nw}$$

For $z_G = 0$, moment coefficient around mass center is:

$$C'_{mw/G} = C'_{mw/O} - \frac{x_G}{D} C'_{Nw} = -\left(\frac{x_G + R}{D}\right) C'_{Nw}$$

For $x_G = -0.26D$:

$$C'_{m/G} = -\left(\frac{R}{D} - 0.26\right) C'_N = -0.94 C'_N = -0.0816 \sin(2\bar{\alpha})$$

$$\Rightarrow C'_{m\alpha/G} = \frac{\partial C'_{m/G}}{\partial \alpha} = -0.163 \cos(2\bar{\alpha})$$

Thus the capsule is statically stable around any trim angle $0 \leq \bar{\alpha} < 33°$. For $z_G \neq 0$ and $\beta = 0$ pitching moment coefficient around GY is:

$$C'_{m/G} = \frac{z_G}{D} C'_A - \left(\frac{x_G + R}{D}\right) C'_N$$

Yawing moment is null.

Using the preceding results, we obtain for $\frac{z_G}{D} = -0.035$ the evolutions which follow for $\beta = 0$ according to incidence α:

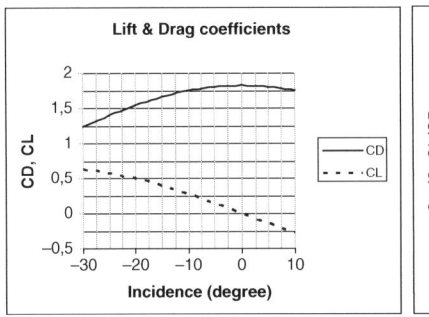
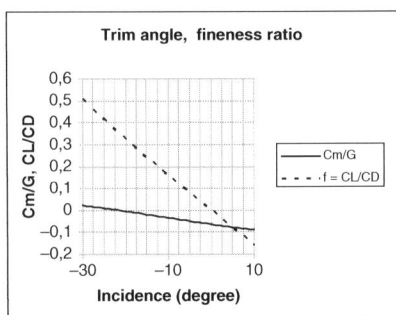

This gives a static trim incidence $-21.5°$, lift coefficient $C_Z = C_L = 0.53$, drag coefficient $C_X = C_D = 1.5$, and fineness ratio $f = 0.35$.

17 Aerodynamics of Apollo Reentry Capsule in Free Molecular Flow

For incidences $\overline{\alpha} < 33°$, the back cover is not impinged by the flow, only the spherical front shield contributes to aerodynamic effects.

According to Sect. 4.4.1, center of pressure is the center of maximum cross section plane. With reference of moments at upstream pole O ($S_{ref} = \pi D^2/4$, $L_{ref} = D$):

$$C_Z = C_L = 0 \,;\quad f = \frac{C_Z}{C_X} = \frac{C_L}{C_D} = 0$$

$$C_X = C_D = 2\cos\overline{\alpha}\,;\quad C_N = C_X \sin\overline{\alpha} = 2\sin\overline{\alpha}\cos\overline{\alpha};\quad C_A = C_X \cos\overline{\alpha}$$

$$C_{m/0} = -\frac{R}{D}(1 - \cos\theta_C)\,C_N;\quad \sin\theta_C = \frac{D}{2R} = \frac{1}{2.4}\,:$$

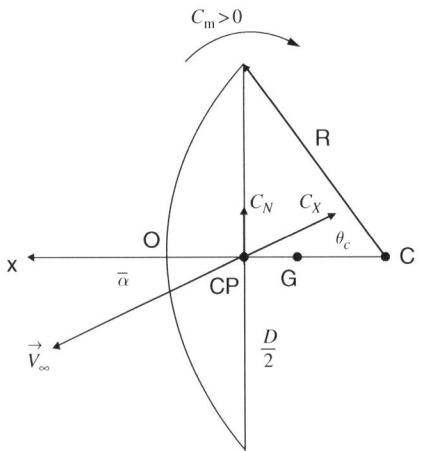

For CG location $\frac{x_G}{D} = -0.26$, we have

$$C_{m/G} = C_{m/O} - \frac{x_G}{D} C_N$$

$$\Rightarrow C_{m/G} = \left[-\frac{x_G}{D} - \frac{R}{D}(1 - \cos\theta_C) \right] C_N = 0.15\, C_N = 0.15 \sin 2\overline{\alpha}$$

Pitching moment coefficient is positive, hence the vehicle is unstable with the heat shield ahead.

$$C_{m\alpha/G} = \frac{\partial C_{m/G}}{\partial \overline{\alpha}} = 0.30 \cos 2\overline{\alpha}$$

18 Aerodynamics of Viking Reentry Capsule in Continuous Flow

According to Sect. "Pressure Coefficients" in Chap. 4, while $S_{ref1} = \pi \cdot R^2$, $L_{ref1} = R$, and θ_m = semiapex angle of tangent cone at the edge of sphere segment, the coefficients of spherical segment (1) are:

$$C_{Aw} = 1 - \sin^4\theta_m - \frac{(1 + 3 \cdot \sin^2\theta_m) \cos^2\theta_m}{2} \cdot \sin^2\overline{\alpha}$$

$$C_{Nw} = \frac{1}{2} \cos^4\theta_m \cdot \sin 2\overline{\alpha}$$

$$C_{mw} = -C_{Nw} \Leftrightarrow x_{CP} = -R$$

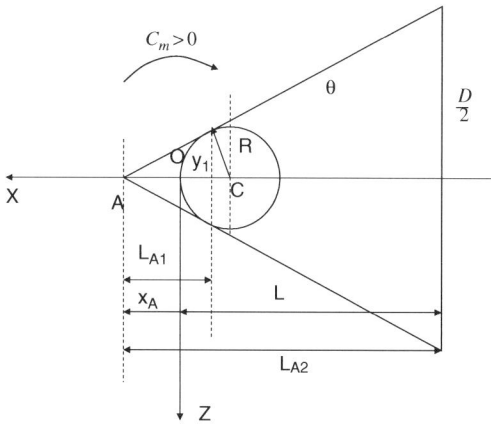

With $\theta_m = 70°$ and $R = 0.25 D$, while multiplying by $\frac{S_{ref1}}{S_{ref}} = \sigma_0^2 = \left(\frac{2R}{D}\right)^2 = \frac{1}{4}$ we obtain coefficients of the cap reported to reference area of vehicle $S_{ref} = \pi \cdot \frac{D^2}{4}$.

$$C'_{Aw1} = \sigma_0^2 C_{Aw1} \ ; \ C'_{Nw1} = \sigma_0^2 C_{Nw1}$$

Pitching moment coefficient around tip O of sphere with reference length D is:

$$C'_{mw/O} = -\frac{R}{D} C'_{Nw1}$$

Coefficients of the truncated cone are obtained by considering the pointed cone (2) with origin A (apex), radius $\frac{D}{2}$, and length L_{A2} such that $\frac{D}{2L_{A2}} = \tan\theta$ and the pointed cone (1) with same origin A, ends at junction of sphere with the conical trunk part, of length $L_{A1} = \frac{R}{\sin\theta} - R\sin\theta$ and maximum radius $y_1 = R\cos\theta$.

Coefficients of truncated cone (2) with $S_{ref} = \pi \cdot \frac{D^2}{4}$ of vehicle are:

$$C'_{Aw2} = \left(1 - \sigma_1^2\right)\left[2 \cdot \sin^2\theta + (1 - 3 \cdot \sin^2\theta)\sin^2\overline{\alpha}\right]$$

$$C'_{Nw2} = \left[1 - \sigma_1^2\right]\cos^2\theta \sin 2\overline{\alpha}$$

with

$$\sigma_1 = \frac{L_{A1}}{L_{A2}} = \frac{y_1}{\left(\frac{D}{2}\right)} = \frac{2R\cos\theta}{D}$$

The coefficient of moment around apex of a pointed cone with reference L_A is:

$$C_{mw/A} = -\frac{2}{3} \cdot \sin 2\overline{\alpha}$$

And for a different reference length D:

$$C'_{mw/A} = C_{mw/A}\left(\frac{L_A}{D}\right) = -\frac{1}{3}\frac{\cos\theta}{\sin\theta}\sin 2\overline{\alpha}$$

The coefficient of moment of the conical part (2) around the apex, with vehicle diameter as reference length is:

$$C'_{mw2/A} = C'_{mw/A}\left[1 - \sigma_1^3\right] \cdot \sin 2\overline{\alpha} = -\frac{1}{3}\frac{\cos^2\theta}{\sin\theta}\left[1 - \sigma_1^3\right]\sin 2\overline{\alpha}$$

The apex of the pointed cone is located at $x_A = OA = \frac{R}{\sin\theta} - R$ from tip O of vehicle.

The coefficient of moment of the conical part (2) around O is thus:

$$C'_{mw2/O} = C'_{mw2/A} + \frac{x_A}{D}C_{Nw2}$$

The coefficients of the vehicle are finally:

$$C_{Aw} = C'_{Aw1} + C'_{Aw2}$$
$$C_{Nw} = C'_{Nw1} + C'_{Nw2}$$
$$C_{mw/O} = C'_{mw1/O} + C'_{mw2/O}$$

The vehicle pitching moment coefficient around G is:

$$C_{mw/G} = C_{mw/O} - \frac{x_G}{D} C_{Nw}$$

We obtain the numerical results which follow:

$$C_{Aw} = 1.769 - 1.65 \sin^2 \overline{\alpha}$$
$$C_{Nw} = 0.1144 \sin 2\overline{\alpha}$$
$$C_{m/O} = -0.1191 \sin 2\overline{\alpha}$$

The center of pressure $\frac{x_{CP}}{D} = \frac{C_{mw/0}}{C_{Nw}} = -1.041$ is independent of incidence. Static margin is constant and positive:

$$SM = \frac{x_G - x_{CP}}{D} = 0.811 > 0$$

$$\frac{z_G}{D} = 0 \rightarrow C_{mw/G} = -0.0927 \sin 2\overline{\alpha} \; ; \; C_{m\overline{\alpha}w/G} = \left.\frac{\partial C_{mw/G}}{\partial \overline{\alpha}}\right|_0 = -0.185$$

$$\frac{z_G}{D} \neq 0 \rightarrow C_{m/G} = \frac{z_G}{D} C_{Aw} - 0.0927 \sin 2\overline{\alpha} \left(\frac{\sin \alpha \cos \beta}{\sin \overline{\alpha}}\right)$$
$$= \frac{z_G}{D} C_{Aw} - 0.185 \cos^2 \beta \cos \alpha \sin 2\alpha$$

$$C_{n/G} = -0.0927 \sin 2\overline{\alpha} \left(\frac{-\sin \beta}{\sin \overline{\alpha}}\right) = 0.185 \cos \alpha \cos \beta \sin \beta$$

While $\beta = 0$, we obtain the curves which follow:

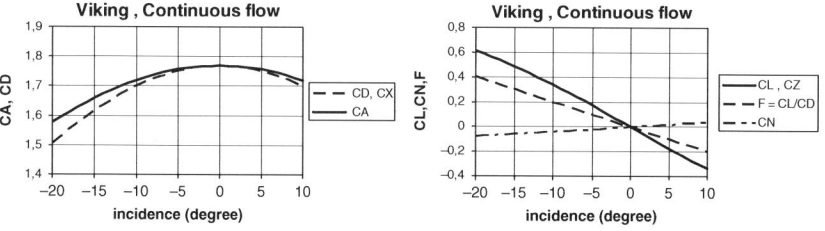

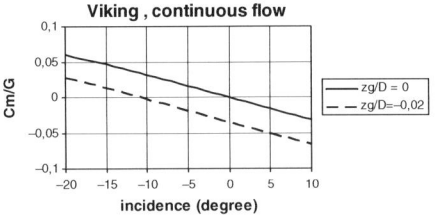

when $\beta = 0$, Yawing moment is null and $\frac{\partial C_{n/G}}{\partial \beta} > 0$. Thus zero sideslip is a stable trim.

Pitching moment curve for $\frac{z_G}{D} = -0.02$ crosses zero at $\alpha = -11°$, which is stable trim because $\frac{\partial C_{m/G}}{\partial \alpha} < 0$. Trim lift coefficient and fineness ratio are respectively $C_L = C_Z = 0.37$; $\frac{C_L}{C_D} = \frac{C_Z}{C_X} = 0.22$.

19 Aerodynamics of Viking Reentry Capsule in Free Molecular Flow

With $S_{ref} = \pi \frac{D^2}{4}$, force coefficients are obtained from simplified theory (Sect. 4.4.1, $\sigma = \sigma' = 1$, Mach ∞):

$$C_Z = C_L = 0; \quad f = \frac{C_Z}{C_X} = \frac{C_L}{C_D} = 0,$$

$$C_X = C_D = 2\cos\overline{\alpha}; \quad C_N = C_X \sin\overline{\alpha} = 2\sin\overline{\alpha}\cos\overline{\alpha}; \quad C_A = C_X \cos\overline{\alpha}$$

The center of pressure is at center of maximum cross section plane, i.e.:

$$x_{CP/A} = -\frac{D}{2\tan\theta} = -0.6369; \quad x_{CP''/O} = x_{CP/A} + R\left(\frac{1}{\sin\theta} - 1\right) = -0.5808$$

$$\frac{x_{CP}}{D} = -0.1659$$

Center of pressure is clearly in front of the center of gravity located at $\frac{x_G}{D} = -0.23$.

Static margin is negative $SM = \frac{x_G - x_{CP}}{D} = -0.064 < 0$.

Pitching moment coefficient around G is positive:

$$C_{m/G} = -\left(\frac{x_G - x_{CP}}{D}\right) C_N = 0.064 \cdot C_N = 0.064 \cdot \sin(2\overline{\alpha})$$

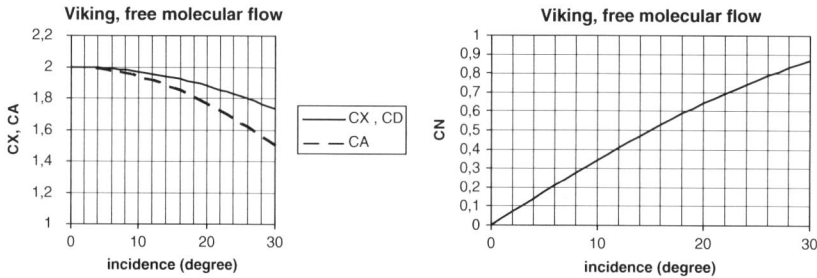

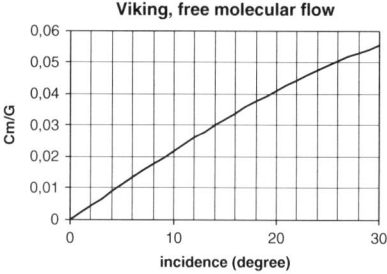

The configuration is unstable, because we have $\frac{\partial C_{m/G}}{\partial \overline{\alpha}} = 0.128 \cos 2\overline{\alpha} > 0$

20 Aerodynamics of Pathfinder Entry Capsule in Intermediate Flow

Erf-Log bridging function is written as:

$$C(Kn) = C_c + \phi(Kn) \cdot (C_m - C_c); \quad \phi(Kn) = \frac{1}{2}\left[1 + erf\left(\frac{\sqrt{\pi}}{\Delta Kn} \ln\left\{\frac{Kn}{Kn_{mi}}\right\}\right)\right]$$

With

$$\Delta Kn = Ln\left(\frac{Kn_m}{Kn_c}\right) = Ln(500)$$

$$Ln(Kn_{mi}) = \frac{Ln(Kn_m) + Ln(Kn_c)}{2} \Rightarrow Kn_m = Kn_{mi} e^{\frac{\Delta Kn}{2}}$$

$$= Kn_{mi}\sqrt{500}; \quad Kn_c = Kn_{mi} e^{-\frac{\Delta Kn}{2}} = \frac{Kn_{mi}}{\sqrt{500}}$$

Knowing $\Phi(Kn_1)$ and $\Delta Kn = 6.2146$ we deduce Kn_{mi} such that:

$$erf\left[\frac{\sqrt{\pi}}{\Delta Kn} Ln\left(\frac{Kn_1}{Kn_{mi}}\right)\right] = 2\Phi(Kn_1) - 1$$

then Kn_m and Kn_c.

Using an approximation of the Erf function [ABR], with $Kn_1 = 0.109$, we obtain:

	Kn_c	Kn_{mi}	Kn_m
C_A	0.0096	0.215	4.8
C_N	0.0060	0.134	3.0
$C_{m/G}$	0.079	1.76	39

According to calculations of reference [PAR] the intermediate regime is shifted toward high Knudsen numbers for stability parameter.

According to exercises 18, 19, and 20 for $2°$ AoA and a CG location $\frac{x_G}{D} = -0.27$, we have the limiting values in continuous and free molecular flow:

Mode	CA	CN	Cm/G (2°)	$C_{m\alpha/G}$ (radian^{-1})
Continuous	1.769	0.00409	−0.00615 (Newton)	−0.176 (Newton)
			−0.00110 (NS)	−0.0315 (NS)
Free molecular	1.998	0.0698	0.00726	0.208

We obtain from these values while using Erf-Log bridging function evolutions according to Knudsen number:

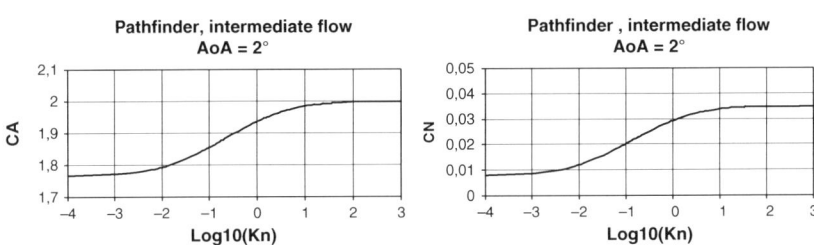

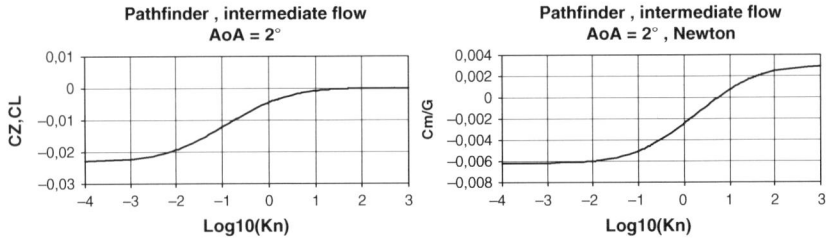

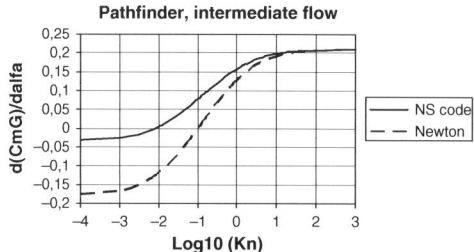

At low AoA, and altitudes corresponding to the beginning of continuous flow, Navier–Stokes code indicates much lower stability than the Newtonian estimate. This is related to the dissociation of CO_2 in Martian atmosphere in the high temperature flow behind the bow shock. Navier–Stokes calculations at lower altitudes agree with Newtonian estimates.

21 Aerodynamics of a Biconic Reentry Vehicle in Continuous Mode

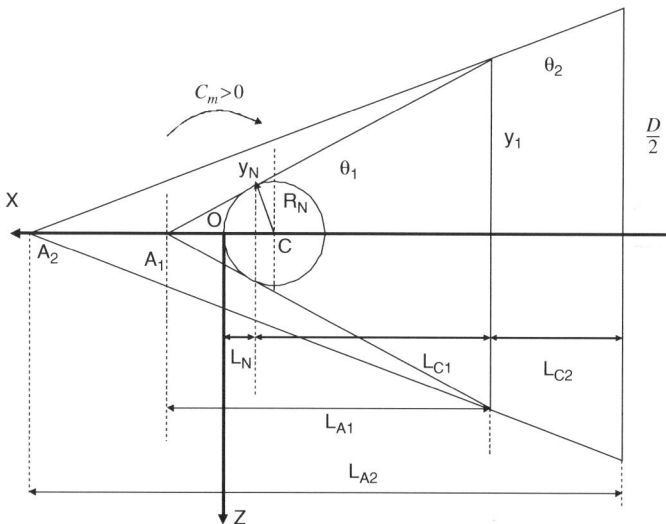

While $D = 0.5\,\text{m}$, $e = \frac{2R_N}{D} = 0.1$, $\theta_1 = 10°$, $\theta_2 = 5°$, and $\lambda = \frac{L_{C1}}{L_{C2}} = \frac{1}{3}$, we obtain:

$$R_N = e\frac{D}{2} = 0.025;\ L_N = R_N(1 - \sin\theta_1) = 0.02489;\ y_N = R_N \cos\theta_1 = 0.02462$$
$$y_1 = y_N + L_{C1}\tan\theta_1 = y_N + \lambda L_{C2}\tan\theta_1;$$

$$\frac{D}{2} = y_1 + L_{C2}\tan\theta_2 = y_N + (\lambda\tan\theta_1 + \tan\theta_2)L_{C2}$$

$$\Rightarrow L_{C2} = \frac{\frac{D}{2} - y_N}{\lambda\tan\theta_1 + \tan\theta_2} = 1.5409; \quad L_{C1} = 0.5136;$$

$$L = L_N + L_{C1} + L_{C2} = 2.0752$$

$$y_1 = 0.1152$$

$$L_{A1} = \frac{y_1}{\tan\theta_1} = 0.6532; \quad L_{A_2} = \frac{D}{2\tan\theta_2} = 2.8575$$

$$x_{A1} = OA_1 = L_{A1} - (L_N + L_{C1}) = 0.1147; \quad x_{A2} = OA_2 = L_{A2} - L = 0.7781$$

Coefficients of a pointed cone with $L_{rf} = L_A$ corresponding to the length and S_{rf} at maximum diameter from apex are:

$$C_{AC}(\theta, \overline{\alpha}) = 2\sin^2\theta + (1 - 3\sin^2\theta)\sin^2\overline{\alpha}; \quad C_{NC}(\theta, \overline{\alpha}) = \cos^2\theta\sin(2\overline{\alpha})$$

$$C_{mC}(\theta, \overline{\alpha}) = -\frac{2}{3}\sin(2\overline{\alpha})$$

Coefficients of the segment of sphere referenced to its radius and base surface are:

$$C_{AS}(\theta, \overline{\alpha}) = 1 - \sin^4\theta - \frac{(1 + 3\sin^2\theta)}{2}\cos^2\theta\sin^2\overline{\alpha}; \quad C_{NS}(\theta, \overline{\alpha}) = \frac{\cos^4\theta}{2}\sin(2\overline{\alpha})$$

$$C_{mS}(\theta, \overline{\alpha}) = -C_{NS}(\theta, \overline{\alpha}) = -\frac{\cos^4\theta}{2}\sin(2\overline{\alpha})$$

By applying formulae of weighted summation (Sect. "Pressure Coefficients" in Chap. 4) we obtain force coefficients of the blunted biconic:

$$C_A = \sigma_0^2 C_{AS}(\theta_1, \overline{\alpha}) + \left(1 - \sigma_1^2\right)\sigma_2^2 C_{AC}(\theta_1, \overline{\alpha}) + \left(1 - \sigma_2^2\right)C_{AC}(\theta_2, \overline{\alpha})$$

$$C_N = \sigma_0^2 C_{NS}(\theta_1, \overline{\alpha}) + \left(1 - \sigma_1^2\right)\sigma_2^2 C_{NC}(\theta_1, \overline{\alpha}) + \left(1 - \sigma_2^2\right)C_{NC}(\theta_2, \overline{\alpha})$$

with

$$\sigma_0 = \frac{2R_N}{D} = e; \quad \sigma_1 = \frac{y_N}{y_1}; \quad \ell_1 = \frac{L_{A1}}{L_{A2}}; \quad \sigma_2 = \frac{2y_1}{D}$$

Total pitching moment coefficient with origin at nosetip O referenced to $\pi\frac{D^2}{4}$ and L_{A2} of complete vehicle from apex is:

$$C_{m/0} = -\frac{R_N}{L_{A2}}\sigma_0^2 C_{NS}(\theta_1, \overline{\alpha}) + C_{mC}(\theta_1, \overline{\alpha})\left(1 - \sigma_1^3\right)\sigma_2^3 \ell_1 + C_{mC}(\theta_2, \overline{\alpha})\left(1 - \sigma_2^3\right)$$
$$+ \frac{x_{A1}}{L_{A2}}C_{NC}(\theta_1, \overline{\alpha})\left(1 - \sigma_1^2\right)\sigma_2^2 + \frac{x_{A2}}{L_{A2}}C_{NC}(\theta_2, \overline{\alpha})\left(1 - \sigma_2^2\right)$$

21 Aerodynamics of a Biconic Reentry Vehicle in Continuous Mode

This pitching moment is then referenced to the real length L of vehicle:

$$C'_{m/O} = \frac{L_{A2}}{L} C_{m/O}$$

We obtain values of coefficients:

$$C_A = 0.03418 + 0.9593 \sin^2 \overline{\alpha}$$
$$C_N = 0.9829 \sin(2\overline{\alpha})$$
$$C'_{m/O} = -0.5664 \sin(2\overline{\alpha}) ; \quad \frac{x_{CP}}{L} = \frac{C'_{m/O}}{C_N} = -0.576$$

alfa	CA	CN	C'm/O	CX	CZ
0	0,0342	0,0000	0,0000	0,0342	0,0000
1	0,0345	0,0343	-0,0198	0,0351	0,0337
2	0,0353	0,0686	-0,0395	0,0377	0,0673
3	0,0368	0,1027	-0,0592	0,0421	0,1007
4	0,0388	0,1368	-0,0788	0,0483	0,1338
5	0,0415	0,1707	-0,0984	0,0562	0,1664

To insure 3% static margin, we must locate the mass center at $\frac{x_G}{L} = \frac{x_{CP}}{L} + 0.03 = -0.546$, i.e., $x_G = -1.133$ m.

At low incidence, we obtain:

$$C_A = 0.03418 + 0.9593 \sin^2 \overline{\alpha} \approx C_{A0} = 0.03418$$
$$C_N = 0.9829 \sin(2\overline{\alpha}) \approx C_{N\alpha} \overline{\alpha} = 1.966 \overline{\alpha}$$
$$C'_{m/G} \approx -\left(\frac{x_G - x_{CP}}{L}\right) C_{N\alpha} \overline{\alpha} = -0.059 \overline{\alpha}$$

With a CG offset $\frac{z_G}{L} = 0.01$, pitching moment coefficient around Gy is:

$$C'_{m/G} \approx -\left(\frac{x_G - x_{CP}}{L}\right) C_{N\alpha} \alpha + \frac{z_G}{L} C_{A0}$$
$$\Rightarrow \alpha_{eq} = \frac{z_G}{x_G - x_{CP}} \frac{C_{A0}}{C_{N\alpha}} \approx 0.0028 \text{ radian}$$

The result is a 0.16° trim angle of attack in the pitching plane.

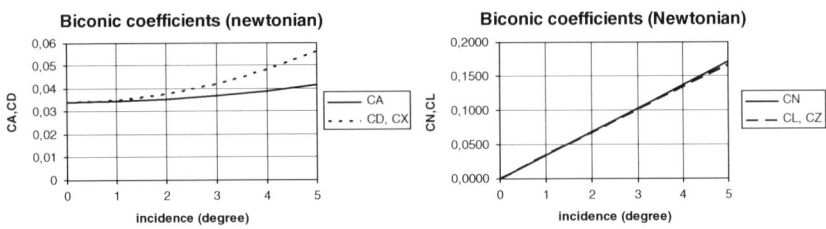

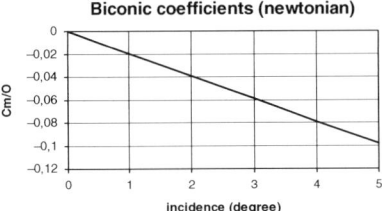

22 Targeting Errors of Ballistic Trajectories

According to (7.49), optimum FPA at sea level is such that:

$$(\cos \gamma_{opt})^2 = \frac{1}{2 - \left(\frac{V_0}{V_c}\right)^2},$$

i.e., $\gamma_{opt} = 37.77°$ with $V_0 = 5000$ m/s and $V_c = 7910$ m/s. According to (7.53), half angular range is $\alpha_{opt} = \frac{\pi}{2} - 2\gamma_{opt}$ and range at sea level P $= 2\alpha r_t = 3215$ Km with $r_t = 6371$ km.

For $\gamma_{opt} \pm 10°$, i.e., 27.77° and 47.77°, range is obtained from (7.37), (7.38), and (7.39). Introducing circular orbit velocity $V_c = \sqrt{\frac{\mu}{r_t}}$ and $x = \left(\frac{V_0}{V_c}\right)^2$, $y = (\cos \gamma_0)^2$ into these relations, we obtain a relation between half angular range and variables x and y:

$$\cos \alpha = \frac{1 - xy}{\sqrt{1 + (x - 2)xy}}$$

It results the respective ranges:
$P_\pm = 2\alpha\, r_t = 2998$ km and 3033 km.

To obtain the sensitivity of the range to errors in V_0 and γ_0, let us square this relation. We obtain $f = [1 + (x - 2)xy](\cos \alpha)^2 - (1 - xy)^2 = 0$. While differentiating f with respect to α, x, and y we obtain expressions of the partial derivative of α with respect to x and y:

$$df = \frac{\partial f}{\partial \alpha}d\alpha + \frac{\partial f}{\partial x}dx + \frac{\partial f}{\partial y}dy = 0,$$

resulting in

$$\frac{\partial \alpha}{\partial x} = -\frac{\partial f}{\partial x}\bigg/\frac{\partial f}{\partial \alpha} \quad \frac{\partial \alpha}{\partial y} = -\frac{\partial f}{\partial y}\bigg/\frac{\partial f}{\partial \alpha}$$

While developing calculations:

$$\frac{\partial \alpha}{\partial x} = \frac{y}{\tan \alpha}\left[\frac{x-1}{1+(x-2)xy} + \frac{1}{1-xy}\right]; \quad \frac{\partial \alpha}{\partial y} = \frac{1}{\tan \alpha}\frac{x}{2}\left[\frac{x-2}{1+(x-2)xy} + \frac{2}{1-xy}\right]$$

The partial derivative of range are:

$$\frac{\partial P}{\partial V_0} = 2r_t \frac{\partial \alpha}{\partial x}\frac{\partial x}{\partial V_0} = 4r_t \frac{V_0}{V_c^2}\frac{\partial \alpha}{\partial x}; \quad \frac{\partial P}{\partial \gamma_0} = 2r_t \frac{\partial \alpha}{\partial y}\frac{\partial y}{\partial \gamma_0} = -4r_t \cos\gamma_0 \sin\gamma_0 \frac{\partial \alpha}{\partial y}$$

For the maximum range trajectory, with given velocity V_0 or x, such that $y = y_{opt} = (\cos\gamma_{opt})^2 = \frac{1}{2-x}$, we verify $\frac{\partial \alpha}{\partial y} = 0$, i.e., an extremum range. With the same hypothesis, $\frac{\partial \alpha}{\partial x} = \frac{1}{2\tan\alpha(1-x)(2-x)} > 0$, i.e., we obtain a range increase with velocity, which is consistent with an optimum FPA. By applying these relations, we obtain for 5000 m/s and for the three FPA:

γ_0 (degree)	37.77°	47.77°	27.77°
$\frac{\partial P}{\partial V_0}$ (km/m/s)	1.64	1.42	1.68
$\frac{\partial P}{\partial \gamma_0}$ (km/degree)	0	−34.6	45

We note the advantage of using a maximum range or minimal energy at a given range, which results in minimum sensitivity to FPA errors (only for the same altitude of initial and final points). We check in addition that a positive FPA error for the trajectory 27.77° corresponds to an increase in range, which is logical because we get closer to the optimum angle. The opposite occurs for angle 47.77°.

Error on initial azimuth: A simple geometrical construction shows that the lateral error is:

$$\delta y = r_t \sin(2\alpha) \sin \delta Az \approx r_t \sin(2\alpha) \delta Az$$

This results in a side error of about 50 km per degree of azimuth for the three trajectories.

23 Stability of Free Rotational Motion of a Satellite with Spin

According to Sect. 7.2.2, $\tan\theta_0 = \frac{I_T \omega_{y0}}{I_x \omega_{x0}}$ results in:

$\theta_0 \sim 31°$, $\dot\varphi_0 = \left(1 - \frac{I_x}{I_T}\right) p_0 = -0.698$ radians /s; $\dot\psi_0 = \frac{I_x p_0}{I_T \cos\theta_0} = 4.396$ radians/s

Initial angular momentum is $H_0 = \sqrt{(I_x \omega_{x0})^2 + (I_y \omega_{y0})^2} = 6106$ N.m.s

Exterior couples being null, angular momentum is constant relative to inertial frames. Exercise 8 gives its rotational energy as a function of the transverse angular rate:

$$E = \frac{1}{2}\left[\frac{H_0^2}{I_x} + I_T\left(1 - \frac{I_T}{I_x}\right)\omega_T^2\right]$$

Initial energy equals

$$E_0 = \frac{1}{2}\left[\frac{H_0^2}{I_x} + I_T\left(1 - \frac{I_T}{I_x}\right)\omega_{y0}^2\right]$$

Thus, dissipation of energy $\dot E < 0$ results in decreasing transverse angular rate:

$$E - E_0 = \dot E \cdot t = I_T\left(1 - \frac{I_T}{I_x}\right)\left(\omega_T^2 - \omega_{y0}^2\right) < 0$$

We obtain:

$$\omega_T^2 = \omega_{y0}^2 + \frac{\dot E \cdot t}{I_T(1 - \frac{I_x}{I_T})} \tag{1}$$

Angular momentum is constant, which enables us to determine roll rate:

$$\omega_x^2 = \frac{H_0^2}{I_x^2} - \left(\frac{I_T}{I_x}\right)^2 \omega_T^2 = \omega_{x0}^2 - \left(\frac{I_T}{I_x}\right)^2 \frac{\dot E t}{I_T(1 - \frac{I_T}{I_x})} \tag{2}$$

Transverse rate ω_T decreases and roll rate ω_x increases.
Then, we obtain the evolution of nutation angle from:

$$\tan\theta = \frac{I_T \omega_T}{I_x \omega_x} \tag{3}$$

The nutation angle tends toward zero, which corresponds to a final trim state where the x axis of the satellite is aligned along $\vec{H}_0$, with roll rate $\omega_x = \frac{H_0}{I_x}$.

The expression of instantaneous energy shows that its minimum allowable value is, $E_{min} = \frac{1}{2}\frac{H_0^2}{I_x}$, reached when $\omega_T = 0$. Thus, dissipation of energy cannot last beyond this moment. In fact this dissipation is related to the deformation of some nonrigid part of the satellite under the effect of the periodic component of motion, dependent on the transverse angular rate vector turning at frequency $-(1 - \frac{I_x}{I_T})\omega_x$ in satellite axes. When $\omega_T = 0$, the amplitude of excitation is null, the rotational motion is uniform and corresponds to a rigid mode, the dissipation of energy becomes null. Expression (1) for $\dot{E} = -10\,mW$ results in canceling transverse angular velocity at 131595 s, i.e., approximately 36 hours 33 minutes. Using graphical software and formulae (1)–(3), we obtain the evolution of nutation.

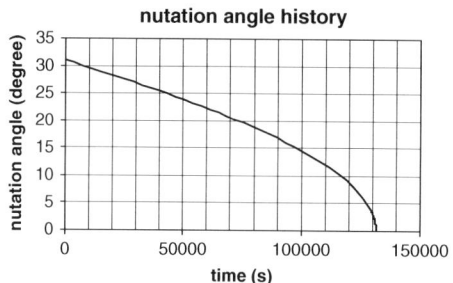

However, the assumption of a constant dissipated power is not physical, because the amplitude of the transverse angular velocity at origin tends toward zero. A more realistic figure would be a power model using a decreasing function of transverse angular rate. This means the convergence time is likely much longer.

24 Allen and Eggers Reentry

The application of Allen's results (Chap. 9) determines velocity versus altitude for the two trajectories. We obtain from results of Chap. 3 on isothermal atmospheres:

- Average temperature T = 239.13 K, reference height 7000 m.
- Sound speed a = 310 m/s and dynamic viscosity is $1.54\,10^{-5}$ (Sutherland Formula).
- Molecular mean-free-path $\lambda_\infty = \frac{8.13\,10^{-8}}{\rho}$

Using graphical software, we obtain evolutions versus altitude of Mach number $\frac{V}{a}$, Knudsen number $\frac{\lambda_\infty}{L_{ref}}$ (independent of the trajectory), and Reynolds number of the vehicle $\text{Re} = \frac{\rho V L_{ref}}{\mu}$ and of the spherical nosetip $R_N = 50$ mm$\text{Re} = \frac{\rho V R_N}{\mu}$:

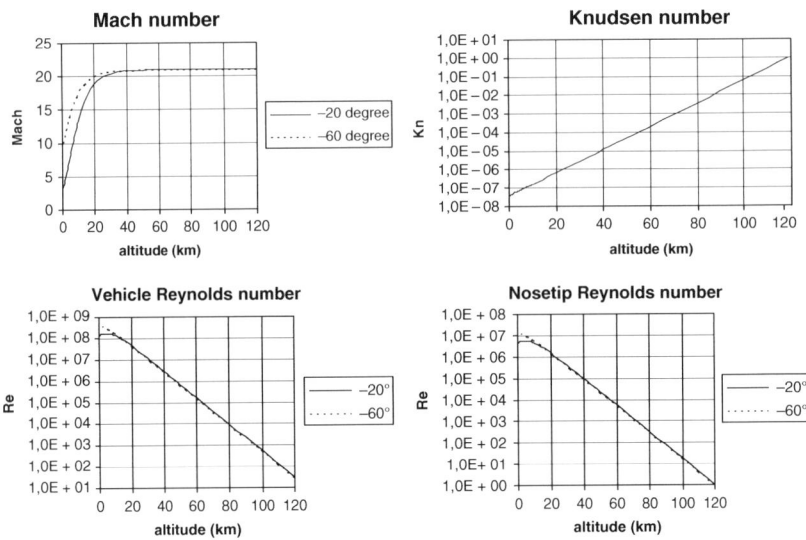

From the assumed values of transition criteria (in practice, they highly depends in material properties, such as roughness and mass loss rate), evolutions of Reynolds number make it possible to define altitudes of beginning of transition, i.e., 47 km for the vehicle and 22 km for nose. Analytical expression of maximum heat flux at stagnation point gives for nose radius 25–100 mm and the two FPA (flux in MW/m², inversely proportional to square root of R_n):

Rn	20°	60°
0,025	67,4	107,2
0,05	47,6	75,8
0,075	38,9	61,9
0,1	33,7	53,6

Altitudes of these maximum heat fluxes are respectively 7.4 Km and 4.6 Km.

25 Mars Atmosphere Measurement

Axial accelerometer measurement is:

$$A_X = -\frac{1}{2}\rho V^2 S_{ref} C_A$$

Assuming low angle of attack, we have:

$$\dot{V} \approx A_X + \varphi \sin \gamma$$
$$V\dot{\gamma} = \varphi \cos \gamma$$
$$\dot{z} = -V \sin \gamma$$

Knowing the estimate of velocity and FPA at beginning of entry (from guidance data), using measured data $A_x(t)$ and the preceding dynamic entry model[1] allows to determine by numerical integration the temporal evolution of velocity V, FPA γ, and altitude z.

Then, using the design model for drag coefficient (for Pathfinder the model was a function of velocity and Knudsen number $Kn = \frac{0.7513 \, 10^{-7}}{\rho D}$), we obtain the estimated density:

$$\hat{\rho} = \frac{-2A_X}{S_{ref} V^2 C_A (V, Kn)}$$

A first iteration is made using the C_A from the Knudsen calculation with previous density model. One can then proceed by iteration using each time Knudsen calculations based on the density resulting from the preceding iteration step.

The order of magnitude of accelerometer measurement for pathfinder is obtained using Allen approximation between 125 and 20 km, then between 25 km and 8 km.

$$V = V_0 e^{-K(\rho - \rho_0)}$$

$$V_0 = 7470, \gamma_0 = 13.6°, \quad K = \frac{H_R}{2\beta \sin \gamma_0}$$

Altitudes	ρ_S (kg.m^{-3})	H_R (km)	K
0–25 km	0.0159	11.049	361.4
25–125 km	0.0525	7.295	238.6

We obtain $V_1 = V(25\,\text{km}) = 4972.4$ and the evolution of accelerometer measurement in earth g's.

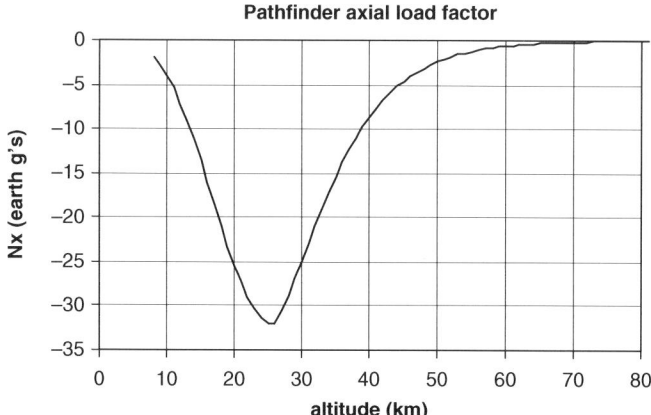

Pathfinder axial load factor

---

[1] In fact, in a similar model, but more exact one, taking into account spherical rotating planet.

In fact this assessment of axial load is overestimated, for Allen's model is very inaccurate for shallow trajectories. Taking into account more exact model, pathfinder maximum load factor was 18 g (to see exercise 34).

The opening of the first parachute of pathfinder occurred at Mach number 1.8 close to the design value at about 8 Km altitude.

26 Entry of Meteorites

Using the approximations of Allen (Chap. 9), with:

$$M = \rho_b V_b = \rho_b \frac{4}{3}\pi R^3; \; S_b = \pi R^2;$$

$$C_X = 1 \Rightarrow \beta = \frac{M}{SC_X} = \frac{4}{3}\rho_b R; \; K = \frac{3H_R}{8\rho_b R \sin \gamma_0}$$

velocity is:

$$\rho = \rho_s \, e^{-\frac{Z}{H_R}} : V \approx V_0 e^{-K\rho}$$

Thus, velocity loss ratio percentage during entry is:

$$100 \frac{V_0 - V}{V_0} = 100\left(1 - e^{-K\rho_s}\right)$$

Heat flux at stagnation point is:

$$\Phi = c\sqrt{\frac{\rho}{R}} V^3 = c\sqrt{\frac{\rho}{R}} V_0^3 e^{-3K\rho}$$

Density and altitude at maximum heat flux are $\rho_m = \frac{1}{6K} \Leftrightarrow Z_m = H_R Ln\,(6K\rho_s)$

- If $Z_m > 0$, then $\Phi_m = c\sqrt{\frac{\rho_m}{R}} V_0^3 e^{-3K\rho_m} = cV_0^3 \sqrt{\frac{1}{6KR}} e^{-\frac{1}{2}} = \frac{2}{3} cV_0^3 \sqrt{\frac{\rho_b \sin \gamma_0}{H_R}}$, maximum heat flux is independent of R.
- If $Z_m \leq 0$, maximum heat flux is at sea level, $Z = 0$, $\rho = \rho_s$, and $\Phi_m = cV_0^3 \sqrt{\frac{\rho_s}{R}} e^{-3K\rho_s}$, which depend on R

Heat quantity per unit area $\frac{dE}{dA}$ received in the vicinity of stagnation point is:

$$p_c = 2\beta \, g \sin \gamma_0; \; p_S = 101325; \; \frac{dE}{dA} \approx cV_0^2 \sqrt{\frac{\pi \beta H_R}{R \sin \gamma_0}} Erf\left[\sqrt{2\frac{p_s}{p_c}}\right]$$

26 Entry of Meteorites

While assuming $\frac{dE}{dA}$ is constant on the front hemisphere, the order of magnitude of the quantity of heat received by the body is:

$$Q \approx 2\pi R^2 \frac{dE}{dA}$$

Rise in temperature of the body, assumed uniform in volume is:

$$\Delta T = \frac{Q}{M\,C_V}$$

If this temperature is lower than $T_V = 2500$ K, we assume the whole mass of the meteorite impacts the ground.
The kinetic energy is:

$$E_{kin} = \frac{1}{2}MV^2 = \frac{4}{3}\pi R^3 \rho_b V_0^2 e^{-2K\rho s}$$

Using graphical software, we obtain:

D (m)	beta	K	dvsv (%)	Zm (km)	Phimax (MW/m²)
0,03	50	140	100,0	49,4	54,0
0,3	500	14	100,0	33,3	54,0
3	5000	1,4	85,7	17,2	54,0
30	50000	0,14	17,7	1,1	53,7
300	500000	0,014	1,9	-15,0	30,4
3000	5000000	0,0014	0,2	-31,2	9,6

D (m)	dE/dS	Q (J)	M (kg)	DT (K)	ECIN (J)	ECIN (kT)
0,03	319074361	451080,751	0,03534292	3190,744	2,4E-163	0,00
0,3	319074361	45108075,1	35,3429174	319,074	3,2E-08	0,00
3	317798064	4492764283	35342,9174	31,780	5,2E+10	0,01
30	202931294	2,8689E+11	35342917,4	2,029	1,7E+15	431
300	72110805,7	1,0194E+13	3,5343E+10	0,072	2,4E+18	6,12E+05
3000	22973354	3,2478E+14	3,5343E+13	0,002	2,5E+21	6,34E+08

We note that meteorites having a diameter about a centimeter and lower are destroyed before the impact. For diameter about 0.1 meter and higher, they impact the ground quasi intact; however, their velocity and kinetic energy are low (they are underevaluated here but when taking account gravity, they reach limit velocity $\sqrt{\frac{2Mg}{\rho_s S\,C_X}}$ about 90 m/s and are dangerous even on a small scale). For a diameter about 1 m, the kinetic energy becomes high, and huge for 10 m. For 30 m diameter, equivalent energy is about 400 KT and corresponds to serious damage by blast effect within a radius of several kilometers. For 300 m of diameter, energy reaches a catastrophic level equivalent to 600 MT of TNT.

27 Normal Load Factor Related to Incidence

The normal aerodynamic force, for a slender cone with half apex angle θ_a is:

$$Z^A = -\bar{q} S_{ref} C_N \approx -\bar{q} S_{ref} C_{N\alpha} \bar{\alpha}; \quad C_{N\alpha} \approx 2 (\cos \theta_a)^2$$

Normal load factor is equal and opposite to apparent normal gravity felt by an observer fixed to the vehicle:

$$N_Z = \frac{Z^A}{mg} = \bar{q} \frac{S_{ref} C_{N\alpha}}{mg} \bar{\alpha}; \quad g_{Z,app} = -N_Z$$

Using Allen's approximation:

$$\bar{q} = \frac{1}{2}\rho V^2 = \frac{1}{2}\rho V_0^2 e^{-2K\rho}; \quad \rho = \rho_S e^{-\frac{Z}{H_R}}; \quad K = \frac{H_R}{2\beta \sin \gamma_0}$$

Ballistic coefficient is $\beta = \frac{m}{S_{ref} C_X} \sim 1.22 \, 10^4$ kg.m^{-2} for $C_x \approx 0.05$

Normal load factor for a constant angle of attack $\bar{\alpha}$ is maximum together with dynamic pressure:

$$\frac{d\bar{q}}{d\rho} = 0 \Leftrightarrow 2K\rho_m = 1 \Leftrightarrow Z_m = H_R Ln(2K\rho_S); \quad \bar{q}_m = \frac{V_0^2}{4Ke}$$

We obtain:

gama0 (degree)	K	Zm (Km)	qbarmax (bars)
20	0,837	5,913	33,2
60	0,331	-0,590	84,1

alfa (degree)	1	3	5
gama0 (degree)	Nz (g)		
20	19,0	56,9	94,9
60	48,1	144,2	240,3

These results illustrate the need to control trim incidence under 20 km altitude.

28 Artefact in Computer Codes

Examination of candidate parameters, for example, axial load factor, shows dependence:

$$N_x = \frac{1}{2}\rho V^2 \frac{S_{ref} C_A}{mg}$$

Here, we must examine evolutions of V, ρ, and C_A.

Expressions of state vector derivatives in the system use models of gravity, atmosphere, and aerodynamic coefficients, which are by construction continuous models (this is mandatory to use a numerical integration algorithm, for example, a Runge Kutta method). Consequently, the first derivatives of the state vector, including velocity derivative $\dot{V}$ are continuous by definition. Thus, the origin of phenomenon can only comes from aerodynamic and atmospheric density models:

- Aerodynamic coefficients are generally modeled by discrete tables as a function of Mach, altitude, and angle of incidence, and the coefficients are linearly interpolated per interval. Thus, the first derivatives of coefficients are discontinuous. However, if the discretization of the tables is well chosen, discontinuities are weak and practically invisible in the evolution of the parameters concerned.
- We are left with the atmosphere. Standard models are generally discretized with a law of temperature having linear evolution by altitude slices. Slope discontinuities in temperature curve induce notable discontinuities of the first derivative of density. This is apparent in the example corresponding to the evolution of the density as a function of altitude for a Martian atmosphere model.

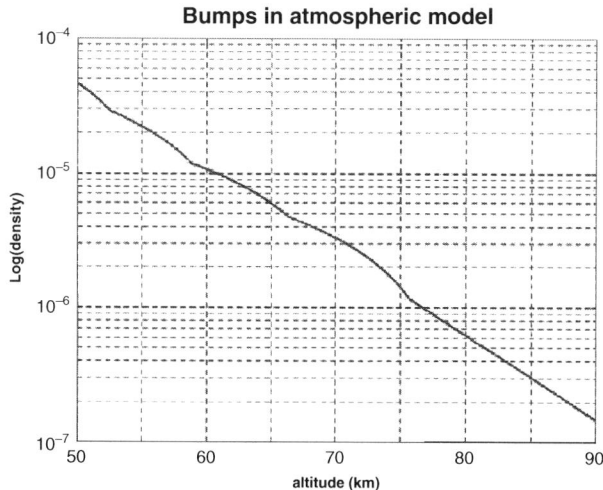

This phenomenon of slope discontinuity in temperature law is not incompatible with nature. For example, in the case of the earth's atmosphere, there is discontinuity at the boundary of low atmosphere with the troposphere near 11 km altitude. Corresponding oscillations are not inevitably all artefacts of calculation. Likely discontinuities of slope induced by modeling are exaggerated compared to reality.

29 Gyroscopic Stabilization of an Entry Capsule

A model of Knudsen number for the Martian atmosphere is found in [PAR]:

$$Kn = \frac{0.7513 \, 10^{-7}}{\rho D}$$

While $\rho = \rho_S e^{-\frac{Z}{H_R}}$ and according to results of exercise 20, we can assess the derivative of pitching moment coefficient $C_{m\alpha/G} = \frac{\partial C_{m/G}}{\partial \bar{\alpha}}$ in an intermediate regime for $\frac{x_G}{D} = -0.25$:

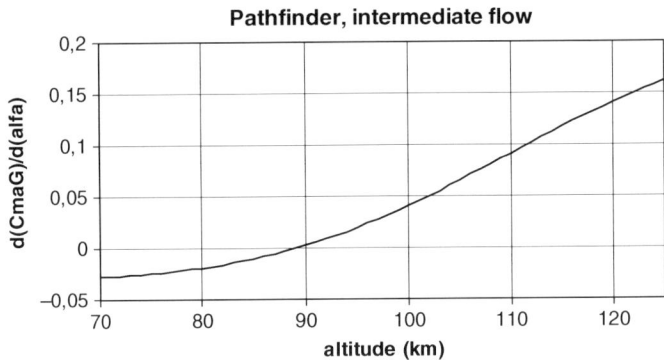

We note that the vehicle is unstable above 87 km altitude.

In the case of a capsule without spin, the incidence motion is divergent, and the probe will start a swing to present the aft cover forward.

Let us examine the effect of spin on incidence behavior. Quasi static approximation of Sect. 10.2 led to approximate evolution of the incidence. In the case of a statically stable vehicle, $C_{m\alpha/G} < 0$, aerodynamic pulsation is defined by $\omega^2 = -\frac{\bar{q} S_{ref} D C_{m\alpha/G}}{I_T}$. Examination of Sect. 10.2 derivations shows that while $C_{m\alpha/G} > 0$, a statically unstable vehicle, approximate evolution remains valid with the same expression of ω^2.

Then we obtain:

$$\frac{\bar{\alpha}}{\bar{\alpha}_0} \approx \left[1 + \left(\frac{\omega^2}{p_r^2}\right)\right]^{-\frac{1}{4}}$$

with $p_r = \mu \frac{p}{2}$; $\mu = \frac{I_X}{I_T}$, where ω^2 is negative in zones of instability and positive outside.

The calculation of evolution of incidence using the preceding expression, with 1° initial AoA at 125 km altitude and roll rate 0.3 to 2 rpm, gives following results:

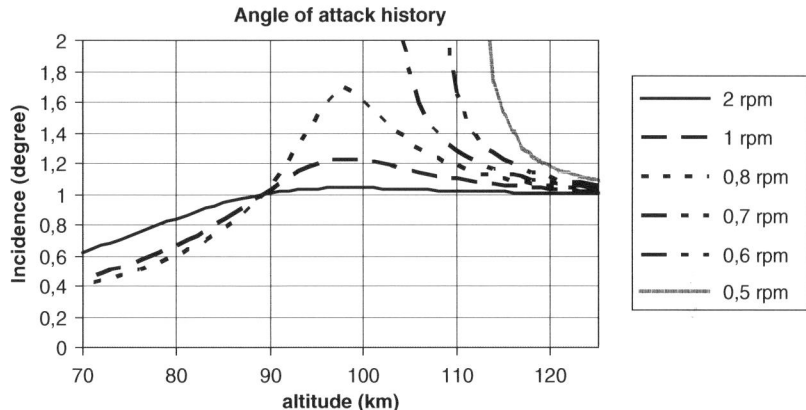

Examination of results shows that using roll rate higher or equal to 1 rpm allows avoiding the divergence for the period of aerodynamic instability. For roll rates lower than 0.8 rpm, we observe a clear divergence of the incidence above 100 km altitude. For increasing roll rates, the divergence decreases, but later convergence is slowed down. Gyroscopic stabilization was widely used on Martian probes (Pathfinder, Beagle 2). In the case of Pathfinder, nominal roll rate was 2 rpm.

30 Effect of Equilibrium Lift on CG Motion

For a conical vehicle 8° and $\bar{\alpha} = 0.1°$; $C_A = 0.05$; Newtonian theory (Sect. 4.3.4.3.4) give:

$$C_{N\alpha} = \left.\frac{\partial C_{Nw}}{\partial \bar{\alpha}}\right|_{\bar{\alpha}=0} \approx 2\cos^2\theta_a \approx 1.96; \quad C_{Z\alpha} = C_{N\alpha} - C_A = 1.91$$

$$C_Z \approx C_{Z\alpha}\bar{\alpha} = 0.0033; \quad f = \frac{C_Z}{C_X} \approx \frac{C_Z}{C_A} \approx 0.066$$

We assume that the lift force is in the vertical plane, directed upward, for a constant fineness and ballistic coefficient.

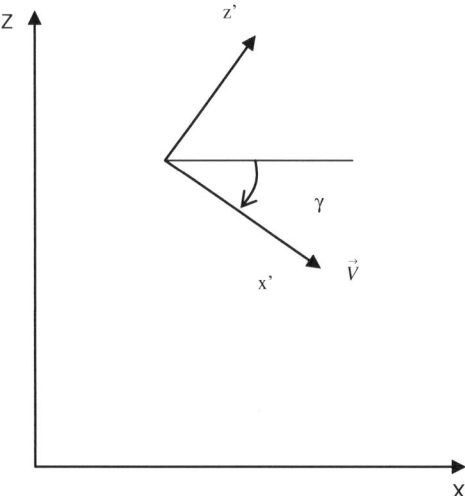

Equations of motion relative to axes x' and z' fixed to relative velocity vector are:

$$\gamma_{X'} = \frac{dV}{dt} = \frac{F_{X'}}{m} = -\bar{q}\frac{S_{ref}C_X}{m} = -\frac{\bar{q}}{\beta}; \quad \gamma_{Z'} = -V\frac{d\gamma}{dt} = f\frac{\bar{q}}{\beta}$$

Evolutions of altitude and air density are given by:

$$\frac{dZ}{dt} = -V\sin\gamma; \quad \frac{d\rho}{dz} = -\frac{\rho}{H_R}$$

While using ρ as the independent variable we obtain:

$$\frac{\rho V \sin\gamma}{H_R}\frac{dV}{d\rho} = -\frac{1}{2\beta}\rho V^2 \quad \Rightarrow \quad \frac{1}{V}\frac{dV}{d\rho} = -\frac{H_R}{2\beta\sin\gamma}$$

$$-\frac{\rho V^2 \sin\gamma}{H_R}\frac{d\gamma}{d\rho} = \frac{f}{2\beta}\rho V^2 \quad \Rightarrow \quad \frac{d\gamma}{d\rho} = -\frac{f H_R}{2\beta\sin\gamma}$$

Second equation gives:

$$\frac{d\cos\gamma}{d\rho} = \frac{f H_R}{2\beta} \quad \Rightarrow \quad \cos\gamma - \cos\gamma_0 = \frac{f H_R}{2\beta}(\rho - \rho_0)$$

Dividing the first equation by the second:

$$\frac{1}{V}\frac{dV}{d\gamma} = \frac{1}{f} \quad \Rightarrow \quad V = V_0 e^{\frac{\gamma-\gamma_0}{f}}$$

30 Effect of Equilibrium Lift on CG Motion

Lateral variation from reference trajectory with no incidence (rectilinear thus $\gamma = \gamma_0$ constant) is:

$$y = -\int_0^t V \sin(\gamma - \gamma_0)\, dt = -H_R \int_{\rho_0}^{\rho} \frac{\sin(\gamma - \gamma_0)}{\sin \gamma} \frac{d\rho}{\rho}$$

For $u = \delta\gamma = \gamma - \gamma_0 \ll 1$ we have:

$$\cos\gamma - \cos\gamma_0 \approx -u \sin\gamma_0 = \frac{f\, H_R}{2\beta}(\rho - \rho_0) \quad \Rightarrow u = -f\, K(\rho - \rho_0)$$

$$\sin\gamma \approx \sin\gamma_0 + u\cos\gamma_0$$

This gives y:

$$y \approx -\frac{H_R}{\cos\gamma_0} \int_0^U \left(\frac{u}{u + \tan\gamma_0}\right) \frac{du}{(u - u_1)} \quad \text{With } u_1 = f\, K\rho_0$$

We obtain by integration:

$$y \approx -\frac{H_R}{\cos\gamma_0} \left[\frac{\tan\gamma_0}{u_1 + \tan\gamma_0} Ln\left(\left|1 + \frac{U}{\tan\gamma_0}\right|\right) + \frac{u_1}{u_1 + \tan\gamma_0} Ln\left(\left|\frac{u_1 - U}{u_1}\right|\right) \right]$$

As $u_1 Ln(u_1)$ tends toward zero with u_1 we have, when $\rho_0 \approx 0$:

$$y \approx -\frac{H_R}{\cos\gamma_0} Ln\left(\left|1 + \frac{U}{\tan\gamma_0}\right|\right) = -\frac{H_R}{\cos\gamma_0} Ln\left(\left|1 - \frac{f\, K}{\tan\gamma_0}(\rho - \rho_0)\right|\right)$$

Numerical application at sea level gives:

$$K = 0.7,\ \gamma_S = 26.1°;\ \delta\gamma = -3.9°,\ V_S = 2131\ \text{m/s},\ y = 953\ \text{m}$$

In the case of reentry with no incidence, we would have according to Allen:

$$V_S = V_0 e^{-K\rho_S} = 2268 m/s, \quad \gamma_S = 30°; \quad \delta\gamma = 0°, \quad y = 0$$

In the case of a vehicle with spin 1.5 Hz at same trim incidence, radius of the helicoids trajectory at the end of the flight (Sect. "Trimmed Phase" in Chap. 14) is:

$$R_N = \frac{|A_{N,eq}|}{p^2}; \quad A_N = \frac{\bar{q}\, S_{ref}\, C_Z}{m} = f\frac{\bar{q}}{\beta}$$

By using same velocity as trajectory with no incidence, we obtain:

$$A_N = 23.6\ \text{m.s}^{-2}, \quad y \sim R_N = 0.26\ \text{m}$$

Effect of spin for a weak trim incidence $0.1°$ is to decrease side offset from 953 m to 0.26 m.

This shows clearly the effectiveness of spin to control effect of asymmetries on trajectory.

31 Effects of Drag Dispersions

Effects on impact velocity, axial load factor and heat flux are assessed using Allen's method (Chap. 9).

$$V_S \approx V_0 e^{-K\rho_S}; \quad K = \frac{H_R}{2\beta \sin \gamma_0} \Rightarrow \frac{dV_S}{V_S} = -K\rho_S \left(\frac{d\rho_S}{\rho_S} + \frac{dK}{K} \right)$$

$$= -Kv_S \left(\frac{dS\, C_X}{S\, C_X} + \frac{d\rho_S}{\rho_S} + \frac{dH_R}{H_R} \right)$$

Maximum load factor, when reached before impact, is at first order independent of C_X and ρ_S, and depends only on V_0, γ_0, and H_R:

$$\frac{dN_X}{N_X} = -\frac{dH_R}{H_R}$$

Maximum heat flux is proportional to $\sqrt{\frac{\beta}{H_R}}$, from whence:

$$\frac{d\Phi_m}{\Phi_m} = -\frac{1}{2} \left(\frac{dS\, C_X}{S\, C_X} + \frac{dH_R}{H_R} \right)$$

To determine the influence on range, we must take account of gravity by using results of Sect. "Flight Path Angle and Downrange Errors" in Chap. 14:

$$\Delta X = -H_R \frac{\cos \gamma_0}{\sin^4 \gamma_0} \frac{\rho_S H_R}{\beta V_0^2} F''(2K\rho_S) \left[\frac{d\rho_S}{\rho_S} + \frac{dH_R}{H_R} + \frac{dS\, C_X}{S\, C_X} \right]$$

with $F''(x) = 1 + \frac{x}{4} + \frac{x^2}{18} + \frac{x^3}{96}$.

$+5\%$ on any of parameters SC_X, H_R or ρ_S gives:

- -11.5% on ground impact velocity V_S i.e. 81 m/s
- -125 m on range

$+5\%$ on SC_X or H_R gives -2.5% on maximum heat flux.
$+5\%$ on H_R gives -5% on maximum axial load factor.

32 Wind Effects

To calculate the trajectory, we use Allen's hypothesis and results, for an observer fixed to the air, at constant velocity relative to earth. At zero AoA, drag is along the relative velocity vector. With a flat earth, no gravity, trajectory relative to the air is rectilinear. For an observer fixed to the air, Allen's results are valid except initial conditions are velocity and FPA relative to the air frame.

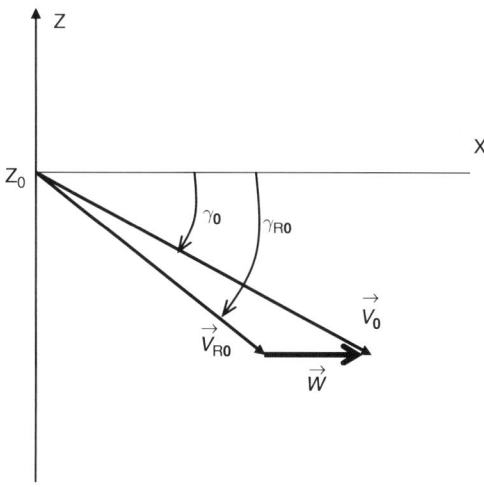

$$V_{R0} \cos \gamma_{R0} = V_0 \cos \gamma_0 - W; \quad V_{R0} \sin \gamma_{R0} = V_0 \sin \gamma_0$$

We have:

$$V_{R0} = \sqrt{V_0^2 + W^2 - 2V_0 W \cos \gamma_0} \approx V_0 \left(1 - \frac{W \cos \gamma_0}{V_0}\right)$$

$$\sin \gamma_{R0} = \frac{\sin \gamma_0}{\sqrt{1 + \left(\frac{W}{V_0}\right)^2 - 2\left(\frac{W}{V_0}\right) \cos \gamma_0}} \approx \left(1 + \frac{W \cos \gamma_0}{V_0}\right) \sin \gamma_0$$

$$\cos \gamma_{R_0} = \frac{\cos \gamma_0 - \frac{W}{V_0}}{\sqrt{1 + \left(\frac{W}{V_0}\right)^2 - 2\left(\frac{W}{V_0}\right) \cos \gamma_0}} \approx \left(1 + \frac{W \cos \gamma_0}{V_0}\right) \cos \gamma_0 - \frac{W}{V_0}$$

Relative velocity evolution according to altitude Z (Z is identical for observer fixed to wind and observer fixed to earth) is given by:

$$V_R \approx V_{R0} e^{-K_R \rho}, \text{ with } K_R = \frac{H_R}{2\beta \sin \gamma_{R0}}$$

Velocity V and FPA γ for observer fixed to earth are such as:

$$V \cos \gamma = V_R \cos \gamma_{R0} + W$$
$$V \sin \gamma = V_R \sin \gamma_{R0}$$

FPA for the motionless observer is not constant, because $\tan \gamma = \frac{\sin \gamma_{R0}}{\cos \gamma_{R0} + \frac{W}{V_R}}$ where V_R depends on altitude (the trajectory is thus not rectilinear)

While denoting $V^* = V_0 e^{-K\rho}$ velocity on reference trajectory with no wind and assuming $\frac{W}{V_R}$ and $\frac{W}{V^*} \ll 1$, we obtain from above results first-order expressions of variations of velocity relative to air and earth relative velocity:

$$V_R - V^* \approx -W \cos \gamma_0 (1 - K\rho) e^{-K\rho}$$
$$V - V^* \approx W \cos \gamma_0 \left[1 - (1 - K\rho) e^{-K\rho} \right]$$

Maximum load factor with wind is determined by the conditions relative to air:

$$N_{X \max} = \frac{V_{R0}^2 \sin \gamma_{R0}}{2g H_R e} \Rightarrow \frac{\Delta N_{X \max}}{N_{X \max}} \approx 2 \frac{\Delta V_{R0}}{V_0} + \frac{\Delta \sin \gamma_{R0}}{\sin \gamma_0} = -\frac{W}{V_0} \cos \gamma_0$$

We obtain in the same way for maximum heat flux:

$$\Rightarrow \frac{\Delta \Phi_{\max}}{\Phi_{\max}} \approx 3 \frac{\Delta V_{R0}}{V_0} + \frac{1}{2} \frac{\Delta \sin \gamma_{R0}}{\sin \gamma_0} = -\frac{5}{2} \frac{W}{V_0} \cos \gamma_0$$

Effect of a tail wind is thus to decrease velocity relative to the air, maximum axial load factor, and maximum heat flux. The opposite occurs for head wind.

Although altitudes Z are identical for the two observers, horizontal ranges since the point of beginning of reentry are different. Denoting X the range for earth-fixed observer and X_R for observer fixed to air:

$$X = W t + X_R, \text{ where } X_R = \frac{Z_0 - Z}{\tan \gamma_{R0}}$$

Range at sea level ($Z = 0, t = T$) for the earth-fixed observer in presence of wind is thus:

$$X = W T + \frac{Z_0}{\tan \gamma_{R0}}$$

32 Wind Effects

where T is total duration of descent from $Z_0 = 120$ km.
Ground range without wind corresponds to $X^* \approx \frac{Z_0}{\tan \gamma_0}$. Range variation induced by wind is thus:

$$\Delta X = W\,T + \frac{Z_0}{\tan \gamma_{R0}} - \frac{Z_0}{\tan \gamma_0} = W\left(T - \frac{Z_0}{V_0 \sin \gamma_0}\right)$$

According to approximate expression of reentry time in Chap. 9 we have:

$$T \approx \frac{Z_0}{V_{R0} \sin \gamma_{R0}} + \frac{H_R}{V_{R0} \sin \gamma_{R0}} K\rho_S \left[1 + \frac{K\rho_S}{4} + \frac{(K\rho_S)^2}{18} + \frac{(K\rho_S)^3}{96}\right]$$

We obtain finally the ground range variation:

$$\Delta X \approx \frac{W\,H_R}{V_0 \sin \gamma_0} K\rho_S \left[1 + \frac{K\rho_S}{4} + \frac{(K\rho_S)^2}{18} + \frac{(K\rho_S)^3}{96}\right]$$

Thus, effect of tail wind is a range increase and head wind a reduction.

For tail wind, numerical calculations give the following results, derived using graphical software (for head wind, only the sign is changed):

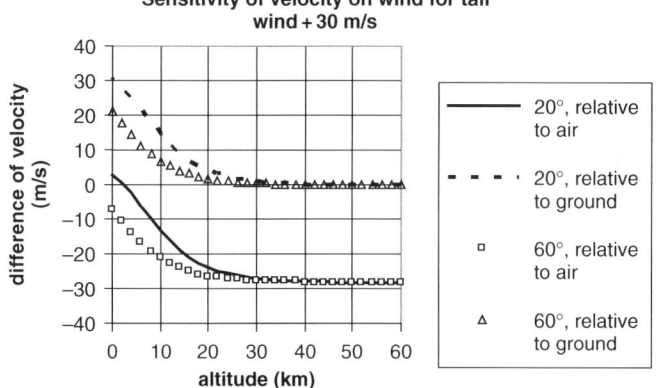

Effects on axial load factor, heat flux, and range:

gam0 (degree)	W (m/s)	DNx/Nx (%)	DPHI/PHI (%)	DX (m)
20	30	-0,47	-1,17	218
60	30	-0,25	-0,63	26
20	-30	0,47	1,17	-218
60	-30	0,25	0,63	-26

These approximate results are in good agreement with more exact three degrees-of-freedom solutions.

33 Effect of a Sharp Stability Variation, Case of a Plane Oscillation

According to methodology, Sect. "Consequences of a Static Stability Variation" in Chap. 14:

Evolution of the incidence,

$$t < 0 \to \xi_0(t) = \theta_0 \sin(\omega_{a0} t + \varphi_0)$$
$$t > 0 \to \xi_1(t) = \theta_1 \sin(\omega_{a1} t + \varphi_1)$$

Energy of rotation,

$$E_0(t<0) = \frac{1}{2} I_T \left[\dot{\xi}_0^2(t) + \omega_{a0}^2 \xi_0^2(t) \right] = \frac{1}{2} I_T \omega_{a0}^2 \theta_0^2$$

$$E_1(t<0) = \frac{1}{2} I_T \left[\dot{\xi}_1^2(t) + \omega_{a1}^2 \xi_1^2(t) \right] = \frac{1}{2} I_T \omega_{a1}^2 \theta_1^2$$

Subtracting the two sets of equations at $t = 0$:

$$E_1 - E_0 = \frac{1}{2} I_T \left[\left(\dot{\xi}_1^2(0) - \dot{\xi}_0^2(0) \right) + \omega_{a1}^2 \xi_1^2(0) - \omega_{a0}^2 \xi_0^2(0) \right]$$
$$= \frac{1}{2} I_T \left(\omega_{a1}^2 \theta_1^2 - \omega_{a0}^2 \theta_0^2 \right)$$

While $\dot{\xi}(t)$ and $\xi(t)$ are continuous functions, even at $t=0$:

$$E_1 - E_0 = \frac{1}{2} I_T \left[\left(\omega_{a1}^2 - \omega_{a0}^2 \right) \xi_0^2(0) \right] = \frac{1}{2} I_T \left(\omega_{a1}^2 \theta_1^2 - \omega_{a0}^2 \theta_0^2 \right)$$

$$E_1 - E_0 = \frac{1}{2} I_T \left[\left(\omega_{a1}^2 - \omega_{a0}^2 \right) (\theta_0 \sin \varphi_0)^2 \right] = \frac{1}{2} I_T \left(\omega_{a1}^2 \theta_1^2 - \omega_{a0}^2 \theta_0^2 \right)$$

Finally:

$$\omega_{a1}^2 \theta_1^2 = \omega_{a0}^2 \theta_0^2 + \left(\omega_{a1}^2 - \omega_{a0}^2 \right) (\theta_0 \sin \varphi_0)^2 \Rightarrow \frac{\theta_1^2}{\theta_0^2} = \frac{\omega_{a0}^2 + \left(\omega_{a1}^2 - \omega_{a0}^2 \right) (\sin \varphi_0)^2}{\omega_{a1}^2}$$

Extreme of amplitudes ratio thus corresponds to:

$$\varphi_0 = 0 \quad \text{or} \quad \varphi_0 = \pi \Rightarrow \frac{\theta_1}{\theta_0} = \frac{\omega_{a0}}{\omega_{a1}}; \quad \varphi_0 = \frac{\pi}{2} \quad \text{or} \quad \varphi_0 = \frac{3\pi}{2} \Rightarrow \frac{\theta_1}{\theta_0} = 1$$

Thus in the case of planar oscillation, unlike motion with spin and circular polarization, amplitude ratio behavior through stability perturbations depends on the phase of motion. In the case of circular motion AoA is constant, as well as potential and total energy. It is not the case when the motion is planar or with elliptic polarization. When stability change occurs at the time of a maximum AoA variations of "potential" and total energy are maximum. When incidence is null, there is no variation of rotational energy. The effect on amplitude depends on the phase of the motion at the time of disturbance.

In the case of plane oscillation, if the disturbance occurs close to zero incidence, total energy remains constant. While stability is decreased, the maximum amplitude of the oscillation increases according to inverse ω_a ratio. If disturbance occurs close to a maximum of incidence, the variation of total energy is maximum (negative) and exactly compensate the effect of stability variation, the amplitude does not vary.

34 Mars Skip Out Trajectories

From Chap. 7, while including radial and normal components of aerodynamic drag in the equations of motion in central gravitation field, we obtain:

$$\ddot{r} = -\frac{\mu}{r^2} + r\dot{\theta}^2 - \frac{F_A \sin \gamma}{m}$$

$$r\ddot{\theta} + 2\dot{r}\dot{\theta} = -\frac{F_A \cos \gamma}{m}$$

$$\frac{F_A}{m} = \frac{\rho V^2}{2\beta}$$

$$V \cos \gamma = r\dot{\theta}$$

$$V \sin \gamma = \dot{r}$$

$$V = \sqrt{\dot{r}^2 + (r\dot{\theta})^2}$$

$$\ddot{r} = -\frac{\mu}{r^2} + r\dot{\theta}^2 - \frac{\rho}{2\beta} V \dot{r}$$

$$\frac{d}{dt}\left(r^2 \dot{\theta}\right) = -\frac{\rho}{2\beta} V r^2 \dot{\theta}$$

Denoting $u = r^2\dot{\theta}$; $V_r = \dot{r}$, we obtain the system of motion equations:

$$\begin{cases} \dot{u} = -\frac{\rho V}{2\beta} u \\ \dot{V}_r = -\frac{\mu}{r^2} + \frac{u^2}{r^3} - \frac{\rho V}{2\beta} \dot{r} \\ \dot{\theta} = \frac{u}{r^2} \\ \dot{r} = V_r \end{cases}$$

The range is $r_p \theta$, with r_p = planet radius.

Initial conditions are $u(0) = r(0) V_0 \cos \gamma_0$; $V_r(0) = V_0 \cos \gamma_0$; $\theta_0 = 0$; and $r(0) = r_p + h_0$.

A numerical solution is obtained with a very simple FORTRAN program, using the fourth order Runge Kutta algorithm shown in Chap. 8.

```
      IMPLICIT REAL*8 (A-H,O-Z)
      character*80 fileout
      DIMENSION X(4),XPRIM(4),xs(4),xi(4)
      common /param/amu,rp,beta,ros25,hr25,ros125,hr125
      common /sorties/r,u,teta,vr,h,ro,v
      external deriv
      ndim=4
      pis2=asin(1.d0)
      pi=2.d0*pis2
      trang=pi/180.d0
      d=6786000.d0
      rp=d/2.d0
      g0=9.81d0
      amu=0.38d0*g0*rp*rp
      ros25=0.01586d0
      ros125=0.05253d0
      hr25=11049.d0
      hr125=7281.d0
      open (10 , file = 'fileout')
      beta=58.8d0
      v0=7479.d0
      h0=130000.d0
      gam0d=-13.71d0
      gam0=gam0d*trang
      r0=rp+h0
      u0=r0*v0*cos(gam0)
      vr0=v0*sin(gam0)
      t=0
      x(1)=r0
      x(2)=u0
      x(3)=0.d0
      x(4)=vr0
      dt=0.1d0
      call deriv(t,x,xprim)
      hkm=h/1000
      qbar=0.5d0*ro*v*v
      an=qbar/beta
      an=an/g0
      range=rp*teta/1000.d0
      write(6,100)t,hkm,vr,v,range,an
      write(10,100)t,hkm,vr,v,range,an
  100 format(6(f11.3,1x))
      ns=0
c***************************************************************
c***************************************************************
      do i=1,2000
      call drk4(T,X,XS,XI,DERIV,XPRIM,DT,NDIM)
      call deriv(t,x,xprim)
      ns=ns+1
      if(ns.eq.10)then
      ns=0
      hkm=h/1000
      qbar=0.5d0*ro*v*v
      an=qbar/beta
      an=an/g0
      range=rp*teta/1000.d0
      write(6,100)t,hkm,vr,v,range,an
      write(10,100)t,hkm,vr,v,range,an
      end if
      if(h.le.8000.d0)then
      go to 2
      end if
      end do
   2  hkm=h/1000
      qbar=0.5d0*ro*v*v
      an=qbar/beta
      an=an/g0
      range=rp*teta/1000.d0
      write(6,100)t,hkm,vr,v,range,an
      write(10,100)t,hkm,vr,v,range,an
      stop
      END
c***************************************************************
      subroutine deriv(t,x,xprim)
      IMPLICIT REAL*8 (A-H,O-Z)
      DIMENSION X(1),XPRIM(1)
      common /param/amu,rp,beta,ros25,hr25,ros125,hr125
      common /sorties/r,u,teta,vr,h,ro,v
      r=x(1)
      u=x(2)
      teta=x(3)
      vr=x(4)
      h=r-rp
      call atmos(h,ro)
      r2=r*r
      tetapoin=u/r2
      vt=r*tetapoin
      vt2=vt*vt
      v=sqrt(vr*vr+vt2)
      a=ro*v/(2.d0*beta)
      upoin=-a*u
      vrpoin=-amu/r2+vt2/r-a*vr
      xprim(1)=vr
      xprim(2)=upoin
      xprim(3)=tetapoin
      xprim(4)=vrpoin
      RETURN
      END
c***************************************************************
      subroutine atmos(h,ro)
      IMPLICIT REAL*8 (A-H,O-Z)
      common /param/amu,rp,beta,ros25,hr25,ros125,hr125
      if(h.ge.25000.d0)then
      ro=ros125*exp(-h/hr125)
      else
      ro=ros25*exp(-h/hr25)
      end if
      return
      end
```

34 Mars Skip Out Trajectories

Using post flight entry condition, numerical results are obtained:

time (s)	altitude (km)	vr (m/s)	velocity (m/s)	range (km)	axial load (g)
0	130,8	-1772,581	7479	0	0
5	122,082	-1714,714	7483,034	35,068	0
10	113,654	-1656,377	7486,942	70,308	0
15	105,519	-1597,58	7490,704	105,714	0,001
20	97,679	-1538,324	7494,266	141,283	0,004
25	90,136	-1478,599	7497,488	177,007	0,011
30	82,893	-1418,358	7500,004	212,88	0,029
35	75,954	-1357,477	7500,901	248,893	0,076
40	69,32	-1295,661	7497,996	285,028	0,188
45	63	-1232,275	7486,33	321,251	0,446
50	57,002	-1166,059	7455,283	357,491	1,007
55	51,347	-1094,746	7383,741	393,602	2,148
60	46,069	-1014,785	7234,034	429,295	4,258
65	41,221	-921,855	6949,894	464,045	7,648
70	36,877	-813,258	6471,523	497,028	12,043
75	33,11	-692,088	5778,199	527,193	16,106
80	29,959	-569,151	4930,856	553,575	18,079
85	27,399	-458,045	4053,699	575,688	17,369
90	25,344	-367,471	3260,077	593,666	14,896
95	23,682	-301,578	2629,178	608,112	11,143
100	22,295	-256,08	2153,515	619,853	8,476
105	21,098	-224,391	1790,304	629,538	6,528
110	20,035	-202,24	1509,145	637,639	5,107
115	19,065	-186,766	1288,115	644,503	4,062
120	18,16	-176,038	1111,656	650,386	3,284
125	17,299	-168,74	968,725	655,476	2,696
130	16,468	-163,951	851,411	659,918	2,245
135	15,656	-161,019	753,98	663,823	1,895
140	14,855	-159,465	672,219	667,275	1,619
145	14,06	-158,934	602,988	670,345	1,4
150	13,265	-159,157	543,913	673,084	1,224
155	12,467	-159,921	493,176	675,539	1,082
160	11,665	-161,057	449,364	677,743	0,966
165	10,856	-162,425	411,365	679,728	0,871
170	10,041	-163,909	378,291	681,518	0,793
175	9,217	-165,408	349,422	683,134	0,729
180	8,387	-166,834	324,164	684,593	0,676
182,4	7,985	-167,469	313,174	685,243	0,655

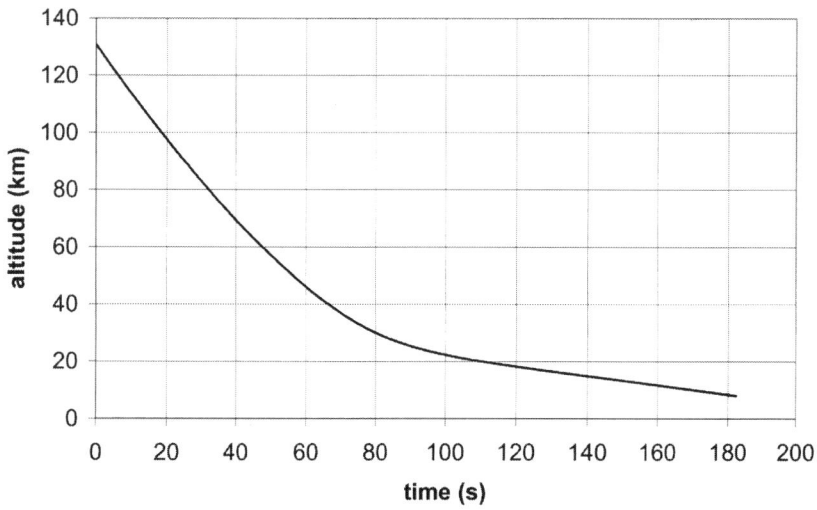

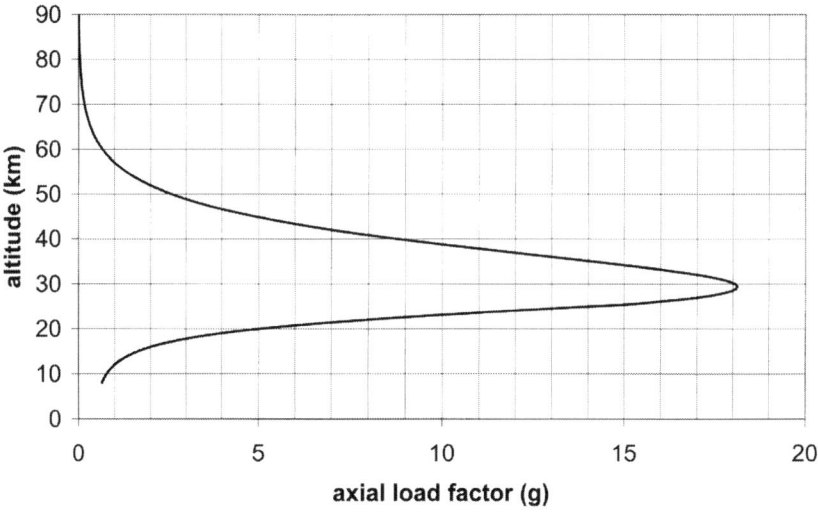

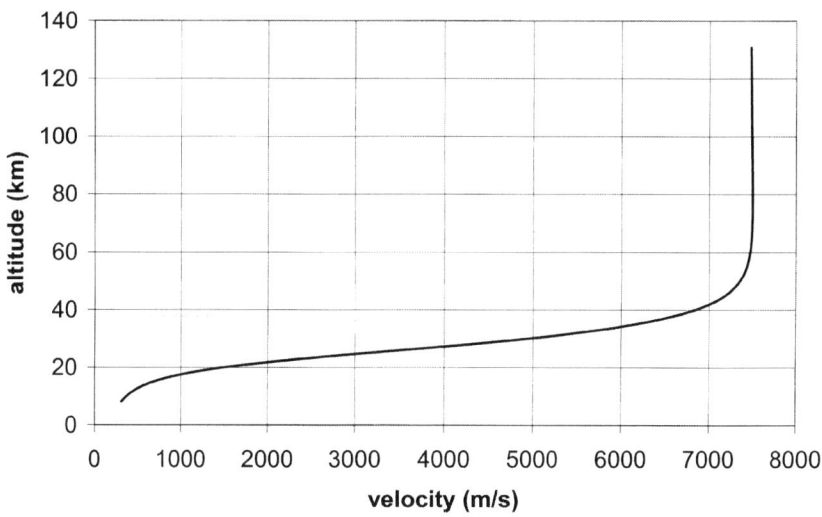

34 Mars Skip Out Trajectories

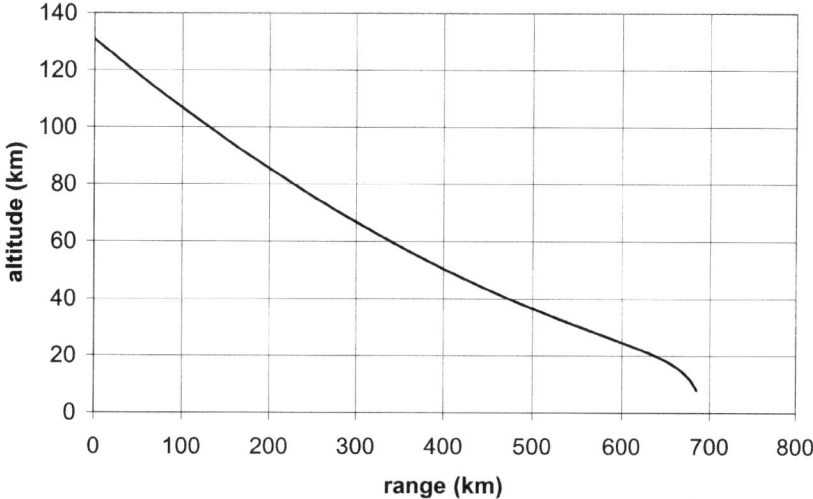

Maximum axial load factor, 18 g is much lower than predicted by Allen solution (32 g) in exercise 25, range and flight time are accordingly much longer. This is obviously related to the curvature of the planet and to the central gravitation field.

Using the program with decreasing flight path angles shows beginning of skip out at $-11.5°$ FPA, as shown in the next figure. The nominal FPA, 14.8°, was chosen with sufficient margin to prevent skip out with regards to navigation uncertainties.

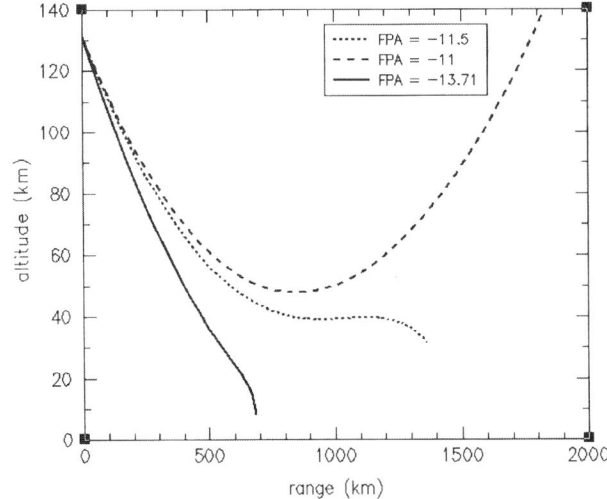

Printed in Great Britain
by Amazon